MAN-INDUCED LAND SUBSIDENCE

Dr. Joseph F. Poland stands at the approximate point of maximum subsidence in the San Joaquin Valley, California. Subsidence of approximately 9.0 m occurred from 1925 to 1977 due to aquifer compaction caused by pumping of ground water. Signs indicate the former elevations of the land surface in 1925 and 1955 respectively. Photo taken December 1976.

REVIEWS IN ENGINEERING GEOLOGY
VOLUME VI

MAN-INDUCED LAND SUBSIDENCE

Edited by
THOMAS L. HOLZER

The Geological Society of America
Boulder, Colorado 80301
1984

© 1984 by The Geological Society of America, Inc.
All rights reserved.

Copyright is not claimed on any material prepared
by government employees within the scope of their
employment.

All material subject to this copyright and included
in this volume may be photocopied for the noncommercial
purpose of scientific or educational advancement.

Published by the Geological Society of America, Inc.
3300 Penrose Place, P.O. Box 9140, Boulder, Colorado 80301

Printed in U.S.A.

Library of Congress Cataloging in Publication Data
Main entry under title:
Man-induced land subsidence.

    (Reviews in engineering geology; v. 7)
    Includes bibliographies.
    1. Engineering geology—Addresses, essays, lectures.
2. Subsidences (Earth movements)—Addresses, essays,
lectures. I. Holzer, Thomas L. II. Series.
TA705.R4 vol. 7 624.1′51 s [624.1′51] 84-16881
ISBN 0-8137-4106-8

Contents

Dedication vii

Preface ix

PART 1. FLUID WITHDRAWAL FROM POROUS MEDIA

*Field-based computational techniques for predicting subsidence
  due to fluid withdrawal* 1
  Donald C. Helm

Subsidence over oil and gas fields 23
  J. C. Martin and S. Serdengecti

Subsidence due to geothermal fluid withdrawal 35
  T. N. Narasimhan and K. P. Goyal

*Ground failure induced by ground-water withdrawal from
  unconsolidated sediment* 67
  Thomas L. Holzer

PART 2. DRAINAGE OF ORGANIC SOIL

Organic soil subsidence 107
  John C. Stephens, Leon H. Allen, Jr., and Ellen Chen

PART 3. COLLAPSE INTO MAN-MADE AND NATURAL CAVITIES

Coal mine subsidence—eastern United States 123
  Richard E. Gray and Robert W. Bruhn

Coal mine subsidence—western United States 151
  C. Richard Dunrud

*Sinkholes resulting from ground-water withdrawal in carbonate
  terranes—an overview* 195
  J. G. Newton

*Mechanisms of surface subsidence resulting from solution
  extraction of salt* 203
  John R. Ege

To Joseph Fairfield Poland, Ph.D., who for more than 50 years dedicated most of his professional career to increasing man's understanding of land subsidence. His curiosity about subsidence began during his graduate education at Stanford University where he enrolled after his graduation from Harvard University in 1929. It was kindled by the first recognized occurrence in the United States of land subsidence caused by ground-water withdrawal in the nearby San Jose area in the Santa Clara Valley, California. This interest came to a head in 1956 when Dr. Poland organized a U.S. Geological Survey research group in Sacramento, California, for the purpose of investigating the mechanics of aquifer systems in order to understand the mechanisms that cause land subsidence. This group, which he directed until his retirement in December 1974, achieved worldwide recognition for its pioneering research.

The esteem and respect for Dr. Poland is reflected in numerous assignments and awards. He was called as an expert witness by the court during the lawsuit brought by the U.S. Navy and the City of Long Beach against Standard Oil Company of California to adjudicate the damaging and costly land subsidence at the Wilmington Oil Field, California. He served as a consultant to UNESCO on the subsidence of Venice, Italy. In 1970 he received the Claire P. Holdredge Award of the Association of Engineering Geologists, a professional society he helped found. In 1972 the O. E. Meinzer Award of the Hydrogeology Division of the Geological Society of America was presented to Dr. Poland and George H. Davis for their paper "Land Subsidence Due to Withdrawal of Fluids."

Dr. Poland has inspired many young scientists with his quiet passion for subsidence research and for his checking and rechecking each hypothesis with field data gathered painstakingly using carefully designed and innovative instruments. Many of these scientists were doctoral candidates from Stanford University and the University of California at Berkeley who wrote their dissertations on topics related to subsidence under the actual, if not official, guidance of Dr. Poland.

Preface

Land subsidence, lowering of the land surface by mass movement, has been caused by man's activities in at least 37 of the 50 states of the United States. It affects an aggregate area of more than 40,000 km^2. The nationwide economic impact of man-induced subsidence is difficult to estimate. Data on damage and loss of property values are not systematically collected. Projections of data from a few sites where economic impact is well documented suggest that total annual cost to the nation is more than $100 million.

Recognizing the vital roles of engineering geologists and hydrogeologists in the prediction, control, and mitigation of man-induced land subsidence, the Engineering Geology and Hydrogeology Divisions of the Geological Society of America cosponsored a symposium on man-induced subsidence at the annual meeting in Atlanta, Georgia, on November 19, 1980. It was entitled "Joseph F. Poland Subsidence Symposium: State-of-the-Art." Ten papers were presented. The enthusiastic response to the symposium prompted nine of the authors to expand their presentations into the papers represented by this volume. Their purpose was to compile comprehensive summaries of our knowledge of the mechanisms of land subsidence in the United States and the techniques that have been developed to predict and mitigate it.

This volume includes nine papers. The papers are arranged into three categories:

Part 1, Fluid Withdrawal from Porous Media, reviews subsidence and associated effects that result from compaction of porous media when pore-fluid pressures are lowered.

Part 2, Drainage of Organic Soil, reviews subsidence that results from draining areas underlain by peat and muck soils.

Part 3, Collapse into Man-Made and Natural Cavities, reviews subsidence that results from man-induced collapse into underground cavities.

Part 1 opens with a paper by Donald C. Helm that describes the range of available field-based techniques for prediction of land subsidence due to fluid withdrawal. Choice of techniques depends on the availability of appropriate field data. The paper by J. C. Martin and S. Serdengecti reviews the fundamentals of reservoir compaction and surface subsidence in oil and gas fields and explains why large vertical surface displacements are exceptional. The third paper by T. N. Narasimhan and K. P. Goyal reviews both surface deformation at areas of geothermal fluid withdrawal worldwide and the state of our knowledge of the physical and thermal mechanisms that cause subsidence. The last paper by the editor, Thomas L. Holzer, summarizes by area the nationwide occurrence of ground failure or rupture caused by ground-water withdrawal from unconsolidated sediment and reviews the state of our knowledge of mechanisms by which failures form.

Part 2 is a paper by John C. Stephens, Leon H. Allen, Jr., and Ellen Chen that reviews worldwide rates and the mechanisms of land subsidence caused by the drainage of peat and muck soils. They also review the methods for prediction and control of organic soil subsidence.

Part 3 opens with two papers on coal mine subsidence. First, Richard E. Gray and Robert W. Bruhn describe subsidence in the eastern United States over both active and abandoned coal mines and review the state of our capability to predict subsidence and associated damage to engineered structures. Second, C. Richard Dunrud describes the mechanics of subsidence at coal fields in the western United States. The paper by J. G. Newton reviews both the mechanisms by which declining water tables induce sinkhole formation and the geologic and hydrologic conditions conducive to collapse. In the last paper John R. Ege reviews the mechanisms by which the solution extraction of salt causes land subsidence, and he documents a case history study of the Grosse Ile subsidence.

Two omissions from the volume require discussion. First, I have consciously omitted a paper specifically focused on subsidence caused by ground-water withdrawal in the United States. Joseph F. Poland and George H. Davis already published this paper in *Reviews in*

Engineering Geology, Volume II (see pages 187–269). New developments in this subject area since their publication are discussed in the papers by D. C. Helm and T. L. Holzer. Second, land subsidence caused by compaction of deposits due to surface wetting was omitted. This subject was treated extensively by Ben E. Lofgren also in *Reviews in Engineering Geology, Volume II* (see pages 271–303).

ACKNOWLEDGMENT

Donald C. Helm provided valuable assistance during both the preparation of the symposium on which this volume is based and the organization of the activities to honor Joseph Poland in conjunction with the symposium. Foremost of these activities was the compilation of Poland's professional papers into a doctoral dissertation. During preparation of the symposium, Helm learned that Poland had completed in the 1930s all of his requirements at Stanford University for a Ph.D. degree except for the dissertation. Helm spearheaded the effort that resulted in the conferring of a Ph.D on Poland at the June 1981 graduation ceremony at Stanford University. Poland's graduate career is the longest one of record known to either Helm or me.

Thomas L. Holzer
Reston, Virginia

Field-based computational techniques for predicting subsidence due to fluid withdrawal

Donald C. Helm*
Lawrence Livermore National Laboratory
University of California
Livermore, California 94550

ABSTRACT

Choice of a predictive technique for land subsidence is based on the availability of appropriate field data. If only the depth and thickness of compressible beds can be estimated, a simple hand calculation is available as a predictive technique and for many purposes is adequate. An example of such a technique uses Schatz, Kasameyer, and Cheney's depth-porosity model. Their depth-porosity model for reservoir compaction is modified in the present paper and is combined with a modified form of Geertsma's nucleus of strain model to form a single predictive technique. The nucleus of strain model accounts for attenuation of vertical movement through the overburden. Based on this combined model, predictions of ultimate vertical subsidence are made at eleven specified locations. At five of these sites in California, Texas, and New Zealand, these predictions are compared to measured subsidence. Agreement between predicted and observed subsidence is good.

If field measurements of water-level fluctuations and the resulting time-dependent compression and expansion of geologic strata are locally available, use of a more refined predictive technique is justified. An example of such a technique uses Tolman and Poland's aquitard-drainage model in conjunction with a field-based method of parameter evaluation (such as Riley's field-based method of estimating specific storage and vertical hydraulic conductivity). Tolman and Poland's conceptual model and Riley's parameter evaluation method have been combined with Helm's one-dimensional finite difference computational scheme to form a powerful time-dependent predictive technique. This technique has been field verified successfully at more than two dozen sites in California and Texas. Use of such a computer code allows prediction of observed time lags in nonlinear compression and expansion of layered sedimentary materials in response to arbitrary sequences of rising and falling water levels.

INTRODUCTION

Several techniques are available for predicting subsidence due to fluid withdrawal. They have been classed by Poland (1984) into three broad categories: empirical, semitheoretical, and theoretical. Empirical methods essentially plot past subsidence versus time and extrapolate into the future based on a selected curve fitting technique. Semitheoretical methods link ongoing induced subsidence to some other measurable phenomenon in the field. Theoretical techniques traditionally use the results of laboratory tests in order to predict subsidence. Essentially, however, theoretical techniques use equations derived from fundamental laws of physics, such as mass balance. Some of the semitheoretical and theoretical methods that are mentioned in Poland's (1984) report are briefly reviewed in Appendix 1.

Two field-based techniques for subsidence prediction are the primary subject of the present paper. Owing to their utility, they

---
*Now with the Commonwealth Scientific and Industrial Research Organization, Division of Geomechanics, Mount Waverly, Victoria 3149, Australia.

deserve special emphasis. These two techniques are called field-based to emphasize their use of data collected in the field rather than in the laboratory. A discussion of the role the undrained overburden plays in predicting subsidence is included. The methods discussed here are not necessarily limited to shallow groundwater systems. They can be applied to subsidence associated with deeper reservoirs, whether the fluid being withdrawn is water, brine, steam, gas, or petroleum.

FIELD-BASED PREDICTIVE TECHNIQUES

The first technique that will be discussed in detail uses a depth-porosity model. This model leads to a semitheoretical method that, unlike others of its type, does not require induced subsidence to have already begun. This is its power; it can give a useful estimate of potential ultimate subsidence at a new undeveloped area where fluid has not yet been withdrawn. It can most effectively be used to estimate potential long-term subsidence. The second technique that will be discussed uses an aquitard-drainage model. This technique is theoretical and, unlike many of its type, does not require the results of laboratory tests. In contrast to the depth-porosity model, this second technique requires compressible beds to be stressed in the field and induced compaction to have already begun. It is a sophisticated model that accounts for nonlinear, nonequilibrium transient effects. It can most effectively be used during the early development of an underground fluid resource. It predicts future transient compression and expansion with precision. What both techniques share and what distinguishes them from most other techniques is that they yield to the user appropriate values for the very parameters required to make subsidence predictions. They presuppose that in situ measurements constitute a type of field laboratory. For this reason we call them field-based techniques for predicting subsidence.

The term "subsidence" pertains to movement of the land surface. This contrasts to the term "compaction" which is used by geologists to denote a decrease in thickness of sediments due to increase in vertical effective stress. The term "consolidation" is used by civil engineers to denote an identical mechanism in the laboratory. Subsidence due to fluid withdrawal is the surface expression of compaction of sediments at depth. Between the land surface and the zone of compaction is material called overburden. Strictly speaking, the depth-porosity and the aquitard-drainage models predict compaction rather than subsidence due to fluid withdrawal.

Overburden material influences land surface movement within an induced subsidence bowl in a complex way. At present the mechanics of movement within the undrained overburden is not sufficiently well understood to warrant inclusion of prediction of its behavior with any compaction model except the less sophisticated depth-pososity model. Later in this paper I shall combine the simpler depth-porosity model appropriately with a model for overburden behavior (namely, a modified nucleus of strain model) in order to estimate maximum ultimate subsidence over the center of a subsidence bowl. The resulting values will be compared to observed subsidence at a number of selected sites.

DEPTH-POROSITY MODEL

The parameter found by using the depth-porosity model is a generalized depth-dependent coefficient of volume change, m_v. This coefficient is used by soil engineers to relate total change in volume ΔV of a column of soil of volume V in response to a small incremental change of effective stress $\Delta p'$, namely,

$$\Delta V = -V\, m_v\, \Delta p', \qquad (1)$$

where ΔV is positive for expansion. The parameter m_v and the skeletal component of the hydrogeologist's nonrecoverable specific storage S_{skv} are related simply through a conversion factor, namely,

$$S_{skv} = \rho_w g\, m_v, \qquad (2)$$

where ρ_w is density of water and g is gravitational acceleration. The distinction between recoverable (elastic) and nonrecoverable specific storage is discussed in detail by Helm (1975, 1976). In many cases the nonrecoverable term is one to two orders of magnitude larger than the recoverable (Ireland and others, 1982).

The soil engineer's coefficient of compressibility a_v is defined by

$$a_v = -\Delta e / \Delta p', \qquad (3)$$

where e ($= V_v/V_s$) is void ratio and the volume of solids V_s is required to remain constant. The parameter m_v is related to a_v through the expression

$$m_v = a_v V_s / V. \qquad (4)$$

V_s is the volume of solids, and V_v is the volume of voids within a specified bulk volume V, namely,

$$V = V_s + V_v. \qquad (5)$$

The material of interest is required to be permeable. Fluid is thereby free to be "squeezed" from the porous system in response to an induced increase in effective stress. This is similar to specifying that we are interested only in interconnected pore spaces. It is assumed by soil engineers that the structure of the skeletal matrix is much more compressible than individual solids or interstitial fluid. For most saturated and permeable sedimentary deposits within 1,000 m of land surface, this can be considered to be a reasonable assumption (see Scott, 1963, p. 180). It should be pointed out that no time lag (of compressional response to stress change) is expressed in equations 1, 2, or 3. These equations refer only to equilibrium states. Hence, the interrelated parameters m_v,

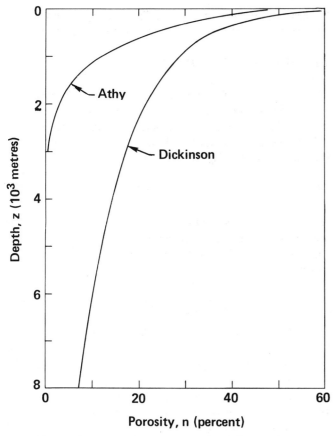

Figure 1. Two examples of decrease of porosity with depth.

a_v, and S_{skv} control the long-term or "ultimate" response of a finite column of material to a specified change in effective stress at a boundary of the column. The very real and observable phenomenon of time lag is discussed later.

Porosity of sedimentary materials is known in general to decrease with depth (Fig. 1). On the basis of empirical data for shale and mudstone, Athy (1930) and Magara (1978) suggested that a relation between porosity and depth can be found from an exponential expression

$$n = n_0 e^{-cz} \qquad (6)$$

under conditions of compaction equilibrium where n (= V_v/V) is porosity at a specified depth z, n_0 is an extrapolation of n to land surface (z = 0), and c is an empirically determined constant. A possible theoretical basis for equation 6 is explored by Schatz (1976). For Athy's data on shale porosities from Oklahoma, n_0 = 0.48 and c = 0.0014 when z is expressed in metres. Schatz and others (1978) suggested using equation 6 with site-specific values of n_0 and c as a method of finding an approximate value of porosity-dependent S_{skv} for compressible sedimentary material including shale, mudstone, sandstone, and clay. They argued that decreasing fluid pressure due to producing an artesian aquifer system has an equivalent long-term effect on porosity that increasing overburden load would have. This latter mechanism is equivalent to lowering a bed to a greater depth in Figure 1. Hence, the curves in Figure 1 are treated by them to represent an in situ stress/strain relation for ultimate equilibrium compaction.

Not all empirical depth-porosity curves follow an exponential relation as expressed by equation 6. A notable exception (Magara, 1978, p. 93) is Dickinson's (1953) data for shale porosities from the Gulf of Mexico coast (Fig. 1). It is found that Dickinson's shale porosities follow a logarithmic relation with depth, namely,

$$\Delta n = - a\, \Delta \ln z, \qquad (7)$$

which implies

$$n = n_{ref} - a \ln (z/z_{ref}), z > 0 \qquad (8)$$

where n_{ref} and z_{ref} are reference values for porosity and depth and a is an empirical constant. If a is interpreted as a type of compression index, equations 7 and 8 approximate the type of stress-strain relation one would anticipate from standard soil mechanics interpretation of laboratory stress/strain data. For Dickinson's field data from the Gulf Coast, a = 0.103 and n_{ref} = 0.05 for an arbitrary reference depth z_{ref} of 10^4 m.

It is now possible to get two depth-dependent theoretical values of S_{skv} based on equations 6 and 8. Under lithostatic conditions, effective stress p' due to submerged weight of overlying material changes with depth z by the gradient

$$\frac{dp'}{dz} = (1-n)(G-1)\rho_w g = \frac{G-1}{1+e}\rho_w g \qquad (9)$$

where G is the specific gravity of solids. Equation 9 is directly applicable to normally pressured strata. The phrase "normally pressured" signifies both that the interstitial fluid pressure within a representative elemental volume is hydrostatic and that the effective stress has never been greater in the past than it is at present.

We shall now discuss the applicability of equation 9 to describe stress gradients within abnormally pressured strata that are found at depth at many locations, notably along the Texas Gulf Coast. Permeable beds with high fluid pressures are called artesian or confined aquifers and in extreme cases are called geopressured or abnormally pressured strata. The semipervious overlying zone is called a caprock or a confining bed. Fluid pressures increase steeply with depth within these semipervious zones. Within the underlying geopressured or artesian zone itself, however, equation 9 can be considered a useful approximation because the local downward gradient of fluid pressure is much less than it is within the overlying confining bed. Although the absolute value of effective stress within a geopressured zone is much reduced, its local vertical gradient across a horizontal plane at depth z can be expressed roughly by equation 9. We shall therefore apply it as a first approximation to interior zones within both normally and abnormally stressed reservoirs, but not within a confining bed.

In accordance with equations 2, 3, and 4, we write

$$S_{skv} = -\frac{V_s \Delta e}{V \Delta p'} \rho_w g = -\frac{1}{1+e}\frac{de}{dp'}\rho_w g$$
$$= -(1-n)\frac{de}{dp'}\rho_w g. \quad (10)$$

Equation 10 can be written

$$S_{skv} = -(1-n)\frac{de}{dn}\frac{dn}{dz}\frac{dz}{dp'}\rho_w g$$
$$= -\frac{1}{1-n}\frac{dn}{dz}\frac{dz}{dp'}\rho_w g, \quad (11)$$

which, in accordance with equation 9, becomes

$$S_{skv} = -\frac{1}{(1-n)^2 (G-1)}\frac{dn}{dz}. \quad (12)$$

It now becomes necessary to determine dn/dz from Figure 1. For an exponential relation of porosity to depth, equation 12 becomes

$$S_{skv} = \frac{c}{(G-1)}\frac{n}{(1-n)^2} \quad (13)$$

where we have used equation 6. For Dickinson's curve, however, equation 12 becomes

$$S_{skv} = \frac{a}{(G-1)(1-n)^2 z} \quad (14)$$

where we have used equation 8.

Figure 2 shows the relation of S_{skv} and depth z in accordance with Athy's data (dashed line) and Dickinson's data (solid line). The dashed line in Figure 2 was found by substituting equation 6 into the right-hand side of equation 13 and assuming that $n_0 = 0.48$ and $c = 0.0014$ m^{-1}. The solid line in Figure 2 was similarly found by substituting equation 8 into the right-hand side of equation 14 for $n_{ref} = 0.05$, $z_{ref} = 10^4$m, and $a = 0.103$.

Symbols in Figure 2 indicate a selection of S_{skv} values determined by other methods. The X's in the upper right-hand corner of Figure 2 represent values of S_{skv} calculated from results (Marsal and Graue, 1969, Table 5, p. 190) of standard laboratory consolidation tests on lake-bed soil samples near Mexico City. This material is known to be highly porous and unusually compressible. The symbols SC, T, and W near some X's represent values of S_{skv} calculated from results of standard laboratory consolidation tests on soil samples taken respectively from the Santa Clara Valley, California (Poland, 1978, written commun), from near Seabrook, Texas (Gabrysch and Bonnet, 1976, Table 3), and a composite of values from the Wilmington oil field, California (Allen and Mayuga, 1969). Subsidence is known to have occurred at all of these locations. The circles represent values of S_{skv} determined from simulating observed compaction and expansion in California by means of a digital computer code (Helm, 1977, 1978). This computer simulation technique uses the aquitard-drainage model and is discussed in the next section.

For many areas it is evident from Figure 2 that within 2000 m of land surface, the curve based on Dickinson's data gives a somewhat high, but reasonable, estimate of S_{skv}. The solid curve in Figure 2 can therefore be easily used as a first approximation of S_{skv}. For example, assume that a confined aquifer has not been developed and hence no field-based compaction records are available. Suppose one knows that the areally extensive aquifer system lies at a depth between roughly 100 and 300 m. Within this 200-m interval there is found to exist about 100 m of fine-grained, compressible interbeds. Hence, the thickness of compressible beds b' is about 100 m, the average depth is about 200

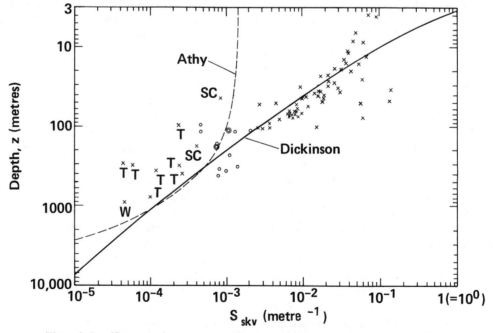

Figure 2. Specific storage for nonrecoverable compaction S_{skv}, as a function of depth z.

m, and in accordance with Dickinson's curve in Figure 2 we can estimate S_{skv} to approximate $10^{-3} m^{-1}$. Using equations 1 and 2, we find

$$\Delta b' / \Delta h \simeq S_{skv} b' = 10^{-3} \times 10^2 = 10^{-1} \quad (15)$$

where for convenience we assume

$$(\Delta b')/b' \simeq -(\Delta V)/V \quad (15a)$$

$$\Delta p' \simeq \rho_w g \Delta h. \quad (15b)$$

Equation 15 tells us, as a first approximation, that for every metre of long-term drawdown Δh, one can expect about 10 cm of ultimate compaction $\Delta b'$. For convenience we use a convention in which $\Delta b'$ is positive for compression, $\Delta p'$ is positive for increase in effective stress, Δp is positive for increase in fluid pressure, and Δh is positive for drawdown (decrease in fluid pressure).

Equation 15 and its use with Dickinson's curve in Figure 2 tacitly require that sedimentary material within a confined aquifer system initially is normally consolidated. By normally consolidated we mean effective stress within a specified bed has never been greater in the past. Holzer (1981) discussed field evidence which indicates that for shallow sediments (namely, z < 300 m) the water levels at several locations had to be drawn down a few metres to a few tens of metres below their initial levels in order for local effective stress values to reach a preexisting maximum (preconsolidation) value. This in situ preconsolidation stress had occurred before people induced any stresses on the system. According to Holzer's analysis, one must be careful in assuming that effective stress has never been greater in the geologic past than it is at present. Within Australia's Latrobe Valley, it was estimated (Helm, 1984) that the in situ preconsolidation stress could not be reached at many localities by the process of lowering potentiometric heads. For the purpose of making a preliminary rough estimate of maximum ultimate subsidence, neglecting geologic preconsolidation would tend to give a high estimate for predicted future subsidence. At locations where natural preconsolidation stresses are unknown or unrecorded, we are forced to make unadjusted use of Figure 2.

Even at a site that actually is normally consolidated, equation 15a states an additional maximizing assumption, namely, that volume strain expresses itself entirely in vertical compression. Note, by way of contrast, that for a confined aquifer system whose volume strain is isotropic, the vertical component of strain is one-third the volume strain. Whenever in situ strain is actually isotropic, use of equation 15 would thereby automatically give an estimate of vertical compression three times too large. Values for S_{skv} in Figure 2 inherently pertain to volume strain, not necessarily to the vertical component.

Equation 15 applies only to situations in hydrodynamic equilibrium. It takes time to reach a new equilibrium state. Skeletal deformation must also be interpreted within a time frame. One

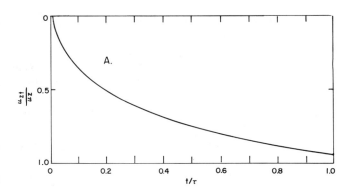

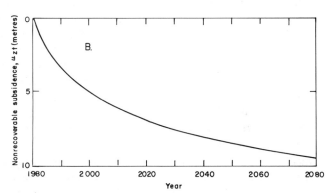

Figure 3. Subsidence versus time. A. Idealized case in accordance with soil mechanics time-consolidation theory. B. Application to a hypothetical site (see equation 15) assuming: (1) u_z = 10 m in response to a sustained constant drawdown of 100 m of water, and (2) τ = 100 years.

would expect a time lag of decades (Helm, 1978) before compaction in the field would reach its ultimate equilibrium value (equation 15) in response to a constant long-term change in stress. Time lag due to hydraulic flow is discussed in the next section.

Methods for approximating a regional distribution of long-term drawdown Δh, which appears in the left-hand side of equation 15, are available (for example, Figueroa-Vega, 1971). Discussion of these methods is beyond the scope of this paper.

TIME LAG

It must be recognized that ultimate subsidence does not occur immediately. Figure 3A indicates at the center of a subsidence bowl (r = 0, z = 0) the relation of transient subsidence u_{zt} to ultimate subsidence u_z as a function of dimensionless time (t/τ) where t is time and τ is a time constant for primary strain (Taylor, 1948). Time constant τ is a function of compressibility or specific storage S_{skv}, a resistance 1/K to fluid flow through interconnected pores, and the square of a characteristic length H. According to time-consolidation theory, τ represents the time it takes saturated permeable and compressible material to reach 93% of ultimate compaction. Figure 3A is an idealized curve based on soil mechanics time-consolidation theory (Taylor, 1948) and requires that an induced stress Δh is applied at an initial instant (t/τ

= 0) and then remains constant. Secondary strain, such as gradual creep, also occurs. For the purposes of this paper, we limit our discussion to primary strain as described above.

Figure 3B demonstrates a hypothetical application of Figure 3A to a site where 10 m of ultimate subsidence is estimated to occur in response to a sustained drawdown Δh of 100 m of water (see equation 15). A time constant τ of 100 years is also assumed. This is a reasonable estimate for τ at some sites. For example, by applying the aquitard-drainage model to observed field data of transient stress and compaction, time constants for confined aquifer systems in the San Joaquin and Santa Clara Valleys, California, were calculated (Helm, 1978) to equal several decades. It must be emphasized that use of Figure 3A gives a gross approximation. After a few years of carefully collected field data of stress and strain at any site, much improved estimates of S_{skv} and τ can be found for that site by using the aquitard-drainage model. This in turn vastly improves predicted values of future subsidence.

Figure 3B is not meant to imply that once a change in fluid pressure is introduced, nonrecoverable compaction proceeds inevitably to an "ultimate" value. Quite the contrary is true. Induced drawdown must be maintained indefinitely in order for an "ultimate" value of compaction or subsidence eventually to occur. If drawdown Δh at any time recovers to above a transient critical value, any further nonrecoverable compaction and subsidence will cease. Predicting this time-dependent critical elevation of head is possible through use of the aquitard-drainage model (Ireland and others, 1982).

AQUITARD–DRAINAGE MODEL

Tolman and Poland (1940) suggested that subsidence in the Santa Clara Valley, California, is caused not simply by declining artesian heads and the resulting compaction of permeable sands, but primarily by the nonrecoverable compaction of slow-draining clay layers within the confined system. This marks the conceptual birth of the aquitard-drainage model.

By linking Terzaghi's (1925) theory of one-dimensional consolidation quantitatively to the aquitard-drainage model in accordance with Tolman and Poland's suggestion, Riley (1969) initiated a field method of parameter evaluation. Helm (1972, 1975, 1976) used Riley's insights and developed a one-dimensional computer code that simulates time-delayed aquitard compression and expansion in response to arbitrary fluctuations of observed hydraulic head. By matching carefully measured vertical strain and developing a parameter calibration technique, Helm (1977, 1978) carried Riley's field method of parameter evaluation a step further. In turn, Freeze (Witherspoon and Freeze, 1972; Gambolati and Freeze, 1973), Narasimhan (Narasimhan and Witherspoon, 1977), Swain (1981, personal commun.), Pollard (Harris Galveston Coastal Subsidence District, 1982), and Helm (1984) have expanded upon the computational side of Helm's approach by borrowing his one-dimensional method for calculating recoverable and nonrecoverable aquifer-system compaction and expansion and combining it with regional fluid-flow models.

Using Tolman and Poland's aquitard-drainage model to estimate in situ parameter values by means of a one-dimensional computer code has led directly to a powerful field-based predictive technique (Helm, 1978) for land subsidence caused by water-level fluctuations within a confined aquifer system. The aquitard-drainage model requires knowing not only the mathematical "ultimate" response to an increment of stress but also the time-rate of compression. The parameters found by using the aquitard-drainage model are site-specific average values of specific storage S_{skv} and the vertical component of hydraulic conductivity K. These parameters control the time-delayed response to stress of compressible interbeds or clay lenses within a confined saturated system. For such a system, time lag is caused by slow drainage as water is "squeezed out" of compressible interbeds. The time constant τ is evaluated to equal

$$\tau = S_{skv} (H/2)^2/K \qquad (16)$$

where H is the thickness of a doubly draining interbed.

In the aquitard-drainage model (Fig. 4) the confined aquifer system contains two basic types of porous material: (1) a group of fine-grained interbeds, each of which is completely surrounded by (2) a hydraulically connected system of coarse-grained material. The fine-grained interbeds (aquitards) are considered much less permeable than the interconnected coarse-grained portion of the confined aquifer system. The term "aquitard" is restricted in this paper to refer only to slow-draining inclusions (such as clay

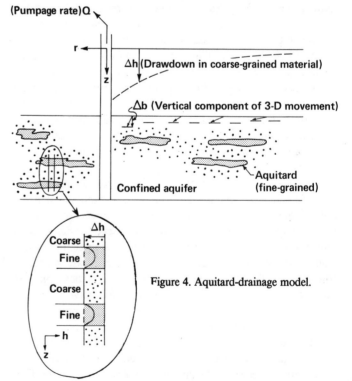

Figure 4. Aquitard-drainage model.

lenses) within an aquifer. Such inclusions are not of regional extent. They are conceptually distinct from slow-draining beds that are regionally extensive, and serve as a confined aquifer system's upper or lower boundary. In other words, we are not referring to a caprock, confining bed, or semi-confining bed. The aquitard-drainage model conceptually attributes the time delay between stress change in the aquifer and compaction in the aquitard to the vertical component of fluid flow from one idealized material (aquitard) to another (aquifer) within the two-material system itself. The difference in compressibility between the two idealized materials is assumed to be similar to the difference in structural compressibility between clay and sand, namely, one to three orders of magnitude (Scott 1963, p. 180). The slow vertical drainage from highly compressible aquitards to the less compressible aquifer material serves a rheological role in this model that is somewhat similar to a viscous "dashpot" in the viscoelastic reservoir model of Corapcioglu and Brutsaert (1977). This latter model is innovative and treats the entire complex aquifer system as a single idealized undifferentiated material. The viscoelastic reservoir model is discussed in Appendix 1.

On the basis of appropriate site specific field data, the aquitard-drainage model predicts (1) "residual" nonrecoverable compaction within the confined system, (2) time-dependent in situ preconsolidation pressure, and (3) a time constant τ for the confined system that is being stressed.

Residual compaction is the part of calculated ultimate compaction that at a specified time has not yet occurred. It can be expected to occur only if the system is not unloaded. Preconsolidation pressure is essentially the maximum effective stress that a volume element has experienced in the past. For a sequence of strata in the field this definition can be translated to represent a critical depth to water at which nonrecoverable compaction is stopped during the unloading phase (rising water levels) and is triggered during the reloading phase (declining water levels) of a specified unloading-reloading cycle. In an area of active fluid production this critical depth is usually shallower than the maximum observed depth to water.

Because the term "H" in equation 16 represents the thickness of an individual doubly draining interbed or an average thickness of a series of such beds, it contrasts to the term "b'" of equation 15 which represents the cumulative thickness of several compressible interbeds within a confined aquifer system. Ultimate compaction is a function of specific storage S_{skv} and total cumulative thickness b' in equation 15. Time lag, however, is a function not only of S_{skv} but also vertical hydraulic conductivity K and a weighted average thickness H of a series of compressible beds. Let us illustrate this distinction. Consider 100 m of compressible interbeds that consist of ten beds of 10 m each. Now consider 100 m of the same material composed of twenty beds of 5 m each. Note from equation 15 that ultimate compaction $\Delta b'$ would be identical in both cases for an identical stress. However, according to equation 16, the time constant in the second case would be one-quarter of what it would be in the first case. Helm (1975, 1976, 1977, 1978) discussed in detail the inverse problem

of evaluating S_{skv} and K that is associated with this distinction between b' and H. Residual nonrecoverable compaction (see item (1) above) is associated with b', whereas time constant τ (see item (3) above) is associated with H.

It is appropriate here to distinguish in passing between the cumulative thickness b' of compressible interbeds and the total thickness b of the entire aquifer system that contains these interbeds, such that $b \geq b'$. For the cruder models, namely, the depth-porosity model and the nucleus of strain model, it is generally assumed that b' equals b and that a reservoir contains only one material. For the aquitard-drainage model, this single-material assumption need not be made.

Examples of use of the aquitard-drainage model will now be mentioned. According to my earlier definition, residual compaction is the amount of nonrecoverable compaction or subsidence one can expect to occur in the future in addition to what has already occurred if water levels are held constant at their previous maximum depths. By simulating field compaction and expansion at 8 sites in the Santa Clara Valley and 7 sites in the San Joaquin Valley, California, Helm (1978, Table 2) estimated that in 1978 residual compaction in the Santa Clara Valley ranged from a minimum of 0.52 m at one site to a maximum of 2.53 m at another. Among the analyzed sets of field data collected in the San Joaquin Valley, calculated residual compaction ranged from a minimum of 0.85 m at one site to a maximum of 9.85 m at another. In the early 1970s the critical depths to water at which nonrecoverable compaction or subsidence would cease were calculated site by site. They ranged from a few metres above observed past maximum depths to water at some sites to several tens of metres above maximum depths at other sites. Time constants also were estimated from site-specific field data. In the Santa Clara Valley, τ was calculated to range from a minimum of 13 years at one site to a maximum of 125 years at another. In the San Joaquin Valley, τ was calculated to range from 5 years at one site to 1350 years at another. For large time constants, one needs longer calibration periods in order to reach a high degree of confidence in estimated parameter values.

Figures 5 and 6 illustrate the use of a one-dimensional computer simulation based on the aquitard-drainage model. Twelve years of detailed continuous measurements between 1959 and 1972 were available. Using the measured stress curve shown in the upper graph of Figure 5 as known input values, parameter values within the mathematical model were calibrated in order to make calculated compaction (dotted line in the lower graph of Figure 5) be as close an approximation to observed compaction (solid line in the lower graph of Figure 5) as possible. The value τ was found from the final estimated parameter values to equal 96 years. Using these values and the field-based input stress curve shown in the upper graph of Figure 6, a "predicted" compaction curve was calculated for the period 1921 to 1974 when cruder measurements were made. This prediction is shown by the solid line in the lower graph of Figure 6. The brackets on the calculated curve in Figure 6 indicate the time interval used for calibration (Fig. 5). Observed compaction values are based on measured

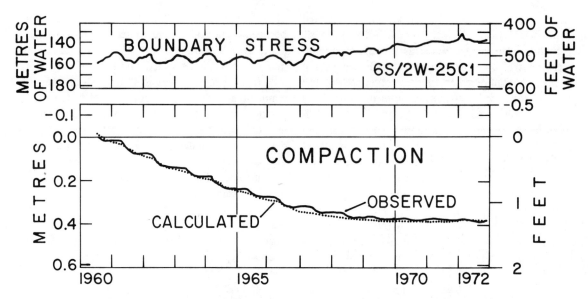

Figure 5. Simulation of compaction based on water-level data for well 6S/2W-25C1 (1960–72) and compaction data observed in well 6S/2W-24C3.

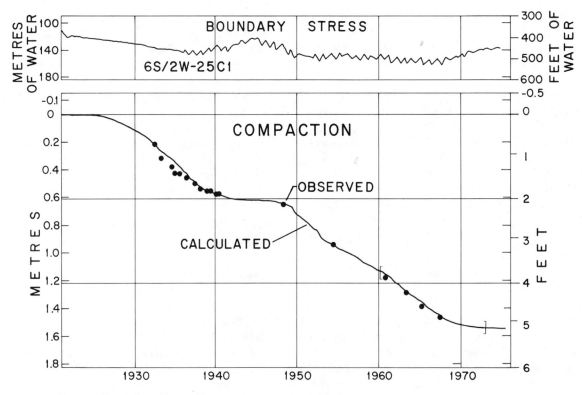

Figure 6. Simulation of compaction based on water-level data for well 6S/2W-25C1 (1921–75) and a portion of subsidence measured at bench mark J111.

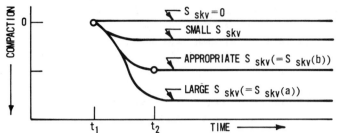

Figure 7. Hypothetical computer runs using a large value for K [= K(a)] to simulate a known compaction value.

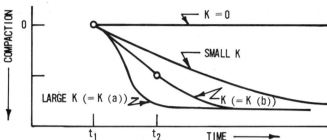

Figure 8. Hypothetical computer runs using a large value for S_{skv} [= $S_{skv}(a)$] to simulate a known compaction value.

subsidence at bench mark JIII and are plotted as solid circles in the lower graph of Figure 6. The excellent agreement between predicted and observed compaction in Figure 6 confirms the parameter values found during the calibration process (Fig. 5). This result increases one's confidence in the site-specific residual compaction, preconsolidation stress, and time constants that are estimated from this field-based computational procedure.

One can generate confidence levels in one'c choice of parameter values by defining an error surface ϵ [= $\epsilon(K, S_{skv})$] where ϵ is defined by [observed compaction (t) – calculated compaction (K, S_{skv}, t)] and then minimize ϵ by adjusting values of K and S_{skv}. A simplified trial-and-error calibration procedure is illustrated below. The first step in searching for appropriate parameter values is to select arbitrarily a value for hydraulic conductivity K and nonrecoverable specific storage S_{skv} for an initial computer run. Let us label this initial set $(K(a), S_{skv}(a))$. A computer compaction curve will be generated by the model in response to known or otherwise specified changes in boundary stress. If, for example, one finds that the calculated compaction curve gives a value for net cumulative compaction between two selected dates (t_1 and t_2) larger than the observed values, one can decrease K or S_{skv} or both for the next trial computer run(s).

For a few hypothetical subsequent runs let us keep K (= K(a)) constant and decrease only S_{skv}. For an extreme case, zero-valued S_{skv} (representing incompressible intergranular structure) will obviously give a calculated compaction curve smaller than the observed compaction. It is possible to find a value of S_{skv} (namely, $S_{skv}(b)$ in Fig. 7) for our initial arbitrary K(a) that will match the net observed compaction between t_1 and t_2. The two circles in Figures 7, 8, and 10 represent hypothetical observed values of compaction at $t = t_1$ and t_2.

Let us now keep the hypothetical initial value of S_{skv} (= $S_{skv}(a)$ constant and decreases K (Fig. 8) on subsequent computer runs. Eventually we shall have two possible pairs of parameters that adequately account for observed net compaction between t_1 and t_2, namely, $(K(a), S_{skv}(b))$ from Figure 7 and $(K(b), S_{skv}(a))$ from Figure 8. These two pairs of values represent two points on a "zero-error" curve (Fig. 9) between t_1 and t_2. The zero error curve represents an infinite number of combinations of K and S_{skv} that satisfy the hypothetical simulation criterion we posed of matching only two time-compaction data points.

When more observed time-compaction data are available than merely two values (circles in Figs. 7 and 8), we can eliminate many pairs of coefficients (e.g., those outside the bracketed segment of the zero-error curve in Fig. 9) because they do not give a reasonable approximation of the slope of an observed time-compaction curve between t_1 and t_2 (see the dashed line in Fig. 10). For sufficient field data, candidate values may be found to fall within a small area on the (K, S_{skv}) plane. This gives high confidence that the parameter values within this small area are valid and can be used successfully for long-term (namely, decades) predictions. Helm (1977) has discussed and illustrated the application of the "zero-error" curve method (Fig. 9) to three sets of actual field data from the Santa Clara Valley, California.

The aquitard-drainage model has been used to simulate field data from the San Joaquin and Santa Clara Valleys, California, and Houston-Galveston area, Texas. As mentioned before, the resulting parameter values (Helm, 1978, Table 1) yield time con-

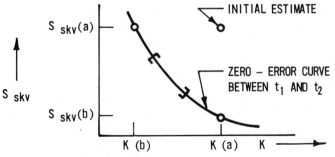

Figure 9. Hypothetical zero-error curve.

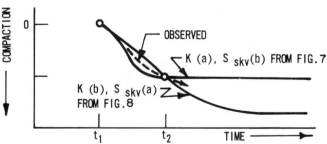

Figure 10. Hypothetical case when rate of compaction is observed.

stants in California that range from several years to centuries (Helm, 1978, Table 2). Comparable values for time constants can be calculated from Texas data (Pollard and others, 1979, Exhibits IV-3 through IV-6).

The aquitard-drainage model was initially used when time-dependent pressure changes were specified as boundary conditions on aquitards at a site of interest rather than when a regional discharge rate Q is specified. It has been successfully field tested and combined with a three-dimensional ground-water flow digital model to predict subsidence in the Houston, Texas, area (Harris Galveston Coastal Subsidence District, 1982) and in the Latrobe Valley, Australia (Helm, 1984). This allows Q to be specified in a two-step predictive technique.

Recently (Helm, 1984), the computer code has been extended to include the calculation of the compression and expansion of (1) semi-confining beds between two distinct aquifer systems as well as aquitards (clay lenses) within any single system (2); overconsolidated material as well as normally consolidated material; and (3) unconfined aquifer systems as well as confined aquifer systems. These extensions of the model have been successfully applied to five locations in the Latrobe Valley, Australia.

COMBINATION OF DEPTH-POROSITY AND NUCLEUS OF STRAIN MODELS

The emphasis of this paper turns now to developing a straightforward method of estimating the vertical component of ultimate subsidence. What this method may lack in precision, it makes up for in simplicity of application. This method consists of combining the depth-porosity model (Fig. 1) and a modification of the nucleus of strain model (Fig. 11), which will be described in the following section, and applying the resulting method to make initial estimates of ultimate subsidence at the center of possible future subsidence bowls.

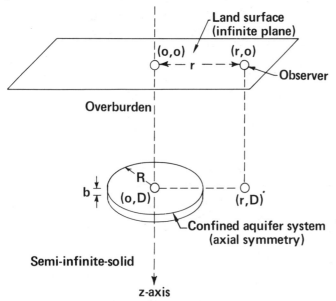

Figure 11. Nucleus of strain model.

A fundamental feature of the nucleus of strain model is that the role of the undrained (unpumped) overburden is taken into consideration. Incorporating this model into a predictive technique allows the compaction models of the previous sections to become subsidence models.

INFLUENCE OF MATERIAL WITHIN THE UNPUMPED OVERBURDEN

Subsidence due to fluid withdrawal is the expression at land surface of the compression at depth of a stressed artesian aquifer system. Material within the intervening unpumped overburden may possibly play a role in mitigating land surface effects. It is assumed that no fluid is withdrawn from the overburden. Geertsma (1957, 1973) has in effect discussed quantitatively the role of the overburden. His equation for ultimate vertical displacement u_z directly over the center of a disk-shaped reservoir with a uniform pressure drop (Fig. 11) is

$$u_z(0,0) = -2(1-\nu)c_m b \Delta p \left\{ 1 - \left[C/(1+C^2)^{1/2} \right] \right\} \quad (17)$$

where

- ν is Poisson's ratio;
- b is reservoir thickness (assumed to equal the cumulative thickness b' of compressible beds);
- Δp is pressure drop (assumed to equal $-\Delta p'$);
- C is D/R;
- D is depth to compressing bed(s); and
- R is radius of the hydraulically stressed system (reservoir).

For highly compressible elastic porous material, Geertsma's coefficient of uniaxial compaction, c_m, becomes

$$c_m = (1-2\nu)(1+\nu)/E(1-\nu) \quad (18)$$

where E is Young's modulus.

Equation 17 should be used with caution for the following reasons. If reservoir fluid is removed to a distant place, bedrock underlying a reservoir may move upward very slightly due to unloading. For a structurally compressible reservoir, such upward movement is small compared to the cumulative decrease of porosity within the reservoir. Gravitational body forces require the upper boundary of a compressing reservoir to move downward for the system to remain contiguous. Unfortunately, Geertsma neglects body forces in his analysis, with the unrealistic result that under some circumstances the base of his idealized reservoir mathematically moves upward a significant amount in response to a decrease of porosity within the overlying reservoir. This physically unlikely upward movement can mathematically nearly equal the total compaction of the stressed system. Hence, there may mathematically be negligible downward movement of the top of Geertsma's compressing reservoir. Correspondingly, under these mathematical circumstances there would mathematically be negligible subsidence at land surface.

Whenever the upper boundary of a compacting aquifer system moves downward in response to compression of underlying sediments, Geertsma (1973) pointed out that according to his model the volume through which this upper surface moves is preserved at land surface. When the vertical movement of the base of a confined aquifer system can be considered negligible, land subsidence, according to Geertsma, volumetrically equals the total loss in volume of the confined aquifer system. This condition, I assert, describes the general case. In a later section, equation 17 will be modified by requiring that due to weight of overlying material the base of a reservoir is constrained from moving upward significantly.

The areal distribution of subsidence is influenced by the ratio of depth D to radius R (Fig. 11) of the depressured confined aquifer system. The effect of compression of an aquifer system with a large D/R ratio may be spread over a large area at land surface and hence minimize the vertical component of volumetric subsidence. For most ground-water systems the D/R ratio is sufficiently small that the effect of spreading can be ignored. This implies the direct applicability of a depth-porosity model as described earlier. Later in this report I shall discuss systems with a large D/R ratio.

The conceptual model used by many investigators, including Geertsma, can be called a half-space model (Fig. 11). The earth is represented as a homogeneous, isotropic, semi-infinite elastic (or poro-elastic) medium. Land surface is represented as a flat upper surface that is free to move. Neither calculated surface movement nor topographic relief affects the essential flatness of the idealized surface. A depressed zone at depth D below land surface with a center at radial distance r from an observer on land surface is represented by an idealized spherical tension center by Carrillo (1949), by vertical pincers by McCann and Wilts (1951), and by a radially symmetric group of strain nuclei by Geertsma (1957). These various representations are conceptionally similar. Gambolati (1972) has discussed the major conceptual distinctions between the tension-center representation (Carrillo, 1949; McCann and Wilts, 1951) and the strain-nucleus representation (Mindlin and Cheng, 1950; Sen, 1950; Geertsma, 1957, 1966, 1973; Finol and Farouq Ali, 1975). Briefly, Carillo's tension center concept tacitly requires a reservoir to be infinitely compressible within an elastic half-space. Geertsma's nucleus of strain concept requires that the reservoir's compressibility equals the compressibility of the surrounding elastic half-space. Finite heterogeneity between the compressing aquifer system (reservoir) and the surrounding half-space was introduced to the model by Gambolati (1972).

HYDRAULIC HETEROGENEITY OF PERMEABLE MATERIAL

The question of finite heterogeneity deserves some comment. According to the half-space model, a confined aquifer system by definition yields fluid to a discharging well and experiences a pressure loss. The surrounding material, including the overburden, by definition does not yield fluid. In the process of yielding fluid, porosity within a compressible confined aquifer system decreases. The surrounding half-space does not experience any net decrease in porosity due to fluid loss. This distinction is analogous to the distinction in soil mechanics between drained and undrained compression (Lambe and Whitman, 1969, p. 423f). Decrease in porosity as a source for fluid discharge inherently introduces an extra component S_{skv} into a compressibility term S_s for a confined aquifer system. This extra component does not appear in a corresponding compressibility term for the undrained surrounding half-space.

To clarify this point, specific storage S_s for a pressure drop within a saturated confined system can be expressed in terms of three components,

$$S_s = S_{skv} + S_{sw} + S_{sg} \quad (19)$$

where S_{skv} pertains to decrease in porosity associated with fluid withdrawal and where S_{sw} and S_{sg} pertain to individual constituent properties. S_{sw} equals the product $n\rho_w g \beta_w$ where β_w is the compressibility of interstitial water. S_{sg} is correspondingly a coefficient that represents the expansion of individual solid grains.

Let us contrast the compression or expansion of a nondraining saturated half-space whose net porosity does not change significantly and the compression of a confined aquifer system that lies within the half-space. Only S_{sw} and S_{sg} of equation 19 are appropriate to use for the surrounding half-space. Their sum equals a bulk compressibility of an equivalent solid. The sum of S_{skv}, S_{sw}, and S_{sg}, however, is appropriate to use for estimating availability of fluid from the discharging aquifer system. Flow of underground fluid determines an apparent heterogeneity of compressibility between a compressing confined aquifer system and the surrounding nondraining half-space. Flow controls when and where an S_{skv} component (Fig. 2) should be invoked. This apparent material heterogeneity due to in situ hydraulic effects is distinct from standard differences in material properties that are tested and recorded in the laboratory. In the laboratory, reservoir and nonreservoir material may behave identically to similar stress and flow patterns. Apparent heterogeneity of permeable material that is due merely to fluid flow from reservoirs is of crucial importance to understanding and modeling in situ reservoir behavior. Some otherwise fine studies of subsidence models (for example, Miller and others, 1980a) fail to make this important distinction between reservoir and nonreservoir behavior. Hydraulic heterogeneity, as discussed above, should not be confused with the more standard concept of material heterogeneity, which dictates what value of S_{skv} should be used for a specified draining layer.

ATTENUATION OF VERTICAL DISPLACEMENT

In light of the foregoing series of discussions, I shall use the depth-porosity model in conjunction with equation 15 to estimate an ultimate vertical downward movement $\Delta b'$ of the upper boundary of an idealized compressible reservoir. Geertsma's model will be used to approximate the degree to which vertical displacement is attenuated between the upper boundary of an idealized reservoir and land surface.

Because of the existence of gravitational forces, I shall not use Geertsma's estimate of the absolute value of vertical displacement of the base of a reservoir ($z = D+b/2$). Geertsma's theory also neglects hydraulic heterogeneity. This is additional reason not to use his estimate of the absolute value of the vertical displacement of the top of a reservoir ($z = D-b/2$). These two criticisms, however, do not apply to estimating strain within the overburden from which no fluid is withdrawn. Geertsma's is currently the best and most commonly used method for calculating the difference between the vertical displacement of the land surface ($z = 0$) and the moving top of a compacting reservoir ($z = D-b/2$).

Equation 15 independently approximates the ultimate vertical displacement of the top of an idealized reservoir at $z = D-b/2$. Geertsma's theory will be modified somewhat in order to estimate an ultimate value of subsidence at the center of a subsidence bowl.

Equation 17 is an expression of a more general equation that was developed by Geertsma (1973). Assuming axially symmetric movement, Geertsma (1973, p. 49) found for $r = 0$, $0 < z < D$ that

$$u_z(0,z) = -c_m b \Delta p$$
$$\left\{ \frac{C(Z-1)}{2[1+C^2(Z-1)^2]^{1/2}} - \frac{(3-4\nu)C(Z+1)}{2[1+C^2(Z+1)^2]^{1/2}} \right.$$
$$\left. + \frac{CZ}{[1+C^2(Z+1)^2]^{3/2}} + 2(1-\nu) \right\} \quad (20)$$

where Z is z/D. At land surface ($z = 0$), equation 20 reduces to equation 17. Theoretical attenuation A of u_z through the overburden at $r = 0$ can be expressed directly from equation 20 as

$$A \equiv u_z(0, D-b/2) - u_z(0,0)$$
$$= -c_m b \Delta p \left\{ \frac{2C(1-\nu)}{(1+C^2)^{1/2}} - \frac{B}{2(1+B^2)^{1/2}} \right.$$
$$\left. - \frac{(2C-B)(3-4\nu)}{2[1+(2C-B)^2]^{1/2}} + \frac{C-B}{[1+(2C-B)^2]^{3/2}} \right\} \quad (21)$$

where B is $b/2R$. The modified estimate of subsidence $u_{z\,mod}(0,0)$ suggested in this paper is simply the difference between $\Delta b'$ of equation 15 and A of equation 21, namely,

$$u_{z\,mod}(0,0) = \Delta b' - A. \quad (22)$$

A coning factor, E, can now be defined as the ratio of land subsidence to reservoir compaction over the center of an idealized reservoir. Geertsma (1973, p. 49) pointed out that in all cases the compaction Δb at the center of his idealized reservoir equals $-c_m b \Delta p$. Our modification only requires that the base of the reservoir remains essentially fixed due to gravitational forces and, for computational convenience, that $\Delta b = \Delta b'$ of equation 15. With this in mind, we find from equations 21 and 22 that

$$E \equiv u_{z\,mod}(0,0)/\Delta b$$
$$= 1 - \frac{2C(1-\nu)}{(1+C^2)^{1/2}} + \frac{B}{2(1+B^2)^{1/2}}$$
$$+ \frac{(2C-B)(3-4\nu)}{2[1+(2C-B)^2]^{1/2}} - \frac{C-B}{[1+(2C-B)^2]^{3/2}} \quad (23)$$

Note from equation 23 that for a deeply buried reservoir ($C \to \infty$) whose radius is much larger than its thickness ($B \to 0$), the coning factor E approaches 0.5 as a limiting minimum value regardless of the value of Poisson's ratio ν.

Poisson's ratio as it appears in equation 23 pertains to strain within the overburden from which by definition no fluid is withdrawn. For undrained saturated material, Poisson's ratio is commonly considered to equal 0.5 (Lambe and Whitman, 1969, p. 455). Assuming this value for ν allows equation 23 to simplify to

$$E = 1 - \frac{C}{(1+C^2)^{1/2}} + \frac{B}{2(1+B^2)^{1/2}}$$
$$+ \frac{2C-B}{2[1+(2C-B)^2]^{1/2}} - \frac{C-B}{[1+(2C-B)^2]^{3/2}} \quad (23a)$$

Equation 23a can be solved to yield the coning factor E ($\equiv u_{z\,mod}(0,0)/\Delta b$) as a function of normalized depth C ($\equiv D/R$) and normalized half-thickness B ($\equiv b/2R$) of a hypothetical disc-shaped reservoir. Figure 12 shows the resulting theoretical attenuation of vertical displacement directly over the center of a subsidence bowl. This upward attenuation is assumed to be caused by the spreading of a constant volume of subsidence due to the presence of nondraining overburden material. Figure 12 indicates that there is less attenuation for near-surface and thicker reservoirs. Note that even if a reservoir lies at the land surface with no intervening overburden, the depth D of the midplane of the reservoir cannot be less than the half-thickness b/2 of the reservoir itself. This physical constraint requires that $B \leq C$. Because Geertsma's theory was developed for a disc-shaped reservoir, there exist limits to reservoir thickness b beyond which the theory is not intended to be applicable. In Figure 12 we have in effect required $0 \leq b \leq R$ which directly implies that $0 \leq B \leq 0.5$.

ILLUSTRATION OF PRELIMINARY PREDICTION OF INDUCED LAND SUBSIDENCE

It is now possible to estimate ultimate subsidence $u_{z\,mod}(0,0)$ at sites where reasonable guesses can be made of reservoir depth D, radius R, thickness b, and long-term change in stress Δh. The previously discussed modification to the depth-porosity and half-space models can be combined to yield the relation

$$u_{mod}(0,0)/\Delta h = E S_{skv} b \quad (24)$$

where equation 15 and the left-hand identity of equation 23 have

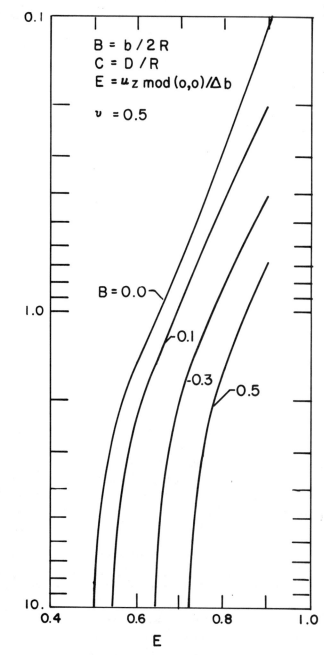

Figure 12. Coning factor E over the center of a subsidence bowl versus normalized depth C and normalized half-thickness B of a disc-shaped reservoir in accordance with equation 23a.

TABLE 1. SITE SPECIFIC RESERVOIR ASSUMPTIONS

Site	Fluid (with assoc. water)	Area (km^2)	Thickness, b (km)	Depth, D (km)
Wilmington	Oil	29	0.34	0.91
Baldwin Hills	Oil	8	0.15	0.38
Chocolate Bayou	Oil and Gas	90	0.2	3.5
Salton Sea	Hydrothermal	54	2.0	2.1
Brawley	Hydrothermal	27	1.5	2.4
Heber	Hydrothermal	42	2.0	--
East Mesa	Hydrothermal	28	1.1	2.3
Border	Hydrothermal	3	0.6	--
Geysers	Hydrothermal (steam)	--	2.0	--
Austin Bayou	Geopressured	10	0.9	4.5
Wairakei	Hydrothermal	15	0.6	0.75

which (Geysers and Wairakei) are currently being developed and have recorded a history of subsidence (Crow, 1979; Pritchett and others, 1978). Geysers is the only steam-dominated system. Others are liquid-dominated. Three sites (Wilmington, Baldwin Hills, and Chocolate Bayou) describe oil fields that also have a history of production-related subsidence. One site (Austin Bayou) describes a geopressured field. The listed assumption for the California hydrothermal sites are taken directly from U.S. Geological Survey Circulars 726 and 790 (Renner and others, 1976; Brook and others, 1979). Wairakei and Austin Bayou assumptions are essentially from U.S. Department of Energy reports (Pritchett and others, 1978; Miller and others, 1980b) and Mercer and others (1975). Wilmington, Baldwon Hills, and Chocolate Bayou data are from Allen and Mayuga (1969), Lee (1977), and Grimsrud and others (1978).

Depth information D from Table 1 is repeated in Table 2. Values of D in Table 2 that do not appear in Table 1 are my approximations. Instead of using equation 6 with site-specific depth-porosity data, as suggested by Schatz and others (1978), it is simpler for the purposes of this paper to use equations 6 and 8 with the classic data of Athy and Dickinson on Figures 1 and 2. Based on equation 14 (solid curve [Dickinson] in Fig. 2), estimates of nonrecoverable specific storage S_{skv} are listed in Table 2. Equation 13 (dashed curve [Athy] in Fig. 2) may give a more reasonable value for S_{skv} for normally consolidated reservoirs buried more deeply than 1000 m. For such reservoirs, in Table 2 the Athy values for S_{skv} is added in parentheses. An exception to this rule of thumb for deeply buried reservoirs is possibly a deeply buried geopressured reservoir, such as Austin Bayou. Such a reservoir may not be normally consolidated. This means it may be more porous and compressible than hydrostatic reservoirs at similar depths. Hence, for geopressured reservoirs, as well as for shallow reservoirs, it may be appropriate to use equation 14 (solid curve [Dickinson] in Fig. 2). It may be noted that Dickin-

been used. Similar to Geertsma, we make a rough approximation that reservoir thickness b equals the total thickness b' of compressible beds. In accordance with standard terminology (Poland and others, 1971), equation 24 depicts an estimated ultimate specific subsidence.

Table 1 lists assumed reservoir geometry at eleven selected locations. Eight of the first nine are located in California. The third and tenth are located in Texas, and the eleventh is in New Zealand. Seven sites describe hydrothermal reservoirs, two of

TABLE 2. SITE SPECIFIC CALCULATED PARAMETERS

Site	Depth, D (km)	S_{skv} (m^{-1})	Radius, R (km)	C	B	E
Wilmington	0.91	1.3×10^{-4}	3.0	0.3	0.06	0.82
Baldwin Hills	0.38	4.2×10^{-4}	1.6	0.2	0.05	0.87
Chocolate Bayou	3.5	2.4×10^{-5} (3.0×10^{-6})	5.3	0.66	0.02	0.72
Salton Sea	2.1	4.4×10^{-5} (2.2×10^{-5})	4.1	0.51	0.24	0.84
Brawley	2.4	3.7×10^{-5} (1.5×10^{-5})	2.9	0.83	0.26	0.78
Heber	2.0	4.7×10^{-5} (2.5×10^{-5})	3.7	0.54	0.27	0.84
East Mesa	2.3	4.0×10^{-5} (1.7×10^{-5})	3.0	0.77	0.18	0.76
Border	2.0	4.7×10^{-5} (2.5×10^{-5})	0.98	2.0	0.31	0.71
Geysers	3.0	2.9×10^{-5} (6.0×10^{-6})	5.6	0.54	0.18	0.80
Austin Bayou	4.5	1.8×10^{-5} (7.3×10^{-7})	1.8	2.5	0.25	0.66
Wairakei	0.75	1.7×10^{-4}	2.2	0.34	0.14	0.85

son's (1953) depth-porosity field data are taken from geopressured sites.

Table 2 indicates approximate values for radius R of idealized reservoirs along with calculated values for C ($\equiv$ D/R) and B ($\equiv$ b/2R). Subsituting these values into equation 23a gives an estimated value of E. The results are listed in Table 2.

Table 3 gives estimates of ultimate subsidence $u_{z\ mod}(0,0)$ at eleven selected sites due to compression of idealized reservoirs in response to an arbitrarily specified long-term drawdown Δh of 100 m (= 328 ft). These estimates are found by calculating ultimate specific subsidence, $u_{z\ mod}(0,0)/\Delta h$, in accordance with equation 24 and using values listed in Tables 2 and 3. The values in parentheses are based on Athy's curve; the other values are based on Dickinson's curve.

The reader is reminded that the specific subsidence values in Table 3 are calculated with the assumption (see equation 15a) that there is no horizontal component of volume strain within a reservoir. Wherever horizontal strain occurs in the field, one would expect an empirically based specific subsidence value to be smaller than the maximum value listed in Table 3. This is the reason for estimating a range of values for ultimate subsidence in Table 3 rather than estimating a single larger value. If actual volume strain within a reservoir is more nearly isotropic than entirely vertical, then the lower estimated value in Table 3 can be considered a more realistic approximation for the vertical component of ultimate subsidence.

At five sites, namely, Wilmington (Allen and Mayuga, 1969), Baldwin Hills (Lee, 1977), Chocolate Bayou (Grimsrud and others, 1978), the Geysers (Lofgren, 1978), and Wairakei (Pritchett and others, 1978, 1979), measurements have been made of induced subsidence. These examples furnish an opportunity to test predictions based on the depth-porosity model. Table 4 summarizes predicted ultimate subsidence at these five sites in response to specified pressure drops based on field observations of induced stress change. In making these predictions, isotropic strain was assumed. This assumption automatically makes the estimated values of ultimate specific subsidence in Table 4 one-third of their upper-bound values which appear in

TABLE 3. PRELIMINARY ESTIMATES OF ULTIMATE SUBSIDENCE AT ELEVEN SELECTED SITES

Site	Ultimate specific subsidence, $u_{z\ mod}(0,0)/\Delta h$	Range of ultimate subsidence for Δh = 100 m of water (m)
Wilmington	3.6×10^{-2}	$1.2 \rightarrow 3.6$
Baldwin Hills	5.5×10^{-2}	$1.8 \rightarrow 5.5$
Chocolate Bayou	3.5×10^{-3} (4.3×10^{-4})	$0.1 \rightarrow 0.4$ ($0.01 \rightarrow 0.04$)
Salton Sea	7.4×10^{-2} (3.7×10^{-2})	$2.5 \rightarrow 7.4$ ($1.2 \rightarrow 3.7$)
Brawley	4.3×10^{-2} (1.8×10^{-2})	$1.4 \rightarrow 4.3$ ($0.6 \rightarrow 1.8$)
Heber	7.9×10^{-2} (4.25×10^{-2})	$2.6 \rightarrow 7.9$ ($1.4 \rightarrow 4.2$)
East Mesa	3.3×10^{-2} (1.4×10^{-2})	$1.1 \rightarrow 3.3$ ($0.5 \rightarrow 1.4$)
Border	2.0×10^{-2} (1.1×10^{-2})	$0.7 \rightarrow 2.0$ ($0.4 \rightarrow 1.0$)
Geysers	4.6×10^{-2} (9.6×10^{-3})	$1.5 \rightarrow 4.6$ ($0.3 \rightarrow 1.0$)
Austin Bayou	1.1×10^{-2} (4.3×10^{-4})	$0.4 \rightarrow 1.1$ ($0.01 \rightarrow 0.04$)
Wairakei	8.7×10^{-2}	$2.9 \rightarrow 8.7$

TABLE 4. PREDICTION OF ULTIMATE SUBSIDENCE FOR ISOTROPIC STRAIN BASED ON THE DEPTH-POROSITY MODEL

Site	Ultimate specific subsidence		Stress, Δh (metres of water)		Subsidence (m) Predicted	Observed
Wilmington	1.2×10^{-2}	x	930	=	11.2	9 (1926-68)
Baldwin Hills	1.8×10^{-2}	x	150	=	2.7	1.5 (1925-63)
Chocolate Bayou	1.4×10^{-4}	x	7000	=	0.98	0.37 (1943-73)
Salton Sea	1.2×10^{-2}					
Brawley	5.9×10^{-3}					
Heber	1.4×10^{-2}					
East Mesa	4.7×10^{-3}					
Border	3.6×10^{-3}					
Geysers	3.2×10^{-3}	x	130	=	0.42	0.15 (1973-77)
Austin Bayou	3.6×10^{-3} (1.5×10^{-4})					
Wairakei	2.8×10^{-2}	x	246	=	6.9	4.5 (1964-74)

Table 3. Observed subsidence at these five sites is listed for comparison. Dates are added in parentheses in order to emphasize that observed subsidence has not necessarily reached its ultimate value. Stress changes are listed in equivalent changes in a hydraulic head. Dickinson's values for S_{skv} are used for shallow and geopressured reservoirs. Otherwise, Athy's values are used.

Observed subsidence values listed in Table 4 for the Chocolate Bayou site are the most uncertain. Actually, 0.55 m (1.8 ft) of total subsidence was observed at Chocolate Bayou between 1943 and 1973. Only 0.37 m (1.2 ft) of this total has been attributed to producing oil and gas (Brimsrud and others, 1978) from deep strata. The remainder is attributed to ground-water withdrawals from shallow aquifers at the same site.

Similar use of Tables 3 and 4 gives the following initial predictions of ultimate subsidence at the center of an idealized subsidence bowl in response to a hypothetical sustained pressure drop of 100 m of water: 1.2 m at Salton Sea, 0.6 m at Brawley, 1.4 m at Heber, 0.5 m at East Mesa, 0.4 m at Border, and between 0.02 and 0.4 at Austin Bayou. The estimated range at Austin Bayou reflects a state-of-the-art uncertainty as to anticipated compressibility values of geopressured reservoirs.

CONCLUSIONS

Equation 15 shows that in order to minimize compaction $\Delta b'$, one can most easily control drawdown Δh. The value of S_{skv} is given by nature. To estimate this value at a specific site is a primary task of prediction. It cannot be controlled by mitigating action. The value of b' cannot be easily controlled either, because it represents the thickness of beds that serve as a source of the fluid being withdrawn.

Two techniques for predicting subsidence have been described. One is the depth-porosity method, and the second is the aquitard-drainage method. Both have been field tested. Each is recommended to be used for different purposes. In order to approximate ultimate subsidence in a new area yet to be exploited, the depth-porosity method is appropriate. In order to estimate time-dependent aspects of compaction and partial expansion in response to complex scenarios of stress changes, the aquitard-drainage method is appropriate.

Discretion should be used in applying values of S_{skv} from the depth-porosity method (Fig. 2) to a site of interest as we have done in Tables 3 and 4. These values can be used most beneficially as a guide in preliminary planning and for giving initial estimates of possible ultimate behavior. Although the initial predictions listed in Table 4 are reasonable, they should be used only when site-specific field data are not available. It is advisable to collect and analyze carefully designed and measured field data at those sites that are considered potentially susceptible to subsidence. Collection of appropriate time-dependent field data of a stressed aquifer system is described by Lofgren (1970). Analysis of these data which are to be used with the aquitard-drainage prediction technique is described by Riley (1969) and Helm (1975, 1976).

If, for some reason, hypothetical worst-case preliminary estimates of ultimate subsidence at a site of interest are found from the depth-porosity technique to be unacceptable, then more realistic site-specific predictions based on the aquitard-drainage model can be made as a local follow-up. To use this model, one needs to start producing the reservoir. Field measurements of stress change and the resulting skeletal movement are of paramount importance for precise predictions of future land surface deformation at a site of interest. Precision and confidence in prediction increase greatly with appropriate and intelligently gathered field data used in conjunction with the aquitard-drainage model.

For pressures greater than an initial in situ preconsolidation value, about 90% of any subsidence that has actually occurred can generally be considered nonrecoverable. Inducing a stress on a reservoir does not necessarily commit the location to experiencing the entire mathematically "ultimate" subsidence (see Fig. 3). During an initial stressing phase, subsidence within a liquid-dominated system occurs slowly. During a stress recovery phase, however, subsidence ceases quickly. Whenever pressure drops are controllable, anticipated future subsidence can be stopped almost at will. Like drawdown, subsidence can be controlled.

ACKNOWLEDGMENT

I am indebted to Joe Poland not only for his research in subsidence but also for his friendship. I know of no kinder nor gentler man more dedicated to science.

Work was performed under the auspices of the U.S. Department of energy by the Lawrence Livermore National Laboratory under contract NO. W-7405-ENG-48.

APPENDIX 1. OTHER PREDICTIVE TECHNIQUES

The main body of this paper describes two predictive techniques that are called the aquitard-drainage and the depth-porosity techniques. This appendix is a review of some additional techniques.

Analysis of field measurements is a key to understanding physical processes in situ. The plotting of correlated field data is basic. the challenge is to discover which field data are causally connected.

Yamamoto (Poland, 1984) pointed out that Wadachi (1940) made a pioneering comparison (Fig. 13) of the observed rate of subsidence in Tokyo and the observed depth to water. This is an early example of a semitheoretical technique as defined in the Introduction to this paper. Wadachi's suggested relation based on Figure 13 is

$$du_z/dt = c\,(h_0 - h) \tag{25}$$

where c is an empirical constant and h_0 is a critical value of hydraulic head h where subsidence ceases. The concept of h_0 is

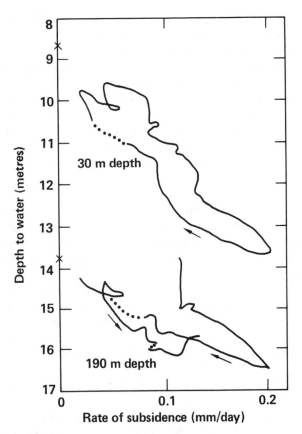

Figure 13. Wadachi's correlation of water levels and observed rates of subsidence, Osaka, Japan (from Poland, 1984).

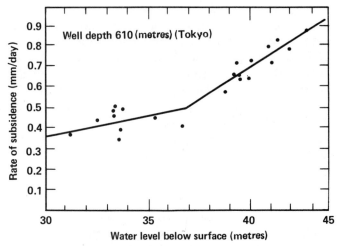

Figure 14. Yamamoto's correlation of water levels and observed rates of subsidence, Tokyo, Japan (from Poland, 1984).

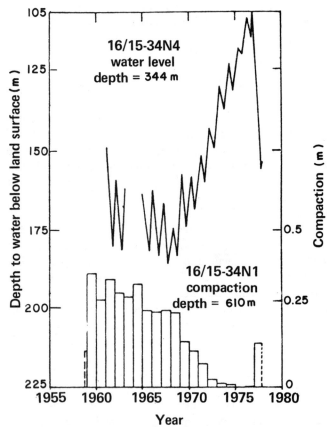

Figure 15. Water-level fluctuations and measured rates of compaction as functions of time near Cantua Creek, California (from Lofgren, 1979).

associated with the concept of transient preconsolidation stress discussed earlier in the section Aquitard-Drainage Model. Equation 25 tacitly presupposes subsidence to be a viscous phenomenon. Yamamoto (Poland, 1984) indicated that a curve with two characteristic slopes (Fig. 14) more closely associated rate of subsidence to water levels. These two characteristic slopes may be related to Holzer's (1981) discussion of geologic preconsolidation stress. Others (for example, Lofgren, 1979) have chosen to plot both measurements of water levels and rates of subsidence more primitively as empirical functions of time (Fig. 15) rather than as functions of one another (Figs. 13 and 14). Corapcioglu (1977; Brutsaert and Corapcioglu, 1976; Corapcioglu and Brutsaert, 1977) has developed a theoretical predictive method that is consistent with the viscoelastic line of evidence indicated by equation 25 and Figures 13 through 15. Corapcioglu assumes that the compacting interval behaves like a single undifferentiated material with viscoelastic properties of stress and strain. Applying Corapciolu's viscoelastic reservoir model to actual field data has indicated that to find time constants that fit both short-term (months) and long-term (decades) compaction data requires either introducing to the model an additional viscoelastic parameter (Kasameyer, 1979, written commun.) or putting no upper limit on long-term or ultimate compaction. The latter alternative was chosen by Corapcioglu and Brutsaert (1977). This implies that over the decades, a compacting interval can theoretically reach less than zero thickness. The first alternative may be physically more realistic even if mathematically it would be more unwieldy.

A second fruitful way to plot field data is to compare depth to water with cumulative subsidence or cumulative compaction (Figs. 16 and 17) rather than with rate of subsidence (Figs. 13,

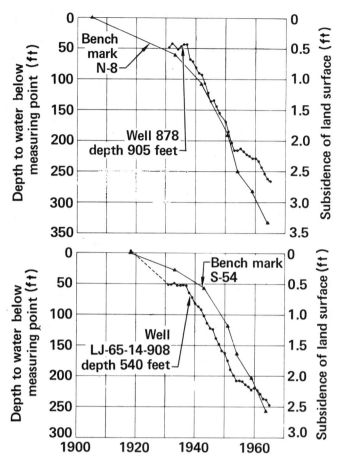

Figure 16. Subsidence of bench marks and water-level declines in nearby wells as functions of time, Houston-Galveston area, Texas (from Gabrysch, 1969).

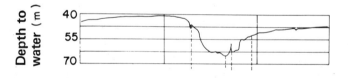

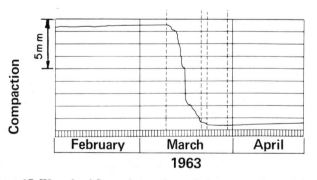

Figure 17. Water-level fluctuations and cumulative compaction as functions of time south of Tulare, California (from Lofgren, 1970).

14, and 15). This is a second general type of semitheoretical approach. For long periods of continuing pressure drops, the relation is striking (Fig. 16). When water levels recover, however, one does not see a comparable recovery of earlier compaction (Fig. 17). Figure 17 is a short-term detailed plot of a portion of the data shown in Figure 18. Figure 18 shows stress versus compaction (strain) as an alternative causal interpretation to the stress versus rate of subsidence (strain rate) of Figures 13 and 14. Riley (1969) analyzed the stress/strain curves in Figure 18 using the aquitard-drainage model. Using the same conceptual model and the theory of primary consolidation from soil mechanics, Helm (1975, 1976) successfully simulated the compaction curve (430–760 ft in Fig. 18) by using the change in applied stress curve (Fig. 18) as input. This is an example of a theoretical technique as defined in the Intruduction. The aquitard-drainage model treats slow-draining aquitards mathematically in a way somewhat analogous to the way a viscous dashpot is used in a viscoelastic rheological model. The main difference is that for the aquitard-drainage model the confined aquifer system is assumed to be heterogeneous, containing two homogeneous saturated materials, namely, clay and sand. The idealized clay lense in the aquitard-drainage model is assumed to have a nonrecoverable quasi-elastic property. In addition, both sand and clay are assumed to have recoverable elastic properties. In the viscoelastic rheological model, the confined aquifer system is assumed to contain one homogeneous material.

Plots of some other correlated field data are worth mentioning. These are examples of additional semitheoretical techniques. One is the influence of clay content and another is the influence of production rate. For the special case of ongoing long-term drawdown (Fig. 16), the ratio of observed subsidence to drawdown gives a type of storage coefficient. Gabrysch (1969) has plotted this ratio against percent of clay (Fig. 19). The interesting result is that greater subsidence per unit pressure decline is associated with a higher clay content. This tends to corroborate laboratory data and Tolman and Poland's (1940) field model that slow-draining fine-grained aquitards are more compressible than the surrounding coarse-grained material.

Aquitard compaction is a major source of stored water. However, as Lofgren (1979) has appropriately emphasized, Figures 17 and 18 indicate that when water of compaction is once used, it is essentially nonrecoverable. Roughly 90% of the decrease of pore volume, where water is stored, is permanent.

Castle and others (1969) have demonstrated a linear relation between the vertical component of subsidence and the cumulative volume of fluid production (Fig. 20). Kumai and others (1969) showed a similar linear relation between rate of vertical subsidence and rate of production. The difference in subsidence at two points within a subsidence bowl is shown in Figure 20. The center of the bowl at Huntington Beach subsides more than a peripheral point.

If, instead of the vertical component, the volume of subsidence is plotted against the cumulative volume of fluid production, the local variations mentioned above in the vertical

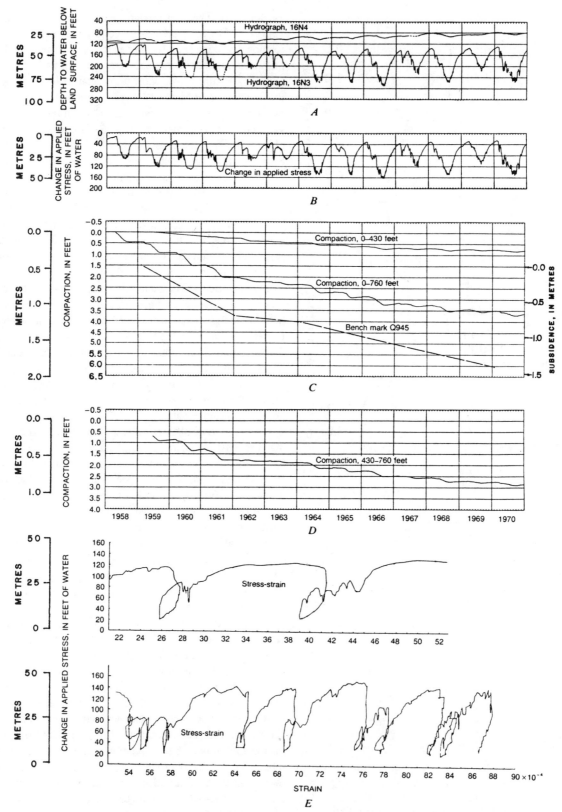

Figure 18. Hydrographs, change in applied stress, compaction, and subsidence as functions of time and the correlation of change in applied stress and compaction per unit thickness (strain) south of Tulare, California (from Poland and others, 1975).

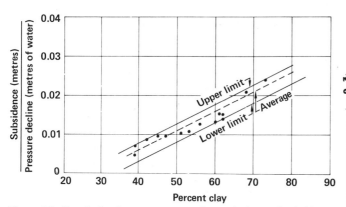

Figure 19. Correlation between percent clay and observed subsidence per unit water-level decline, Houston-Galveston area, Texas (from Gabrysch, 1969).

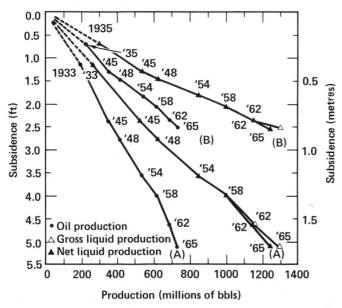

Figure 20. Cumulative oil, gross-liquid, and net-liquid production from Huntington Beach oil field, California, plotted against subsidence at bench marks located (A) near the center of subsidence and (B) midway up the southeast limb of the subsidence bowl (from Castle and others, 1969).

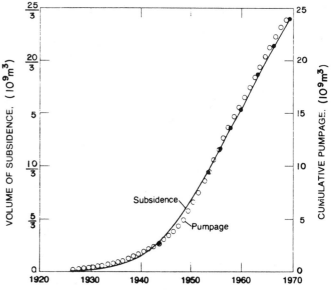

Figure 21. Cumulative volumes of subsidence and pumpage, 1926 to 1969, Los Banos–Kettleman City area, California. Points on subsidence curve indicate times of leveling control (from Poland and others, 1975).

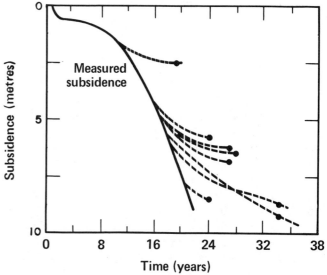

Figure 22. Predictions made for the subsidence associated with developing the Wilmington oil fields, California (from Christian and Hirschfeld, 1974).

component of subsidence drop out. Poland and others (1975) have demonstrated a dramatic correlation between the cumulative volumes of subsidence and fluid production (Fig. 21). Throughout the entire 43-year period (1926–1969), the source of one-third of all the water pumped from confined aquifers in the Los Banos–Kettleman City area, California, was the permanent decrease in pore volume of underlying sediments. This decrease manifested itself at land surface as nonrecoverable subsidence.

Throughout the years there have been several attempts to predict subsidence associated with pumping the Wilmington oil fields, California, (Christian and Hirschfeld, 1974). The dashed lines on Figure 22 indicate a number of these attempts. Two of these lines represent predictions made in the 1950s by Terzaghi himself (Pacific Fire Rating Bureau, 1958) using laboratory time-consolidation data and his epoch-making theory of consolidation. This is an example of a theoretical technique. According to John Martin (1978, personal commun.), Terzaghi considered using a predictive technique similar to the depth-porosity method described earlier in this report, which can be considered a semitheortical technique. He rejected the latter idea, however, because it gave what at the time was believed to be too large an estimate of ultimate subsidence. We see from Table 4 that at Wilmington the observed maximum subsidence in 1968 of 9 m was not much less than the predicted ultimate subsidence of 11.2

m using the depth-porosity model for a drawdown of 930 m. Subsidence had not yet reached a true ultimate value in 1968 when its continuation was successfully arrested by fluid injection.

Figure 22 indicates that predicting subsidence based on parameter values found from standard laboratory tests may be less than helpful. Elsewhere in the world, however, this laboratory-based method of prediction has been used with better results according to L. Carbognin (1980, personal commun.) of Venice, Italy, G. Figueroa-Vega (1979, personal commun.) of Mexico City, Mexico, and the State Electricity Commission of Victoria, Australia (Helm, 1984). According to J. F. Poland (1978, personal commun.), stress-dependent parameter values estimated from carefully performed and analyzed laboratory tests on samples collected from the Santa Clara Valley, California, can quantitatively account for less than half of the subsidence that has already occurred. Although excellent and extensive published laboratory data and analyses of samples from compacting beds throughout the San Joaquin Valley, California, have been available for many years (Johnson and others, 1968), they have not yet been used successfully to predict subsidence. R. Gabrysch (1978, personal commun.) reported that careful interpretation of laboratory analyses of samples from the Houston-Galveston area of Texas had not yet led to a reliable method of subsidence prediction.

Lewis and Screfler (1978) developed a sophisticated coupled flow/strain model and applied it to simulate the subsidence history of Venice, Italy. Gambolati and others (1974) applied an uncoupled flow/strain model to simulate the same subsidence history. For practical reasons, the Harris Galveston Subsidence District (1982) decided to follow an uncoupled approach. They fine-tuned the regional ground-water flow model only at points where subsidence would be calculated. They modeled subsidence only at these predetermined locations and interpolated between them. Their work represents the most ambitious field-based predictive study yet made.

In addition to the computer models already mentioned, a number of other theoretical methods exist that employ digital computers. Most have not yet been applied to field data. Only a few representative ones can be mentioned here. Papadopulos and others (1975) used Hantush's (1960) theory of leaky aquifers to estimate possible subsidence due to producing geopressured fields rimming the Gulf of Mexico. Herrera and others (1974) applied Herrera's (1970; Herrera and Rodarte, 1973) theory of leaky aquifers to predict subsidence near Mexico City. Safai and Pinder (1980), similar to Verruijt's (1969) fine work, used Biot's (1941) elastic theory of three-dimensional consolidation to derive their equations of deformation. Bear and Corapcioglu (1981) have introduced some interesting computational simplifications to Biot's theory. Lewis and Schrefler (1978) and Garg and others (1975) included nonlinear creep in their rheological models. Sandhu (1968) applied a theory of mixtures to the time-consolidation problem. Helm (1982, 1983) derived an equation of motion for solid particles during transient conditions of fluid flow.

The state of the art of taking field measurements at depth has not kept pace with the ability of computational geohydrologists to postulate theories of three-dimensional deformation. Inexpensive and precise techniques for measuring horizontal and vertical movement of the land surface are available (Lofgren, 1978) and are continuing to be improved. Measurements of cumulative vertical movement between land surface and an anchor buried at depth are sufficiently advanced (see Figs. 17 and 18) as to be nearly routine. The corresponding ability to measure compaction at depth, which was developed during the 1950s (Lofgren, 1970), made possible the testing of the more sophisticated and reliable predictive techniques described in this paper.

A critical problem at present is the lack of a reliable and inexpensive method for measuring cumulative horizontal movement at depth and how that horizontal movement migrates to land surface. The amount of horizontal movement at any one point may be small and hence difficult to measure, but the cross-sectional area involved may be so large that such movement may represent a large volume of nonrecoverable loss in storage. The state of the science of predicting horizontal movement today is similar to the state of the science in the 1950s for predicting vertical subsidence (exemplified by Fig. 22). The importance of horizontal movement for ground-water management and oil-reservoir management is underappreciated. Damage associated with horizontal movement and fissuring is costly. To have a reliable predictive technique for transient horizontal movement would be invaluable. Until theories that include horizontal skeletal displacement at depth are checked against appropriate field data, their predictive capability cannot be claimed to have been satisfactorily tested.

REFERENCES CITED

Allen, D. R., and Mayuga, M. N., 1969, The mechanics of compaction and rebound, Wilmington Oil Field, Long Beach, California, U.S.A., in Tison, L. J., ed., Land subsidence, Volume 2: International Association of Scientific Hydrology Publication 89, p. 410–423.

Athy, L. F., 1930, Density, porosity, and compaction of sedimentary rocks: American Association of Petroleum Geology Bulletin, v. 14, p. 1–24.

Bear, J., and Corapcioglu, M. Y., 1981, Mathematical model for regional land subsidence due to pumping: Water Resources Research, v. 17, p. 937–958.

Biot, M. A., 1941, General theory of three-dimensional consolidation: Journal of Applied Physics, v. 12, p. 155–164.

Brook, C. A., Mariner, R. H., Mabey, D. R., Swanson, J. R., Guffanti, M., and Muffler, L.J.P., 1979, Hydrothermal convection systems with reservoir temperatures ⩽90°C.: Assessment of geothermal resources of the United State—1978: U.S. Geological Survey Circular 790, 163 p.

Brutsaert, W., and Corapcioglu, M. Y., 1976, Pumping of aquifer with viscoelastic properties: American Society of Civil Engineers, Journal of the Hydraulics Division, v. 102 (HY11), p. 1663–1675.

Carrillo, Nabor, 1949, Subsidence in the Long Beach–San Pedro area: Stanford Research Institute, Stanford, California, p. 67–79, 227–242.

Castle, R. O., Yerkes, R. F., and Riley, F. S., 1969, A linear relationship between liquid production and oil-field subsidences, in Tison, L. J., ed., Land subsidence, Volume 1: International Association of Scientific Hydrology Publication 88, p. 162–173.

Christian, J. T., and Hirschfeld, R. C., 1974, Subsidence of Venice: Predictive difficulties: Science, v. 185. p. 1185.

Corapcioglu, M. Y., 1977, Mathematical modeling of leaky aquifers with rheological properties, *in* Johnson, A. I., ed., Land subsidence: International Association of Scientific Hydrology Publication 121, p. 191–200.

Corapcioglu, M. Y., and Brutsaert, W., 1977, Viscoelastic aquifer model applied to subsidence due to pumping: Water Resources Research, v. 13, p. 597–604.

Crow, N. B., 1979, An environmental overview of geothermal development: The Geysers-Calistoga KGRA, Volume 4, Environmental geology: Lawrence Livermore National Laboratory Publication No. UCRL-52496, v. 4, 72 p.

Dickinson, G., 1953, Geological aspects of abnormal reservoir pressures in Gulf Coast Louisiana: American Association of Petroleum Geologists Bulletin, v. 37, p. 410–432.

Figueroa-Vera, German E., 1971, Influence chart for regional pumping effects: Water Resources Research, v. 7, p. 209–210.

Finol, A., and Farouq Ali, S. M., 1975, Numerical simulation of oil production with simultaneous ground subsidence: Society of Petroleum Engineers Journal, v. 15, p. 411–422.

Gabrysch, R. B., 1969, Land-surface subsidence in the Houston-Galveston region Texas, *in* Tison, L. J., ed., Land subsidence, Volume 1: International Association of Scientific Hydrology Publication 88, p. 43–54.

Gabrysch, R. K., and Bonnet, C. W., 1976, Land-surface subsidence at Seabrook, Texas: U.S. Geological Survey Water-Resources Investigation 76-31, 53 p.

Gambolati, G., 1972, A three-dimensional model to compute land subsidence: International Association of Scientific Hydrology, v. 17, p. 219–226.

Gambolati, G., and Freeze, R. A., 1973, Mathematical simulation of the subsidence of Venice, 1, Theory: Water Resources Research, v. 9, p. 721–733.

Gambolati, G., Gatto, P., and Freeze, R. A., 1974, Mathematical simulation of the subsidence of Venice, 2, Results: Water Resources Research, v. 10, p. 563–577.

Garg, S. K., Blake, T. R., Brownell, D. H., Jr., Nayfeh, A. H., and Pritchett, J. W., 1975, Simulation of fluid-rock interactions in a geothermal basin: Systems, Science and Software Final Report No. SSS-R-76-2734, La Jolla, California, 63 p.

Geertsma, J., 1957, The effect of fluid pressure decline on volumetric changes of porous rocks: Transactions of American Society of Mechanical Engineers, AIME, v. 210, p. 331–340.

—— 1966, Problems of rock mechanics in petroleum production engineering: Proceedings of First Congress of the International Society of Rock Mechanics, Lisbon, v. 1, p. 585–594.

—— 1973, A basic theory of subsidence due to reservoir compaction, the homogeneous case: Verhandelingen der Koninkljke Nederlands Geologisch Mijhbouw, v. 28, p. 43–62.

Grimsrud, G. P., Turner, B. L., and Frame, P. A., 1978, Areas of ground subsidence due to geofluid withdrawal: Berkeley, California, Lawrence Berkeley Laboratory Publication No. 8618.

Hantush, M. S., 1960, Modification of the theory of leaky aquifers: Journal of Geophysical Research, v. 65, p. 3713–3725.

Harris Galveston Coastal Subsidence District, 1982, Water management study phase two and supplement No. 1: Houston, Texas, 68 p., 160 exhibits.

Helm, D. C., 1972, Simulation of aquitard compaction due to changes in stress [abs.]: EOS (American Geophysical Union Transactions), v. 53, p. 979.

—— 1975, One-dimensional simulation of aquifer system compaction near Pixley, California, 1) Constant parameters: Water Resources Research, v. 11, p. 465–478.

—— 1976, One-dimensional simulation of aquifer system compaction near Pixley, California, 2) Stress-dependent parameters: Water Resources Research, v. 12, p. 375–391.

—— 1977, Estimating parameters of compacting fine-grained interbeds within a confined aquifer system by a one-dimensional simulation of field observations, *in* Johnson, A. I., ed., Land subsidence: International Association of Scientific Hydrology Publication 121, p. 145–156.

—— 1978, Field verification of a one-dimensional mathematical model for transient compaction and expansion of a confined aquifer system: Verification of mathematical and physical models in hydraulic engineering, *in* Proceedings, 26th Hydraulics Division Specialty Conference, College Park, Maryland: American Society of Civil Engineers, p. 189–196.

—— 1982, Conceptual aspects of subsidence due to fluid withdrawal, *in* Narasimhan, T. N., ed., Recent trends in hydrogeology: Geological Society of America Special Paper 189, p. 103–139.

—— 1983, Analysis of sedimentary skeletal deformation in a confined aquifer and the resulting drawdown, *in* Rosenshein, J., and Bennett, G. D., eds., Groundwater hydraulics: American Geophysical Union Monograph 9, p. 29–82.

—— 1984, Latrobe Valley subsidence predictions, The modelling of ground movement due to groundwater withdrawal: Joint report of the Fuel Department and the Design, Engineering, and Environment Department, State Electricity Commission of Victoria, Melbourne, Australia (in press).

Herrera, I., 1970, Theory of multiple leaky aquifers: Water Resources Research, v. 6, p. 185–193.

Herrera, I., and Rodarte, L., 1973, Integro-differential equations for systems of leaky aquifers and applications: Water Resources Research, v. 9, p. 995–1005.

Herrera, I., Alberro, J., Leon, J. L., and Chen, B., 1974, Analisis de asentamientos para la construccion de los lagos del plan Texcoco: Publication 340, Instituto de Ingenieria, Universidad Nacional Autonoma de Mexico, 113 p.

Holzer, T. L., 1981, Preconsolidation stress of aquifer systems in areas of induced land subsidence: Water Resources Research, v. 17, p. 693–704.

Ireland, R. L., Poland, J. F., and Riley, F. S., 1982, Land subsidence in the San Joaquin Valley, California, as of 1980: U.S. Geological Survey Open-File Report 82-370, 129 p.

Johnson, A. I., Moston, R. P., and Morris, D. A., 1968, Physical and hydrologic properties of water-bearing deposits in subsiding areas in central California: U.S. Geological Survey Professional Paper 497-A, 71 p.

Kumai, H., Sayama, M., Shibaski, T., and Uno, K., 1969, Ground sinking in Shiroiski Plain, Saga Prefecture, Japan, *in* Tison, L. J., ed., Land subsidence, Volume 2: International Association of Scientific Hydrology Publication 89, p. 645–657.

Lambe, T. W., and Whitman, R. V., 1969, Soil mechanics: New York, John Wiley and Sons, Inc., 553 p.

Lee, K. L., 1977, Calculated horizontal movements at Baldwin Hills, California, *in* Johnson, A. E., ed., Land subsidence: International Association of Scientific Hydrology Publication 121, p. 299–308.

Lewis, R. W., and Schrefler, B., 1978, A fully coupled consolidation model of the subsidence of Venice: Water Resources Research, v. 14, p. 223–230.

Lofgren, B. E., 1970, Field measurement of aquifer-system compaction, San Joaquin Valley, California, U.S.A., *in* Tison, L. J., ed., Land subsidence, Volume 1: International Association of Scientific Hydrology Publication 88, p. 272–284.

—— 1978, Monitoring crustal deformation in the Geysers–Clear Lake geothermal area, California: U.S. Geological Survey Open-File Report 78-597, 19 p.

—— 1979, Changes in aquifer-system properties with ground-water depletion, *in* Saxena, S. K., ed., Evaluation and prediction of subsidence: Proceedings, Engineering Foundation Conference, January 1978, Pensacola Beach, Florida, American Society of Civil Engineers, p. 26–46.

Magara, Kinji, 1978, Compaction and fluid migration: New York, Elsevier Scientific Publishing Co., 319 p.

Marsal, R. J., and Graue, R., 1969, The subsoil of Lake Texcoco, (in) Nabor Carrillo: The subsidence of Mexico City and Texcoco Project: Secretaria de hacienda y credito publico fidaciaria, Mexico, p. 167–202.

McCann, G. D., and Wilts, C. H., 1951, A mathematical analysis of the subsidence in the Long Beach–San Pedro area: Internal Report, California Institute of Technology, 117 p.

Mercer, J. W., Pinder, G. F., and Donaldson, I. G., 1975, A Galerkin finite-element analysis of the hydrothermal system at Wairakei, New Zealand: Journal of Geophysical Research, v. 80, p. 2608–2621.

Miller, I., Dershowitz, W., Jones, K., Myer, L., Roman, K., and Schauer, M., 1980a, Detailed report on tested models: Companion Report 2 to simulation of geothermal subsidence (LBL-10571): Berkeley, California, Lawrence Berkeley Laboratory Publication No. LBL-10837, 197 p.

—— 1980b, Case study data base: Companion Report 3 to simulation of geothermal subsidence (LBL-10571): Berkeley, California, Lawrence Berkeley Laboratory Publication No. LBL-10839, 66 p.

Mindlin, R. D., and Cheng, D. H., 1950, Thermoelastic stress in the semi-infinite solid: Journal of Applied Physics, v. 21, p. 931.

Narasimhan, T. N., and Witherspoon, P. A., 1977, Numerical model for land subsidence *in* Johnson, A. I., ed., Land subsidence: International Association of Scientific Hydrology Publication 121, p. 133–144.

Pacific Fire Rating Bureau, 1958, Subsidence in Long Beach-Terminal Island-Wilmington, California, (Internal report), 38 p.

Papadopulos, S. S., Wallace, R. H., Jr., Wesselman, J. B., and Taylor, R. E., 1975, Assessment of onshore geopressured-geothermal resources in the Northern Gulf of Mexico Basin, *in* White, D. E., and Williams, D. L., eds., Assessment of geothermal resources of the United States—1975: U.S. Geological Survey Circular 723, p. 125–146.

Poland, J. F., 1972, Subsidence and its control, *in* Cook, T. D., ed., Underground waste management and environmental implications, Memoir No. 18: Houston, Texas, American Association of Petroleum Geologists, p. 50–71.

——editor, 1984, Guide to the study of land subsidence due to ground-water withdrawal: United Nations Educational, Scientific, and Cultural Organization, Paris, (in press).

Poland, J. F., Lofgren, B. E., and Riley, F. S., 1971, Glossary of selected terms useful in studies of the mechanics of aquifer systems and land subsidence due to fluid withdrawal: U.S. Geological Survey Water Supply Paper 2025, 9 p.

Poland, J. F., Lofgren, B. E., Ireland, R. L., and Pugh, R. G., 1975, Land subsidence in the San Joaquin Valley, California, as of 1972: U.S. Geological Survey Professional Paper 437-H, 78 p.

Pollard, W. S., Holcombe, R. F., and Marshall, A. F., 1979, Subsidence cause and effect, Harris-Galveston Coastal Subsidence District Phase 1-A Study: McClelland Engineers, Inc., Houston, Texas, 2 vols.

Pritchett, J. W., Rice, L. F., and Garg, S. K., 1978, Reservoir engineering data: Wairakei geothermal field, New Zealand: Systems, Science and Software Report No. SSS-R-78-3597, La Jolla, California, 359 p.

—— 1979, Summary of reservoir engineering data, Wairakei geothermal field, New Zealand: Berkeley, California, Lawrence Berkeley Laboratory Publication No. LBL-8669, 25 p.

Renner, J. L., White, D. E., and Williams, D. L., 1976, Hydrothermal convection systems, assessment of geothermal resources of the United States—1975: U.S. Geological Survey Circular 726, 155 p.

Riley, F. S., 1969, Analysis of borehole extensometer data from central California, *in* Tison, L. J., ed., Land subsidence, Volume 2: International Association of Scientific Hydrology Publication 89, p. 423–431.

Safai, N. M., and Pinder, G. F., 1980, Vertical and horizontal land deformation due to fluid withdrawal: International Journal for Numerical and Analytical Methods in Geomechanics, v. 4, p. 131–142.

Sandhu, R. S., 1968, Fluid flow in saturated porous elastic media, [Ph.D. thesis]: Berkeley, University of California, 105 p.

Schatz, J. F., 1976 Models of inelastic volume deformation for porous geologic materials, *in* Cowin, S. C., and Carroll, M. M., eds., The effect of voids on material deformation, Volume 16: American Society of Mechanical Engineers, p. 141–170.

Schatz, J. F., Kasameyer, P. W., and Cheney, J. A., 1978, A method of using in situ porosity measurements to place an upper bound on geothermal reservoir compaction: Proceedings, Invitational Well Testing Symposium, 2nd, Lawrence Berkeley Laboratory Publication No. LBL-8883, p. 90–94.

Scott, R. F., 1963, Principles of soil mechanics: Palo Alto, California, Addison-Wesley Publishing Co., 550 p.

Sen, B., 1950, Note on the stresses produced by nuclei of thermoelastic strain in a semi-infinite elastic solid: Applied Mathematics Quarterly, v. 8, p. 635.

Taylor, D. W., 1948, Fundamentals of soil mechanics: New York, John Wiley and Sons, 700 p.

Terzaghi, K., 1925, Principles of soil mechanics: IV, Settlement and consolidation of clay: Engineering News-Record, New York, McGraw-Hill, v. 95, p. 874–878.

Tolman, C. F., and Poland, J. F., 1940, Ground-water, salt-water, infiltration, and ground-surface recession in Santa Clara Valley, Santa Clara County, California: American Geophysical Union Transactions, p. 23–34.

Verruijt, A., Elastic storage of aquifers, 1969, *in* De Wiest, ed., Flow through porous media: New York, Academic Press, p. 331–376.

Wadachi, K., 1940, Ground sinking in west Osaka (second report): Disaster Prevention Research Institute Report No. 3.

Witherspoon, P. A., and Freeze, R. A., 1972, The role of aquitards in multiple-aquifer system, Penrose Conference of the Geological Society of America, 1971: Geotimes, v. 17, p. 22–24.

MANUSCRIPT ACCEPTED BY THE SOCIETY APRIL 18, 1984

Subsidence over oil and gas fields

J. C. Martin
Chevron Oil Field Research Company
P.O. Box 446
La Habra, California 90631

S. Serdengecti
Harvey Mudd College
Claremont, California 91711

ABSTRACT

Most oil and gas reservoirs experience only small amounts of compaction and surface subsidence. Significant subsidence due to production of hydrocarbons has been observed over some oil and gas fields. This paper presents a review of the fundamentals of reservoir compaction and surface subsidence over oil and gas fields and explains why large-scale subsidence is rare. A new method of estimating maximum potential subsidence is presented and used to analyze the subsidence over oil and gas fields in Louisiana. Large-scale compaction and subsidence are evidently associated with inelastic behavior of the reservoir rock and in some cases of the surrounding rock. No reliable methods have been established for predicting either the transition from elastic to inelastic reservoir rock behavior or large-scale reservoir compaction and subsidence.

The preconsolidation effect evident in some field data introduces an additional difficulty in understanding and predicting subsidence. Various possible causes for the effect include uplift and erosion, temporary pressure drop or an increase in horizontal stresses prior to hydrocarbon production, effects of loading rate on reservoir rock, and changes from elastic to inelastic behavior of the reservoir rock or surrounding formations.

INTRODUCTION

Subsidence due to oil and gas production was first observed in the late 1910s when the ground above the Goose Creek oil field in Texas began to subside (Pratt and Johnson, 1926). Evidence of subsidence over one of the Bolivar Coast oil fields in Venezuela appeared in the late 1920s (van der Knaap and van der Vlis, 1967), and evidence of subsidence appeared over the Wilmington oil field in Long Beach, California, in the 1930s (Gilluly and Grant, 1949). More recently, significant subsidence has been reported over the Groningen gas field in the Netherlands (Schoonbeek, 1976); over the Inglewood, Buena Vista Hills, Santa Fe Springs, and Huntington Beach oil fields in California (Castle and others, 1970); and over several oil and gas fields in the USSR (Ilijn, 1977). Subsidence due to production of gas and water has been reported from the Po Delta area in northeast Italy (Poland and Davis, 1969; Schrefler and others, 1977) and from the Niigata area on the northwest coast of the island of Honshu, Japan (Poland and Davis, 1969).

Poland and Davis (1969) presented a review of oil and gas field subsidence in the United States. Almost 9 m of subsidence at Wilmington is the largest amount reported; it was treated by Allen (1968), Allen and Mayuga (1970), Castle and others (1970), Coxe (1949), Frame (1952), Gilluly and Grant (1949), Harris and Harlow (1947), McCann and Wilts (1951), Mayuga (1970), Mayuga and Allen (1970), Miller and Somerton (1955), Pierce (1970), Poland and Davis (1969), Shoemaker (1955), Shoemaker and Thorley (1955), Winterburn (1943), and Yerkes and Castle (1970). Subsidence in the Bolivar coastal fields was treated by Nuñez and Escojido (1977), van der Knaap and van der Vlis (1967), and Merle and others (1975). Subsidence in the Groningen gas field was treated by Boot (1973), Geertsma

(1973b, Geertsma and van Opstal (1973), van Kesteren (1973a, 1973b; de Loos (1973), Schoonbeek (1976), and Teeuw (1973). Erickson (1977) discussed subsidence control over the Beverly Hills (east) oil field in California.

The objectives of this paper are to present (1) a review of the fundamentals of subsidence over oil and gas fields, (2) new material on the mechanical behavior of reservoir rock, (3) a method of estimating the maximum subsidence, (4) results of an analysis of the maximum subsidence over oil and gas fields in Louisiana, and (5) a discussion of subsidence prediction and control.

Early attempts to predict ultimate subsidence at Wilmington were disappointing (Allen and Mayuga, 1970). Results presented in this paper indicate that large-scale subsidence is a result of special conditions, such as large pressure declines in shallow, thick, highly compressible reservoirs. When these rare conditions do occur, the ultimate subsidence is very difficult to predict in the early stages. Estimates can be made, however, of the ultimate subsidence for oil and gas reservoirs in general. Some surface subsidence probably occurs over all oil and gas reservoirs that experience pressure decline; however, subsidence has been detected over only a few of the tens of thousands developed oil and gas fields. Results of a study of the oil and gas reservoirs in Louisiana presented herein indicate that for most the potential subsidence is insignificant.

The Wilmington, Goose Creek, Inglewood, and Bolivar coastal oil fields have experienced large amounts of subsidence and other phenomena associated with reservoir compaction. Deformation at Wilmington included a series of abrupt earth movements in the formations overlying the compacting reservoirs. These movements sheared off many well casings (Mayuga and Allen, 1970) and were accompanied by significant earthquakes (Richter, 1958; Kovach, 1974). Differential horizontal earth movement did much damage to surface structures including buildings, bridges, railroad tracks, and pipelines. In the Bolivar coastal fields, reservoir compaction caused damage to pipe in the producing interval (Kennedy, 1961). In the Goose Creek field, abrupt earth movements and surface faults occurred around the edges of the field, accompanied by earthquakes. Vertical displacement across one of the faults amounted to 20 cm (Pratt and Johnson, 1926). These phenomena occur in very few oil and gas fields, suggesting that very few fields have significant reservoir compaction and surface subsidence. Important factors in oil-field subsidence are the reservoir pressure, thickness, and width-to-depth ratio and the mechanical rock properties of the reservoir and surrounding formations. The mechanical properties are by far the most difficult factors to determine. Evidently they depend upon many parameters, such as the in situ stress distribution and its history, rock type, porosity, grain geometry, mineralogy, and cementation.

The results presented in this paper explain why large-scale subsidence is rare and indicate that large-scale subsidence is associated with two types of in situ rock failure. Internal rock failure involves grain fracturing, crushing, and rearranging. Associated with this are large reductions in porosity and reservoir thickness which can induce large shear stresses in the rock surrounding and overlying the reservoir. These stresses may cause the second type of in situ failure, fractures or faults. In situ rock failures associated with large-scale subsidence and reservoir compaction can make the prediction of ultimate subsidence very difficult.

DISCUSSION

Reservoir Compaction

Figure 1 illustrates a simplified fluid-filled model of a compacting reservoir, in which the reservoir rock has been replaced with springs. The total earth stresses acting on the reservoir are balanced by the pore pressure, P_i, and the effective vertical and horizontal earth stresses, σ_v and σ_h, which are represented by the loads on the springs. As the fluid pressure declines, the fluid supports less of the earth stresses. This loss of support is taken up by an increase in the effective stresses, σ_v and σ_h, and by induced stresses in the formations surrounding the reservoir. The increase in effective stress causes the springs to compress and the reservoir to compact as illustrated in Figure 1(b). If the springs are very weak, the formations surrounding the reservoir will absorb most of the loss of support. In the case of a pressurized cavity containing no rock, the surrounding formations absorb all of the loss of support resulting from a pressure decline. In a mine where the cavity pressure is atmospheric, the surrounding rock must absorb the loss of support from both the fluid pressure and the effective stress on the rock.

Roberts (1970) presented evidence that reservoir sands may compact inelastically. Holzer (1981), Nuñez and Escojido (1977), Merle and others (1975), and Kennedy (1961) presented evidence that initially some aquifers and oil reservoirs behave as if the producing formation is relatively incompressible and becomes significantly more compressible after large pressure drops have occurred. In the model, this behavior corresponds to some of the springs breaking as the fluid pressure declines.

There is a fundamental difference between the effective stresses in the rocks inside and outside a reservoir compacting due to a fluid-pressure decline. Inside the reservoir, rock is compressed, and outside the reservoir, rock is subjected to tensile and shearing stresses.

Most reservoirs in which fluid production has caused significant compaction consist of a sequence of sands and shales. It is not always clear how much each lithology contributes to the reservoir compaction. Nuñez and Escojido (1977) stated that data obtained from radioactive bullets in wells indicate that the sands in the Bolivar coastal fields are responsible for the subsidence. Allen and Mayuga (1970) indicated that approximately two-thirds of the compaction at Wilmington occurred in the sands.

Laboratory tests on sands, reported by Roberts and de Souza (1958), indicate that sands may be more compressible than clays under simulated reservoir conditions. Evidently clays undergo extensive mineralogical changes in the burial process as they are transformed from surface clays to shales associated with

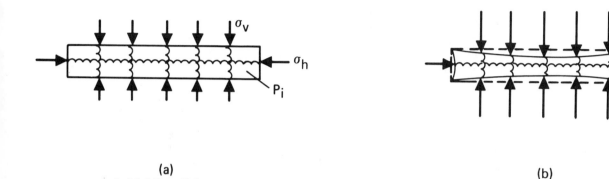

Figure 1. Simplified model of a compacting reservoir.

oil and gas reservoirs (Hower and others, 1976). These mineralogical changes may have a strong influence on the compaction characteristics of shales.

In this paper it is assumed that the time lag between the fluid-pressure decline and the formation compaction can be neglected. This condition is expected for reservoirs in which the sand compacts. If there are interbedded shales present they must either be sufficiently thin so that their compaction keeps up with the pressure decline or their compaction is negligible. If any shales are thick enough to require significant time for their internal fluid pressures to equalize, this time lag should be included in compaction and subsidence calculations by accounting for the effects of pressure transients in the shales.

Preconsolidation Effect

Some aquifers and oil reservoirs behave as if they had been subjected to higher effective stresses in the past than those existing at the beginning of the fluid-pressure decline (Holzer, 1981; Nuñez and Escojido, 1977; Merle and others, 1975; Kennedy, 1961). This could have been caused by uplift followed by erosion, although the geologic evidence does not always support this explanation (van der Knaap and van der Vlis, 1967). Other possible explanations include: (1) a temporary drop in fluid pressure prior to the pressure decline caused by hydrocarbon production; (2) time effects related to the rate of loading of the reservoir rock; (3) an increase in the horizontal stresses prior to the pressure decline; (4) a change in the reservoir rock from elastic to inelastic behavior, not related to previous stresses or rates of loading; and (5) a change from elastic to inelastic behavior in the formations surrounding the reservoir. These possibilities are discussed in the following paragraphs.

A previous drop in the reservoir fluid pressure could induce a preconsolidation effect. In this case the effective stress on the reservoir rock would increase as the fluid pressure decreased and then decrease as the pressure increased.

Laboratory tests that we performed on sandstones that have undergone permanent deformation reveal a preconsolidation effect related to the rate of loading. Similar results have been obtained on shales (Bjerrum, 1967). Samples loaded at a slow rate often respond to a large increase in the rate of loading with a small amount of deformation over a significant increment of load. After this increment the samples resume essentially the same rate of deformation with load as they had before the change in rate of loading. The rate of loading in a compacting reservoir is many times greater than the rate over geologic time. Thus, the preconsolidation effect could be related to a change in rate of loading.

Figure 2 presents the results of a laboratory test that we conducted on a sand pack which illustrates a preconsolidation effect caused by a change in the rate of loading. The sample was loaded axially in steps from $13.8 \times$ MPa to $55.2 \times$ MPa while being radially stressed to maintain a constant sample diameter. From $13.8 \times$ MPa to 34.5×10^{-6} MPa, the average rate of axial loading was 0.079 kPa/s. At this point the rate of loading was increased to 2.3 kPa/s. Above $34.5 \times$ MPa the sample behaved as if it had been preconsolidated $5.0 \times$ MPa. This demonstrates that varying the rate of loading can induce preconsolidation effects in laboratory tests of rock samples.

A preconsolidation effect can also be induced in laboratory samples by an increase in the radial stress which corresponds to

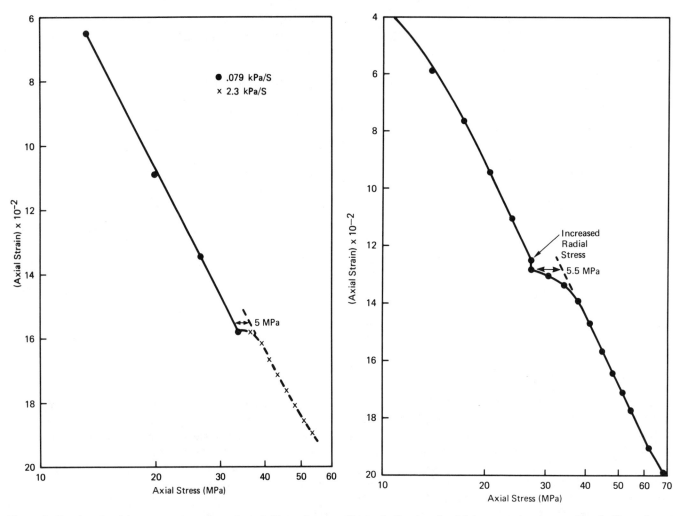

Figure 2. Results of a laboratory test on a sand pack illustrating a preconsolidation effect caused by an increase in the rate of loading.

Figure 3. Results of a laboratory test on a sand pack illustrating a preconsolidation effect caused by an increase in radial stress.

the in situ horizontal stress. Figure 3 presents the axial strain versus axial stress obtained by us from a laboratory test on a sand pack loaded to 27.6 × MPa while maintaining a no-radial-strain condition. The radial stress was then increased to the hydrostatic value, and the length decreased as indicated in the figure. The axial load was then increased under a no-radial-strain condition. The sand pack behaved as if it had been preconsolidated approximately 5.5 × MPa. These results suggest that a preconsolidation effect could be caused by an increase in the horizontal stresses before the reservoir-pressure decline began.

A change from elastic to inelastic behavior may occur that is unrelated to previous stresses or rates of loading. Over geologic time the reservoir rock may have developed a limited capacity to deform elastically due to either physical or chemical phenomena or both. When this limit is exceeded inelastic behavior begins, and the preconsolidation effect would represent the end of elastic behavior. A similar possibility is that for a limited pressure drop the reservoir rock deforms inelastically and the surrounding rock behaves elastically. The end of the preconsolidation effect would thereby be related to the end of the elastic behavior of the overburden.

A delayed response of interbedded shale layers as a result of nonequilibrium fluid pressure is not a satisfactory explanation for the observed preconsolidation effect (Holzer, 1981). Some reservoirs exhibiting this effect contain significant amounts of thin shales. Compaction of these shales should follow the reservoir-pressure decline without appreciable time lag.

Maximum Subsidence

As discussed in the following paragraphs, the maximum surface subsidence in uniformly thick horizontal reservoirs can be expressed as the product of three factors: (1) the one-dimensional compaction, (2) a factor relating to the transfer of stresses from the surrounding formations to the reservoir rock, and (3) a factor relating the maximum reservoir compaction to the maximum surface subsidence. Results will be presented of finite-element calculations performed by us to illustrate the behavior of the last two factors.

The maximum subsidence, S, can be expressed as

$$S = (C_m \Delta ph) \left(\frac{\Delta h_m}{C_m \Delta ph} \right) \left(\frac{S}{\Delta h_m} \right) \quad (1a)$$

or

$$S = \begin{pmatrix} \text{One-dimensional} \\ \text{Compaction} \end{pmatrix} \begin{pmatrix} \text{Stress} \\ \text{Transfer} \\ \text{Factor} \end{pmatrix} \begin{pmatrix} \text{Subsidence} \\ \text{Spreading} \\ \text{Factor} \end{pmatrix} \quad (1b)$$

Note that the right-hand side of equation 1a reduces to S if $C_m \Delta ph$ and Δh_m are cancelled.

The one-dimensional compaction is equal to $C_m \Delta ph$, where C_m is the one-dimensional compaction coefficient (Teeuw, 1973; Geertsma, 1957), Δp is the pressure drop, and h is the reservoir thickness. One-dimensional compaction has been used in soil mechanics for many years to evaluate clay consolidation (Terzaghi, 1943).

The stress transfer, $\Delta h_m / C_m \Delta ph$, is the ratio of the maximum reservoir compaction, Δh_m, to the one-dimensional compaction. This factor is related to the transfer of stresses from the surrounding formations to the reservoir rock. In one-dimensional compaction the weight of the overburden acts as a dead load on the reservoir rock. There is no support of the overburden from the formations surrounding the reservoir. The maximum reservoir compaction, Δh_m, is a result of interaction of the overburden load with the reservoir rock and surrounding formations. Thus, the factor $\Delta h_m / C_m \Delta ph$ indicates the amount of stress transferred to the reservoir rock due to the pressure drop, Δp.

The subsidence spreading factor is a measure of the amount that the maximum reservoir compaction is reflected in the maximum surface subsidence. For shallow reservoirs covering large areas, the maximum subsidence is essentially equal to the maximum reservoir compaction, and the subsidence spreading factor is approximately 1.0. For deep reservoirs covering small areas, the maximum subsidence may be only a small fraction of the maximum reservoir compaction, and the subsidence spreading factor is small. In such cases the subsidence is spread over an area much larger than the reservoir (Geertsma, 1973a).

One-Dimensional Compaction

One-dimensional compaction is small for many reservoirs due to any or all of the three parameters, C_m, Δp, or h, being small. Many reservoirs have small thicknesses and/or small pressure declines. Hard, well-consolidated rocks generally have small values of C_m.

In most cases C_m is the most difficult of the three one-dimensional compaction parameters to determine. Unpublished laboratory data obtained at Chevron Oil Field Research Company indicate that C_m can vary over more than two orders of magnitude, the largest values usually being for the initial loading of high-porosity unconsolidated or semiconsolidated rock samples. Under certain conditions C_m can be obtained from observed compaction. Unfortunately, this requires significant compaction and much more detailed information than is usually available. Values of C_m can be calculated from pressure, and shear sonic velocities from full wave-form acoustic logs. Unfortunately, these values correspond to extremely small and short time-stress changes that are not representative of conditions associated with reservoir compaction.

Perhaps the best way to obtain values of C_m is to measure it on core samples. However, caution should be used in applying these data, since the results are affected by sample disturbance, improper stress conditions, anisotropic and inhomogeneous samples, and experimental error. Few C_m data are available in literature for reservoir rock other than those presented by Teeuw (1973).

In general, C_m can be estimated from the bulk compressibility, C_b, for which there is a larger data base. For well-consolidated rock that behaves elastically, Teeuw (1973) has related C_m to C_b:

$$C_m = \frac{1}{3} \left(\frac{1+\nu}{1-\nu} \right) \left(1-\beta \right) C_b, \quad (2)$$

where ν is Poisson's ratio and β is the ratio of grain compressibility to the bulk compressibility.

For most rocks Poisson's ratio lies between 0.15 and 0.35. Assuming β is sufficiently small that it can be neglected, C_m falls between 0.45 C_b and 0.69 C_b for elastic rocks, which are generally well consolidated. For soft unconsolidated sandstone, where significant grain fracturing and grain rearranging take place, results obtained by G. H. Newman of Chevron Oil Field Research Company indicate that C_m is approximately equal to C_b. For friable, semiconsolidated sandstones C_m generally falls between C_b and the value given by equation 2. Thus, for most sandstones, C_m falls between 0.45 C_b and C_b.

Figure 4 presents values of C_b measured at Chevron Oil Field Research Company, La Habra, California, on sandstone core samples from various parts of the world. These data represent the sample during initial loading, midway between the estimated initial in situ effective stress and the lithostatic stress.

The results presented in Figure 4 indicate that the bulk compressibility, C_b, generally decreases with depth. The unconsolidated and friable samples tend to be more compressible and come from shallow formations. This agrees with the increasing cementation and consolidation with depth observed in most sedimentary basins.

Figure 4 indicates that for sandstones, C_b and hence C_m vary over two orders of magnitude. Most values are much less than the few highest ones. Assuming that reservoirs that have the potential for large-scale compaction have high values of C_m, the values of C_b and C_m for most reservoirs should be much less.

Stress Transfer Factor

We used the finite-element stress analyses computer pro-

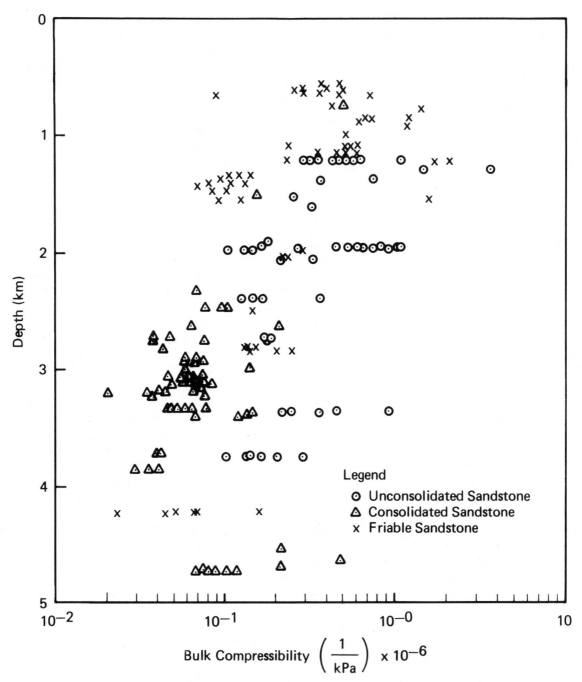

Figure 4. Bulk compressibility, C_b, of sandstone core samples versus depth.

gram, SAP IV (Bathe and others, 1973), to calculate stress transfer and subsidence spreading factors for idealized mathematical models. These models represent a uniformly thin horizontal disc embedded in a poroelastic half-space. The mechanical properties of the reservoir rock may be different from the surrounding rock. Otherwise, all rocks are homogeneous, isotropic, and linearly poroelastic. These models are similar to those used by Geertsma (1973a, 1973b). They are more general in that they have finite thicknesses, and the properties of the reservoir and surrounding rock may be different. The advantages of these mathematical models are their simplicity and the ease with which they can be analyzed. The disadvantages are the limitations imposed by the assumption of homogeneous, isotropic poroelastic rock and the simple geometry of the models.

A finite-element computer program was used in a series of reservoir models to calculate the stress distribution, reservoir compaction, and surface subsidence resulting from a given pressure drop. These models were used to find the stress transfer and subsidence spreading factors. The rock properties were represented by Young's modulus and Poisson's ratio. In most of the

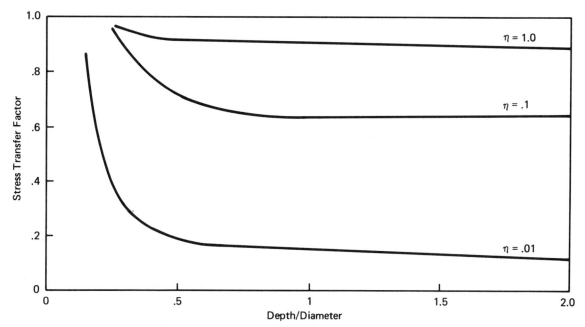

Figure 5. Stress transfer factor versus depth/diameter ratio of circular reservoirs with thickness/diameter ratio of 0.1 and η = 1.0, 0.1, and 0.01.

calculations, Poisson's ratio was 0.25 both inside and outside the reservoir, whereas Young's modulus was smaller inside the reservoir than outside. Thus, in most calculations the reservoir rock was represented as being softer and more compressible than the surrounding rock. This condition was designed to approximate internal failure of the reservoir rock but not of the surrounding rock. The bulk rock compressibility was assumed to be sufficiently greater than the grain compressibility so that the coefficient of fluid pressure $(1-\beta)$, in the equation for effective stress is approximately 1.0.

Figure 5 presents results we obtained for the stress transfer factor as a function of the depth/diameter ratio for a thickness/diameter ratio of 0.1 for three values of η where η is the ratio of Young's modulus inside the reservoir to Young's modulus for the surrounding rock. The stress transfer factor increases to 1.0 as the depth/diameter ratio approaches zero (Geertsma, 1973a, 1973b). This figure indicates that the stress transfer factor is approximately 1.0 for $\eta = 1$ but varies drastically for $\eta = 0.01$ in the depth/diameter ratio range from 0.25 to 0.50. Above 0.5 the stress transfer factor is relatively small (less than 0.20) and decreases slowly with an increasing depth/diameter ratio. In this range the surrounding rock takes on most of the loss of overburden support caused by the reservoir fluid pressure drop.

Finite-element calculations that we performed indicate that shear stresses can be sufficient to cause failure in the surrounding rock. This type of failure should increase the stress transfer factor, since it causes the reservoir rock to support more of the overburden weight.

Reservoir compaction should affect the fluid pressure in the surrounding rock and overburden. These pressure changes will strengthen the rock in some regions but will weaken it in others. A detailed analysis of the effects of rock failure on the stress transfer factor is beyond the scope of this paper. Exploratory calculations indicate that failure should be important in cases of large-scale reservoir compaction. Stress transfer factors calculated on the basis of linear poroelasticity are probably valid for small to moderate amounts of compaction.

Figure 6 presents the variation of the stress transfer factor with the thickness/diameter ratio for $\eta = 0.01$ and for depth/diameter ratios of 0.25, 0.50, and 1.0. This figure indicates that the stress transfer factor can be significantly less than 1.0 for reservoirs with highly compressible rock and thickness/diameter ratios as low as 0.01.

Subsidence Spreading Factor

Geertsma (1973a, 1973b) developed an equation for the subsidence spreading factor for an infinitely thin, horizontal, circular reservoir in uniform, linear, poroelastic half-space. He found for his models the subsidence spreading factor is a function of only the depth/diameter ratio. Our finite-element calculations allow for finite reservoir thicknesses and different mechanical properties inside and outside the reservoir. Our calculations indicate that the subsidence spreading factor is approximately a function of the depth/diameter. In all the calculations the thickness/diameter ratio was equal to 0.1 or less. All the results fall within the shaded region in Figure 7, indicating that the subsidence spreading factor is relatively small for depth/diameter ratios greater than 1.0.

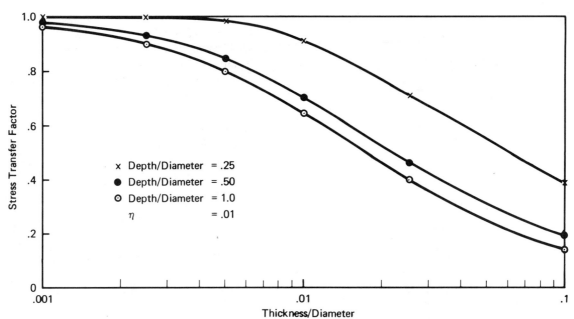

Figure 6. The variation of the stress transfer factor with thickness/diameter ratio for depth/diameter ratios of 0.25, 0.50, and 1.00.

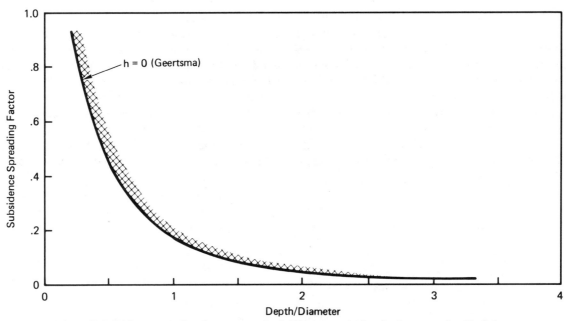

Figure 7. Subsidence spreading factor versus depth/diameter ratio for circular reservoirs. Shaded area contains all of our calculated results for thickness/diameter ratios of 0.10 or less.

Analysis of Field Data From Louisiana

Each of the major factors that determine the maximum subsidence is discussed in the preceding paragraphs. The results imply that for most oil and gas reservoirs the maximum subsidence is small, and that large-scale subsidence requires an unusual set of conditions. In general, the one-dimensional compaction, $C_m \Delta p h$, should be large, and neither the stress transfer factor nor the subsidence spreading factor should be small.

The University of Oklahoma Petroleum Data System was used to investigate the characteristics of oil and gas reservoirs in California, Louisiana, and Texas. Our investigation revealed many similarities in the data from various areas. Unfortunately, only the data from Louisiana were in a form suitable for our analysis. Figure 8 presents a plot of the number of Louisiana reservoirs versus depth expressed as a percent of the total. This

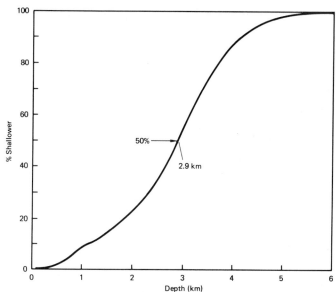

Figure 8. Distribution of Louisiana oil and gas reservoirs with depth.

figure indicates that only about 10% of the reservoirs are shallower than 1 km, almost half are deeper than 3 km, and most are deeper than 2 km.

Figure 9 presents a plot of the number of reservoirs expressed in percent of total versus thickness. This figure indicates that most Louisiana reservoirs are less than 10 m thick.

Figure 10 presents a plot of the number of reservoirs expressed in percent of total versus thickness/diameter ratio for Louisiana reservoirs. The diameter used is that of a circle containing the area of the reservoir as reported in the data file. Obviously a circle can be a poor approximation to the reservoir shape. The significance of Figure 10 is the very small thickness/diameter ratios indicated by the curve. According to this figure most Louisiana reservoirs have a thickness/diameter ratio of less than 0.01. The diameter used can be in considerable error without affecting the overall results.

Figure 11 presents a plot of the number of reservoirs expressed in percent of total versus the depth/diameter ratio for Louisiana reservoirs. This figure indicates that most have depth/diameter ratios greater than 1.0. For most, the subsidence spreading factor is less than 0.20 (Fig. 7).

In summary, most Louisiana reservoirs are deeper than 2 km and have a thickness of less than 10 m, a thickness/diameter ratio of less than 0.01, a depth/diameter ratio greater than 1.0, and a subsidence spreading factor less than 0.20. Therefore, most normally pressured reservoirs should have a maximum possible subsidence of less than 0.02 m. This is based on equation 1 using a bulk compressibility of 0.5×10^{-7} 1/kPa from Figure 4; a maximum pressure drop, Δp, of 19,930 kPa; a thickness of 10 m, for an assumed subsidence spreading factor of 0.2; and an assumed stress transfer factor of 1.0.

A number of the deeper Louisiana reservoirs were initially highly overpressured. For these the maximum possible pressure drop, Δp, can be twice that for normally pressured reservoirs. Thus, considering pressure effects only, most initially overpressured Louisiana reservoirs should have a maximum possible subsidence of less than 0.04 m.

Unfortunately, there are few data on C_m or C_b for rocks from initially highly overpressured reservoirs. These rocks may be more compressible than those from normally pressured reservoirs. Such an effect could be offset by smaller values of the subsidence spreading factor and smaller stress transfer factors due to the overall greater depths of the overpressured reservoirs.

Many oil and gas fields consist of a vertical sequence of reservoirs. Subsidence calculations must include the effects of each reservoir.

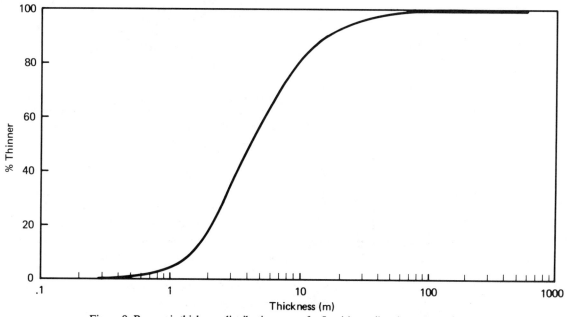

Figure 9. Reservoir thickness distribution curve for Louisiana oil and gas reservoirs.

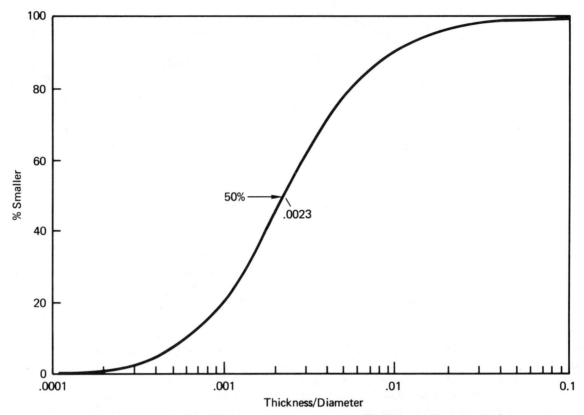

Figure 10. Distribution of Louisiana oil and gas reservoirs with thickness/diameter ratio.

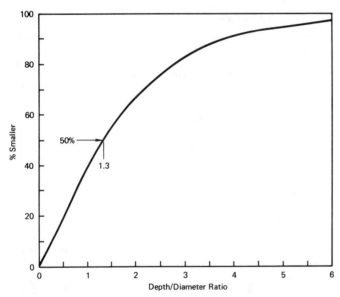

Figure 11. The variation of Louisiana oil and gas reservoirs with depth/diameter ratio.

Subsidence Prediction and Control

Subsidence due to compaction over oil and gas fields can be avoided by injecting fluid to maintain pressure. For strong water-drive reservoirs, subsidence can be avoided by restricting reservoir withdrawals to allow sufficient water influx to maintain reservoir pressure.

For the few reservoirs susceptible to large-scale compaction, as yet there is no proven method of accurately predicting compaction and subsidence. Prediction methods vary from strictly empirical to elaborate models based on principles of soil and rock mechanics. Empirical methods suffer from the lack of well-documented examples to establish their validity. Subsidence models require knowledge of the mechanical rock properties, which are either obtained from laboratory tests on core samples or are deduced from field observations. Tests on core samples are adversely affected by the disturbance the samples receive before they are tested and from the lack of knowledge of the in situ stresses. Matching subsidence models to observed subsidence suffers from the requirements that significant subsidence must take place before the method can be applied and from the difficulties involved in determining mechanical properties from field observations.

Empirical methods of subsidence prediction are given by Miller and Somerton (1955) and Castle and others (1970). Prediction methods based on the principles of soil and rock mechanics include those of McCann and Wilts (1951), Geertsma and Opstal (1973), Finol and Farouq Ali (1975), and Kosloff and others (1980a, 1980b).

At least three factors make subsidence prediction difficult, in addition to the variations in rock properties, fluid pressure, and reservoir geometry. These factors are (1) the preconsolidation effect; (2) the one-dimensional compaction coefficient, C_m, after the preconsolidation effect has been overcome; and (3) the failure characteristics of the rock surrounding and overlying the reservoir. Each of these factors may involve in situ rock failure or its beginning. In general, in situ rock failure is very difficult to predict. Accurately predicting large-scale subsidence, which may depend upon three different aspects of in situ rock failure, can be very difficult, if not impossible using current technology. Subsidence prediction should become easier as the subsidence progresses. Once large-scale subsidence has run its course, it may be relatively simple to adjust parameters to obtain agreement between field data and theoretical calculations.

It may be possible in theory to measure the amount of preconsolidation and the value of C_m on core samples and use this information in predicting subsidence. Unfortunately, reservoirs most likely to have large-scale compaction are likely to have reservoir rock that is weak to unconsolidated, with high porosity. Sample disturbance may destroy the in situ properties of the cores before the laboratory tests begin. In situ stresses, which can strongly affect the rock compression characteristics, are difficult to determine and reproduce in the laboratory.

Because subsidence prediction may be difficult does not mean that it should not be attempted. In making subsidence predictions one should examine and utilize all potentially valuable sources of information, including conventional geology, geophysics, well-logging, surface surveys, collar logs, radioactive bullet logs, well-test results, fluid-pressure surveys, laboratory tests on core samples, and mathematical models.

The approach used should depend upon the conditions of the problem, but in most cases initial efforts should be directed toward determining the one-dimensional compaction, $C_m \Delta ph$. If this proves to be small, the analysis can be terminated. If not, a linear elastic earth-stress model, such as a finite-element model or perhaps an analytic model in a half-space, should be considered. The results of this model should be compared to the subsidence field data, if available, to determine if a suitable fit can be obtained. The results should also be examined for stress conditions that indicate rock failure. If either of these two conditions are not met (that is, no fit to the field data or signs of rock failure), an inelastic model should be considered.

CONCLUSIONS

1. Most oil and gas reservoirs experience only small amounts of reservoir compaction and surface subsidence.
2. There are a number of possible explanations for the preconsolidation effect evident in field data, but the true cause or causes have not been established.
3. Evidently, large-scale oil and gas reservoir compaction and surface subsidence are associated with inelastic behavior of the reservoir rock and in some cases inelastic behavior of the surrounding and overlying rock.
4. As yet no reliable method has been established to predict the transition from elastic to inelastic reservoir rock behavior or accurately predict large-scale reservoir compaction and subsidence.
5. Important factors in oil-field subsidence are reservoir fluid pressure, depth, geometry, and mechanical rock properties (including the mechanical properties of the surrounding and overlying formations). The mechanical rock properties are by far the most difficult to obtain and the factors on which additional research is most needed.

REFERENCES CITED

Allen, D. R., 1968, Physical changes of reservoir properties caused by subsidence and repressuring operations: Journal of Petroleum Technology, January, p. 23–29.

Allen, D. R., and Mayuga, M. N., 1970, The mechanics of compaction and rebound, Wilmington Oil Field, Long Beach, California, U.S.A.: Proceedings of the Tokyo Symposium on Land Subsidence, published jointly by the International Association of Scientific Hydrology, Braamstraat 61 (rue des Ronces), Gentbrugge (Belgium), and Unesco, Place de Fontenoy, 75 Paris-7^e, p. 410–423.

Bathe, K., Wilson, E. L., and Peterson, J. E., 1973, SAP IV—A structural analysis program for static and dynamic response of linear systems: Earthquake Engineering Research Center, University of California, Berkeley, Report no. EERC 73-11.

Bjerrum, L., 1967, Engineering geology of Norwegian normally-consolidated marine clays as related to settlements of buildings: Geotechnique, v. 17, no. 2, p. 94–99.

Boot, R., 1973, Level control surveys in the Groningen Gasfield: Verhandelingen van het Koninklijk Nederlands geologisch mijnouwkundig Genootschap, v. 28, p. 105–109.

Castle, R. O., and others, 1970, A linear relationship between liquid production and oil-field subsidence: Proceedings of the Tokyo Symposium of Land Subsidence, published jointly by the International Association of Scientific Hydrology, Braamstraat 61 (rue des Ronces), Gentbrugge (Belgium), and Unesco, Place de Fontenoy, 75 Paris-7^e, p. 162–173.

Coxe, L. C., 1949, Long Beach Naval Shipyard endangered by subsidence: Civil Engineer, v. 19, no. 11, p. 44–47, 90.

de Loos, J. M., 1973, In situ compaction measurements in Groningen observation wells: Verhandelingen van het Koninklijk Nederlands geologisch mijnbouwkundig Genootschap, v. 28, p. 79–104.

Erickson, R. C., 1977, Subsidence control and urban oil production—a case history, Beverly Hills (East) Oilfield, California: Proceedings of the International Symposium on Land Subsidence, Anaheim, California, International Association of Hydrological Sciences Publication 121, p. 285–297.

Finol, A., and Farouq Ali, S. M., 1975, Numerical simulation of oil production with simultaneous ground subsidence: Transactions of the Society of Petroleum Engineers, v. 259, p. 411–424.

Frame, R. G., 1952, Earthquake damage, its cause and prevention in the Wilmington Oil Field, in summary of operations, California Oil Fields, California: California Department of Natural Resources, Division of Oil and Gas, v. 38, no. 1, p. 5–15.

Geertsma, J., 1957, The effect of fluid pressure decline on volume changes of porous rocks: Transactions of the American Institute of Mining Engineers, v. 210, p. 331.

—— 1973a, Land subsidence above compacting oil and gas reservoirs: Journal of Petroleum Technology, p. 734–744.

—— 1973b, A basic theory of subsidence due to reservoir compaction, the homogeneous case: Verhandelingen van het Koninklijk Nederlands geologisch mijnbouwkundig Genootschap, v. 28, p. 43–62.

Geertsma, J., and van Opstal, G., 1973, A numerical technique for predicting subsidence above compacting reservoirs, based on the nucleus of strain concept: Verhandelingen van het Koninklijk Nederlands geologisch mijnbouwkundig Genootschap, v. 28, p. 63–78.

Gilluly, J., and Grant, U.S., 1949, Subsidence in the Long Beach Harbor Area, California: Geological Society of America Bulletin, v. 60, p. 461–530.

Harris, F. M., and Harlow, E. H., 1947, Subsidence of the Terminal Island-Long Beach Area, California: American Society of Civil Engineers Proceedings, v. 73, no. 8, p. 1035–1072.

Holzer, T. L., 1981, Preconsolidation stress of aquifer systems in areas of induced land subsidence: American Geophysical Union, Water Resources Research, v. 17, no. 3, p. 699–711.

Hower, J., and others, 1976, Mechanism of burial metamorphism of argillaceous sediments 1. Mineralogical and chemical evidence: Geological Society of America Bulletin, v. 87, p. 725–737.

Ilijn, A. S., 1977, The earth surface subsidence at the areas of gas and oil pumping out: Proceedings of the Second International Symposium on Land Subsidence, Anaheim, California, International Association of Hydrological Sciences Publication 121, p. 665.

Kennedy, D.J.L., 1961, A study of failure of liners for oil wells associated with the compaction of oil producing strata [Ph.D. thesis]: University of Illinois, Urbana, Illinois, 251 p.

Kosloff, D., Scott, R. F., and Scranton, J., 1980a, Finite element simulation of Wilmington Oil Field subsidence: I, Linear modeling: Tectonophysics, v. 65, p. 339–368.

—— 1980b, Finite element simulation of Wilmington Oil Field subsidence: II, Nonlinear modeling: Tectonophysics, v. 70, p. 159–183.

Kovach, R. L., 1974, Source mechanisms for Wilmington Oil Field, California, subsidence earthquakes: Seismological Society of America Bulletin, v. 64, no. 3, p. 699–711.

Mayuga, M. N., 1970, Geology and development of California's giant—Wilmington Oil Field: Geology of giant petroleum fields: American Association of Petroleum Geology Memoir 41, p. 158–184.

Mayuga, M. N., and Allen, D. R., 1970, Subsidence in the Wilmington Oil Field, Long Beach, California, U.S.A.: Proceedings of the Tokyo Symposium on Land Subsidence, published jointly by the International Association of Scientific Hydrology, Braamstraat 61 (rue des Ronces), Gentbrugge (Belgium), and Unesco, Place de Fontenoy, 75 Paris-7e, p. 66–79.

McCann, G. D., and Wilts, C. H., 1951, Mathematical analysis of subsidence in the Long Beach–San Pedro area, California: California Institute of Technology, Pasadena, California.

Merle, H. A., Kentie, C.J.P., van Opstal, G., and Schneider, G.M.G., 1975, The Bachaquero study, a composite analysis of the behavior of a compaction drive/solution gas drive reservoir: Society of Petroleum Engineers Paper 5529, presented at the 50th Annual Fall Meeting, Dallas, Texas.

Miller, O. E., and Somerton, W. H., 1955, Operators eye heroic measures to halt Wilmington sinking: Oil and Gas Journal, December 19, v. 54, no. 33, p. 7.

Nuñez, O., and Escojido, D., 1977, Subsidence in the Bolivar Coast: Proceedings of the Second International Symposium on Land Subsidence, Anaheim, California, International Association of Hydrological Sciences Publication 121, p. 257–266.

Pierce, R. L., 1970, Reducing land subsidence in the Wilmington Oil Field by use of saline water: Water Resources Research, v. 6, no. 5, p. 1505–1514.

Poland, J. F., and Davis, G. H., 1969, Land subsidence due to withdrawal of fluids, in Varnes, D. J., and Kiersch, G. E., eds., Reviews in Engineering Geology, Volume II: Geological Society of America, p. 187–269.

Pratt, W. E., and Johnson, D. W., 1926, Local subsidence of the Goose Creek Field: Journal of Geology, v. 34, no. 1, p. 577–590.

Richter, C. F., 1958, Elementary seismology: W. H. Freeman Company.

Roberts, J. E., 1970, Sand compression as a factor in oil field subsidence: Proceedings of the Tokyo Syposium of Land Subsidence, published jointly by the International Association of Scientific Hydrology, Braamstraat 61 (rue des Ronces), Gentbrugge (Belgium), and Unesco, Place de Fontenoy, 75 Paris-7e, p. 368–376.

Roberts, J. E., and de Souza, J. M., 1958, The compressibility of sands: American Society for Testing and Materials Proceedings, v. 58, p. 1269–1277.

Schoonbeek, J. B., 1976, Land subsidence as a result of natural gas extraction in the Province of Groningen: Society of Petroleum Engineers of AIME, SPE Paper 5751, presented at the SPE–European Spring Meeting, Amsterdam, The Netherlands.

Shoemaker, R. R., 1955, Protection of subsiding waterfront properties: American Society of Civil Engineers Proceedings, v. 81, no. 805-1-805.24.

Shoemaker, R. R., and Thorley, T. J., 1955, Problems of ground subsidence: American Water Works Association Journal, v. 47, no. 4, p. 412–418.

Schrefler, B. H., Lewis, R. W., and Norris, V. A., 1977, A case study of the surface subsidence of the Polesine area: International Journal for Numerical and Analytical Methods in Geomechanics, v. 1, p. 377–386.

Teeuw, D., 1973, Laboratory measurement of compaction properties of Groningen reservoir rock: Verhandelingen van het Koninklijk Nederlands geologisch mijnbouwkundig Genootschap, v. 28, p. 19–32.

Terzaghi, K., 1943, Theoretical soil mechanics: New York, John Wiley and Sons, Inc., p. 265–296.

van Kesteren, J., 1973a, The analysis of future surface subsidence resulting from gas production in the Groningen Field: Verhandelingen van het Koninklijk Nederlands geologisch mijnbouwkundig Genootschap, v. 28, p. 11–18.

—— 1973b, Estimate of compaction data representative of the Groningen Field: Verhandelingen van het Koninklijk Nederlands geologisch mijnbouwkundig Genootschap, v. 28, p. 33–42.

van der Knaap, W., and van der Vlis, 1967, On the cause of subsidence in oil-producing areas: 7th World Petroleum Congress, Mexico City, v. 3, p. 85–105.

Winterburn, R., 1943, Wilmington Oil Field, California: California Division of Mines and Geology Bulletin 118, p. 301–305.

Yerkes, R. F., and Castle, R. O., 1970, Surface deformation associated with oil and gas field operations in the United States: Proceedings of the Tokyo Symposium on Land Subsidence, published jointly by the International Association of Scientific Hydrology, Braamstraat 61 (rue des Ronces), Gentbrugge (Belgium), and Unesco, Place de Fontenoy, 75 Paris-7e, p. 55–66.

MANUSCRIPT ACCEPTED BY THE SOCIETY APRIL 18, 1984

Subsidence due to geothermal fluid withdrawal

T. N. Narasimhan
K. P. Goyal*
Earth Science Division
University of California
Lawrence Berkeley Laboratory
Berkeley, California 94720

ABSTRACT

Single-phase and two-phase geothermal reservoirs are currently being exploited for power production in Italy, Mexico, New Zealand, the United States, and elsewhere. Vertical ground displacements of up to 4.5 m and horizontal ground displacements of up to 0.5 m have been observed at Wairakei, New Zealand, that are clearly attributable to the resource exploitation. Similarly, vertical displacements of about 0.13 m have been recorded at The Geysers, California. No significant ground displacements that are attributable to large-scale fluid production have been observed at Larderello, Italy, and Cerro Prieto, Mexico. Observations show that subsidence due to geothermal fluid production is characterized by such features as an offset of the subsidence bowl from the main area of production, time-lag between production and subsidence, and nonlinear stress-strain relationships. Several plausible conceptual models, of varying degrees of sophistication, have been proposed to explain the observed features. At present, relatively more is known about the physical mechanisms that govern subsidence than the relevant thermal mechanisms. Although attempts have been made to simulate observed geothermal subsidence, the modeling efforts have been seriously limited by a lack of relevant field data needed to sufficiently characterize the complex field system.

INTRODUCTION

In many parts of the world geothermal energy is being actively exploited for power generation. Compared to oil and coal, the energy content of a unit mass of geothermal water is relatively small. Hence, power production from geothermal reservoirs, especially those dominated by liquid water, entails the extraction of large volumes of the fluids, leading invariably to the mining of these fluids. This depletion of stored fluid volume is compensated largely by a reduction in the bulk volume of the reservoir with associated reservoir deformation. Abundant field evidence exists to show that the effects of reservoir deformation often propagate to the land surface to be manifested as vertical and horizontal ground displacements. Although the term "subsidence" is suggestive of vertical downward movement of the ground surface, we shall, in this paper, use the term in a more general context to include both horizontal and vertical displacements.

Additionally, the deformations accompanying reservoir depletion may also lead to the activation of movements along preexisting faults, leading to seismic events. The ground displacements, which may often attain magnitudes of several meters, can lead to significant environmental consequences in some areas. For example, vertical movements of only a few feet in some coastal areas, such as in Texas, can lead to flooding and loss of valuable urban or agricultural lands. Abrupt spatial changes in the magnitude of subsidence, on the other hand, can lead to the rupturing of irrigation canals or pipelines. There exists, therefore, a practical desire to exploit the geothermal resource in such a fashion that the deleterious effects of land subsidence are minimal and acceptable. To achieve this end, a proper understanding of the subsidence mechanism is essential so that the consequences of specific exploitation strategies can be foreseen and appropriate ameliorative measures taken.

The purpose of this paper is to assess our current status of knowledge related to subsidence caused by the removal of geo-

---

*Present address: Phillips Petroleum Company, 655 East 4500 South, Salt Lake City, Utah.

thermal fluids. In particular, we shall address the following questions: What are the patterns and magnitudes of subsidence that have been observed in different parts of the world? What are the physical bases that relate fluid withdrawal and ground displacements? What is our current ability to predict land subsidence with the help of mathematical models? And finally, what are the key questions that need to be answered in order to increase our ability to predict subsidence?

We shall begin the paper with a description of case histories relating to geothermal systems from around the world. We shall then describe the physical mechanisms that govern subsidence and examine how these physical mechanisms may be quantitatively analyzed using mathematical models. We shall close the paper with a discussion of the current status of knowledge and the identification of key issues requiring resolution.

FIELD OBSERVATIONS

In general, geothermal systems can be classified into five categories: normal gradient, radiogenic, high heat flow, geopressured, and hydrothermal (DiPippo, 1980). In normal gradient systems the temperature gradient in the earth's crust averages about 30 °C/km. Exploitation of such a system would require one to drill deep in the earth's crust, rendering this resource to be uneconomical at present.

Geothermal energy produced by the radioactive decay of uranium, thorium, and potassium in the earth's crust forms a radiogenic system. Radioactive decay of 1 kg granite can release about one-billionth of a watt of heat. Thus, a fairly large amount of heat energy can be obtained by tapping radiogenic resource of the earth's crust as a whole. However, this energy is quite diffused and a suitable medium may not be readily available to permit its large-scale extraction.

Surface temperatures are principally controlled by conductive flow of heat through solid rocks, by convecting flow in circulating fluids, or by mass transfer in magmas. The conduction-dominated, high-heat-flow areas may be associated with regions in which the crust is abnormally thin, thus allowing the mantle to come into closer proximity to the surface, or in which a large, deep-seated magma chamber is enclosed within the earth's crust. Such areas are often found to have large thermal gradients, sometimes as large as 2 to 4 times the normal gradient as found in the Hungarian Basin (Boldizsar, 1970) where temperature gradients of 40 to 75 °C/km are known. These regions are expected to yield high temperatures at shallow depths. However, such areas may not prove feasible for power production because of the diffused nature of energy contained in them.

The fourth type of geothermal system, the geopressured system, is found in regions where fluid pressures exist in excess of hydrostatic pressure gradient of 9.8 kPa/m (0.433 psi/ft). It is believed that any or all of the following processes are responsible for the existence of a geopressured system: Rapid burial of saturated sediments, with rates of loading exceeding rates of water expulsion; development of osmotic pressure across clay beds; and liberation of water through diagenetic alteration of montmorillonite to illite between temperatures of 80 to 120 °C (Jones, 1969, 1976). Such fields are found along the northern coast of Gulf of Mexico and in many other parts of the world. These fields do not have high temperature gradients, but considerable temperatures are encountered due to great depths ($\simeq 6$ km) involved. Such systems are of economic importance, as they are capable of delivering mechanical energy, thermal energy, and large supplies of methane gas. The Gulf coast of Texas and Louisiana is currently being explored with deep wells to harness this resource.

The last geothermal resource, the hydrothermal type, has been extensively exploited and used for power production, space heating, and other applications throughout the world because of its proximity to the earth's surface and its amenability to energy extraction. The driving heat energy for such systems is supplied at the base of the convection loop. Hydrothermal systems may be subclassified into two types: vapor-dominated and liquid-dominated systems, which differ in the physical state of the dominant pressure-controlling phase. In vapor-dominated systems, pressure is controlled by the steam phase, whereas in the liquid-dominated systems, it is controlled by liquid water. Among the geothermal systems discovered to date, hot-water systems are perhaps 20 times as common as vapor-dominated systems (Muffler and White, 1972). Among the liquid-dominated systems, Wairakei in New Zealand and Cerro Prieto in Mexico are currently producing 140 MW and 180 MW of electric power, respectively. Electric power is also being produced from the vapor-dominated systems such as The Geysers in California, U.S.A. (900 MW), and Larderello in Italy (380 MW).

Of the five categories of geothermal systems, only the geopressured and the hydrothermal systems are currently viable for economic power production. Therefore, we shall limit our discussion of subsidence to hydrothermal and geopressured systems.

In the following section, field observations are presented from several geothermal sites. Attempt is made to emphasize the important features relevant to subsidence. A total of six case histories are discussed. These include: the liquid-dominated systems at Wairakei and Broadlands in New Zealand and at Cerro Prieto in Mexico; the vapor-dominated systems at The Geysers in California and at Larderello in Italy; and the geopressured system at Chocolate Bayou in Texas.

Wairakei, New Zealand

Wairakei is located on the North Island of New Zealand. It is situated on the west bank of the Waikato River and lies 8 km north of Lake Taupo (Fig. 1). This liquid-dominated field occupies an area of 15 km^2 (Grindley, 1965) and extends about 5 km westward from the river over a relatively flat valley underlain by Taupo pumice alluvium. On the west, it is bordered by hills of Wairakei Breccia that rise 90 to 150 m above the valleys and serve as a groundwater recharge area. No boundaries have been indicated toward north and south as evidenced by the behavior of the wells. The structure of this field is controlled by numerous

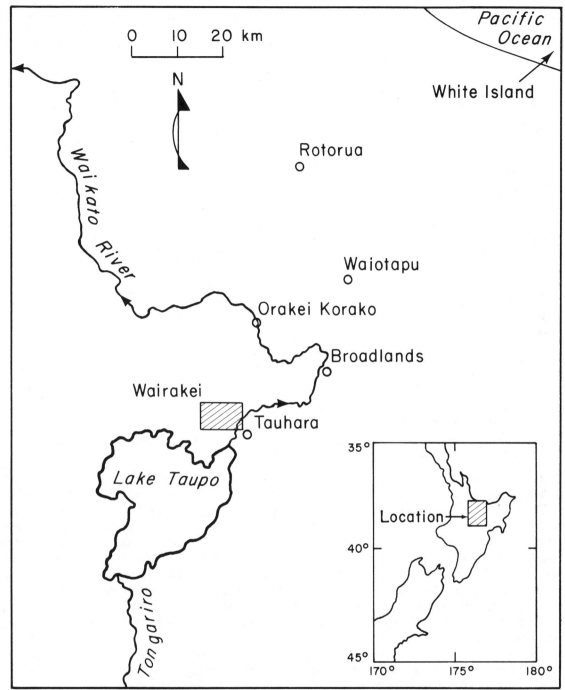

Figure 1. Location of the Wairakei geothermal field (from Mercer et al., 1975).

fractures associated with the Wairakei, Kaiapo, and Upper Waiora faults (Grimsrud et al., 1978). The geology of the Wairakei field is described in Grindley (1965), Healy (1965), and Grange (1937). The reservoir engineering data have been compiled by Pritchett et al. (1978), and subsidence-related studies are reported in Grimsrud et al. (1978), Viets et al. (1979), and Miller et al. (1980a, 1980b). A mixture of steam and water, in a ratio of about 1 to 4 by weight, is yielded by the Waiora Formation which is considered to be the main geothermal reservoir. Above the Waiora lies a relatively impermeable Huka mudstone. The Wairakei ignimbrites, considered to be practically impermeable, underlie the Waiora. The thickness of the Waiora Formation varies from about 366 m (1200 ft) in the west to more than 793 m (2600 ft) in the east. The Huka Falls Formation, a relatively fine grained lacustrine rock, is less than 100 m (300 ft) toward southwest of the main production area and thickens to about 310 m (1000 ft) toward northwest and southeast.

Geothermal fluid production at Wairakei started in early

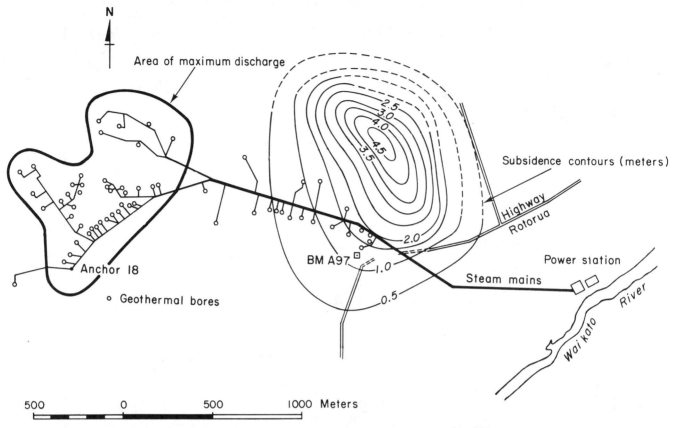

Figure 2. Vertical displacements due to geothermal fluid withdrawal at Wairakei, New Zealand, 1964–1975 (from Stillwell et al., 1975).

1950. The production increased significantly in 1958 with the commencement of power generation. A total of 141 wells were drilled in the field up to 1968 when drilling activity completely ceased. Of these, 65 bores account for about 95% of the total fluid produced from the entire field. It is believed that the reservoir was originally filled with hot water to the base of the Huka Falls Formation before production started. Based on the early exploration measurements, initial temperatures and pressures at the sea level were about 250 °C and 3965 kPag (575 psi g). Data presented by Pritchett et al. (1978) indicate that initial temperatures in the upper part of the reservoir may have been 10 to 40 °C lower than in the deeper parts. Presumably, the hottest fluids were found in areas close to faults and fissures. In the early years of production (1958–1962), recharge to the reservoir was about 10% of the fluid produced. This inadequate recharge led to large pressure drops in the reservoir. For example, pressure drops of the order of 1725 kPa g (250 psi g) were observed in the western production area, and over 2070 kPa g (300 psi g) in the eastern production area. However, recharge rose to about 90% of the fluid produced in the following period, leveling off at a pressure drop rate of less than 69 kPa g (10 psi g) per year. A total of about 1.05 trillion kg (2.33 trillion lb) of fluid had been produced from the Wairakei-Tauhara region as of December 31, 1976. This large-scale extraction of fluids has led to significant ground deformations in and around the Wairakei field.

Initial surface subsidence measurements were made in 1956 on the basis of bench marks established in 1950. Periodic measurements since then have shown that the area affected by subsidence exceeds 3 km^2. Subsidence at Wairakei has been reported by Hatton (1970) and Stilwell et al. (1975), and has subsequently been thoroughly reviewed by Pritchett et al. (1978). Observed vertical subsidence and horizontal ground movements are shown in Figures 2 and 3. As seen in Figure 2, the area of maximum subsidence occurs east of the main production zone, and the maximum subsidence was of the order of 4.5 m between 1964 and 1975. The horizontal movements, accompanying vertical subsidence, are represented by vectors in Figure 3. These vectors point toward the area of maximum subsidence. Also the observed horizontal deformations increase with increasingly vertical settlement. A horizontal movement of about 0.5 m can be observed near the zone of maximum subsidence in Figure 3. A plot of reservoir pressure drop versus subsidence at bench mark A-97 is shown in Figure 4. This figure shows that subsidence at Wairakei is characterized by (1) an off-set subsidence bowl, (2) a linear relation between reservoir pressure and subsidence up to 1963, and (3) a nonlinear relation after 1963.

As discussed elsewhere in this paper, surface and subsurface deformations may be expected to enhance the fault activity and the seismicity of the area. In a recent study Evison et al. (1976) found that both microearthquakes and macroearthquakes were

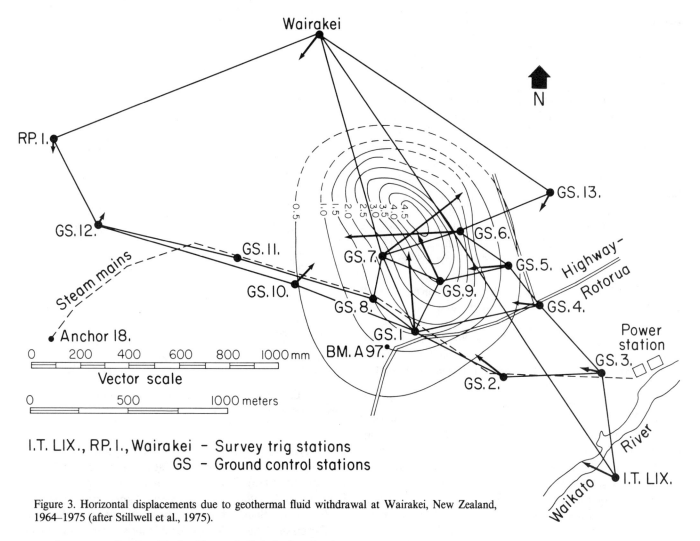

Figure 3. Horizontal displacements due to geothermal fluid withdrawal at Wairakei, New Zealand, 1964–1975 (after Stillwell et al., 1975).

many times more frequent in the Taupo fault belt than in the adjoining basins or in the Kaingaroa Plateau to the east. Nevertheless, to our knowledge no such study exists that specifically relates subsidence with seismicity in the Wairakei area. It is difficult to assert at this time that increased seismicity in the Wairakei field is due to increased subsidence.

At Wairakei spent geothermal fluids with approximately 4400 ppm of dissolved solids are discharged directly into the Waikato River (Defferding and Walter, 1978). Since 1968, no new wells have been drilled in the field, and about 140 MW of power have been steadily produced since then.

The surface deformations in the field have disrupted pipelines carrying steam, cracked drainage canals, and caused the main road to sink by 2 m (Viets et al., 1979). The recurring cost of repair of steam lines might range from $2000 to $10,000 per year. Fixing of drainage canal cost about $250,000.

Broadlands Geothermal Field, New Zealand

The Broadlands geothermal field, located about 28 km northeast of Wairakei, is another liquid-dominated geothermal

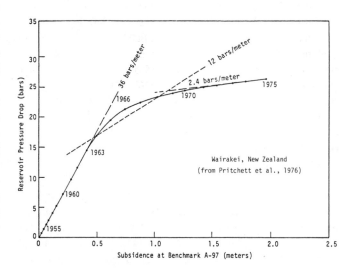

Figure 4. Relation between reservoir pressure drop and subsidence at Wairakei, New Zealand (from Pritchett et al., 1976).

system in New Zealand (Fig. 1). Its behavior, however, appears to be considerably different from that at Wairakei, largely due to the presence of significant amounts of carbon dioxide gas. The New Zealand Electricity Department is expecting to produce 150 MW of electricity from this field by mid 1980s. The first 100-MW unit may be in operation by late 1985. The exploration in this area began in the early 1960s, and drilling activity started in 1965. As of 1977, a total of 32 wells had been drilled, of which only 16 are considered to be good producers (DiPippo, 1980). These are wells BR2, 3, 8, 9, 11, 13, 17 to 23, 25, 27, and 28 (Fig. 5). The depth of the wells in this field vary between 760 and 1400 m, one well (BR15) reaching down to 2418 m. Over the field three thermal anomalies have been recognized (DiPippo, 1980). First is the Broadlands thermal anomaly covering an area of roughly 365-m radius centered around well BR7; second is the Ohaki thermal anomaly with an area of about 550-m radius centered on well BR9, and third is an elongated area between wells BR6 and BR13 which extends about 1220 m in north-south direction. Of these, the Ohaki anomaly is associated with best production.

The geology of the Broadland's area has been extensively studied and reported in Grindley (1970), Browne (1970), Hochstein and Hunt (1970), and Grindley and Browne (1976). The subsurface formations in the descending order include: Recent Pumice alluvium, Huka Falls Formation, Ohaki Rhyolite, the Waiora Formation, Broadlands Rhyolite, Rantawiri Breccia, Rangitaiki Ignimbrites, Waikora Formation, the Ohakuri group, and the Graywacke basement. The thickness of these formations is spatially variable. The Waiora Formation and Rantawiri Breccia are the two main aquifers that provide most of the fluid produced. The formations below the lower aquifer are quite dense and almost impermeable. The Huka Falls Formation and the Ohaki Rhyolite provide confinement to the Waiora aquifer, whereas the Broadlands Rhyolite apparently acts as a boundary separating the two aquifers. The local disruption of alternating permeable and impermeable formations by faults and dikes provides steep channels for fluid flow (Browne, 1970). The Ohaki and the Broadlands faults lie in the respective thermal areas. The lateral extent of the field, as determined from the resistivity surveys (Risk, 1976), is also shown in Figure 5. The resistivity of the region, enclosed by bars, is less than 5 ohm-m, and the resistivity anomaly encloses an area of about 10 km². The boundary between hot and cold ground is essentially vertical down to a depth of at least 3 km.

During a period of 5 years between 1966 and 1971, a total discharge of about 34 billion kg (74 billion lb) of fluid and 4.4×10^{16} J (42 trillion BTU) of heat had been extracted from the Broadlands field. The entire field was almost shut down for over 3 years between August 1971 and December 1974. Initial temperatures of about 260 and 300 °C existed in the upper and the lower reservoirs before exploitation.

The response of the Broadlands geothermal field is noticeably different from that of a conventional liquid-dominated system due to the presence of noncondensible gases, mainly CO_2. The

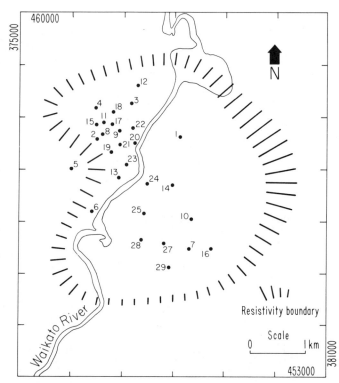

Figure 5. Locations of wells and resistivity boundary at the geothermal field at Broadlands, New Zealand.

partial pressure of gases reduces the boiling point of water by 3 °C at 300 °C and by about 1 °C at a temperature of 260 °C (Macdonald, 1976). Thus, a two-phase region is expected to exist in the reservoir within a depth of about 2 km during the preproduction state. Standard hydrostatic pressures, as defined by Hitchcock and Bixley (1976), existed in the aquifers of the Broadlands system prior to production. Since the commencement of production in early 1966, the reservoir pressures continued to decline until 1971 when the production ceased almost completely. During exploitation it was found that in the Ohaki area to the north the reservoir behaved as a single, interconnected unit, whereas Broadland area to the south was characterized by considerably lower permeability and contained several isolated pockets tapped by individual wells. According to Grant (1977), the communication between the Ohaki bores was also not perfect, as it took about a year for pressure transients to propagate across the Ohaki region. This behavior is quite different from that observed in the Wairakei field where the communication between wells was very good (Pritchett et al., 1978). Another remarkable difference observed between these two fields is in the size of the pressure drop. A pressure drop of as much as 1400 kPa (14 bars) was observed in some wells when exploitation ceased in 1971 in the Broadlands field (Hitchcock and Bixley, 1976). For a comparable amount of fluid withdrawn, the pressure drop observed in the Wairakei field was quite small. Apparently, the presence of small quantities of carbon dioxide has played a major role in determining the response of the Broadlands field. The

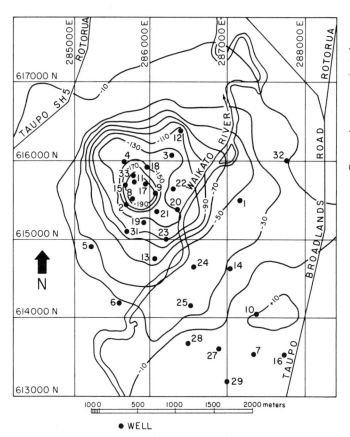

Figure 6. Subsidence at the Broadlands geothermal field between May 1968 and March 1974. Contour values are in millimeters (from Ministry of Works and Development, 1977).

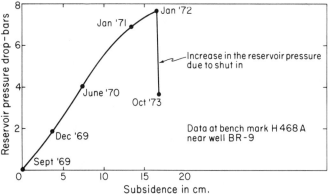

Figure 7. Relation between reservoir pressure drop and subsidence at Broadlands, New Zealand, 1969–1973.

lower effective permeabilty as evidenced at Broadlands can be attributed to the presence of the two fluid phases, each of which impedes the flow of the other (Grant, 1977). According to this same author, the pressure drop in the reservoir was mainly the drop in the partial pressure of the gas, with a little drop in the steam-phase pressure. After the 1971 shut-in, pressures in the field started to build up with a recovery rate of about 173 kPa/yr (25 psi/yr) over a 3-year period (Hitchcock and Bixley, 1976). Grant (1977) thought that this pressure recovery was primarily the recovery of gas pressure, and the amount of pressure increase was the measure of the total amount of CO_2 recharge to the system.

Under these conditions it is reasonable to infer that in the absence of CO_2, the pressure behavior should have been similar to that of Wairakei. Following the pattern of drawdown and recovery, ground subsidence and rebound were also observed at the Broadlands field as described below.

An extensive, precise level network was established in the Broadlands area during the 1967–1969 period. Approximately 500 bench marks were installed over an area of 65 km^2, covering a route distance of about 78 km. Initially, a precise level survey was carried out in May 1968, which was then resurveyed in September 1969 and in March 1974. Local subsidence surveys were conducted in September 1969, December 1969, June 1970, January 1971, January 1972, February 1975, and February 1976. Total vertical subsidence observed between May 1968 and March 1974 is shown in Figure 6, which also includes the recovery in the ground levels during the shut-down period between 1971 and 1974. Thus, the subsidence shown in Figure 6 is, in fact, smaller than the maximum magnitudes observed up to August 1971. For example, a maximum vertical displacement of 220 mm was observed over the period February 1969 to January 1972 compared to 190 mm shown in Figure 6, indicating clearly that a rebound has occurred. It may also be noted by comparing Figures 2 and 6 that unlike at Wairakei, the subsidence bowl in the Broadlands occurs directly over the region of maximum discharge. By using the pressure data from Hitchcock and Bixley (1976) and subsidence data from the Ministry of Works and Development (1977), a plot of reservoir pressure drop versus subsidence has been prepared for bench mark H468A in the vicinity of well BR9 and is shown in Figure 7 for the period September 1969 to October 1973. For this figure, September 1969 is taken as the datum, at which time a certain amount of subsidence and a pressure drop of about 600 kPa (6 bars) was already existing at well BR9 (Hitchcock and Bixley, 1976). A total subsidence of 6 mm took place at H468A between January 1972 and February 1975 (Ministry of Works and Development, 1977). Assuming a linear relation, we have calculated the subsidence for October 1973. Note from Figure 7 that the slope of the curve tends to change between January 1971 and January 1972 and drastically changes beyond January 1972. This is due to the rebound of the ground surface associated with rising reservoir pressures caused by the shut-in of the field in August 1971. Horizontal displacements associated with subsidence have also been observed in the Broadlands field and are shown in Figure 8. As seen in this figure, the maximum movement over a 6-year period (1968–1974) is about 120 mm. A reinjection plan is underway in the Broadlands area to minimize subsidence effects. Reinjection tests were conducted on wells BR7, BR13, BR23, BR33, and BR34 for periods varying from a few weeks to 3 years (Bixley and Grant, 1979). In all cases, water supersaturated with silica

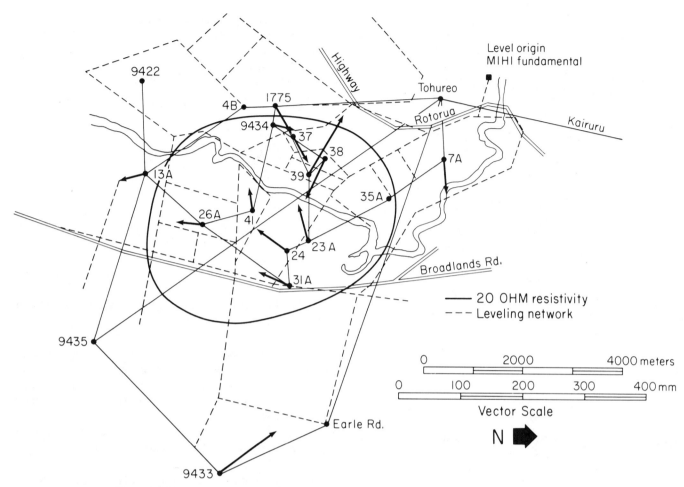

Figure 8. Observed horizontal displacements at the Broadlands geothermal field, New Zealand, 1968–1974 (from Stilwell et al., 1975).

was injected, and it was found that the permeability of the injected formation either increased or remained the same. Thus, reinjection tests are quite encouraging at the Broadlands field, and a full-scale reinjection scheme might be coming forth in the future. The potential problems related to subsidence at this site include flooding by Wairakei rivers and the role of ground deformation in siting power houses and steam lines.

Broadlands geothermal field, which lies in the Taupo-Reparoa basin, has very little microearthquake activity as compared with that in the Taupo fault belt where the activity is about two orders of magnitude higher (Evison et al., 1976). It appears that neither geothermal fluid production nor subsidence has any effect on the microseismicity of the Broadlands area.

Cerro Prieto, Mexico

Cerro Prieto geothermal field is the first liquid-dominated system in North America to be exploited for electric power generation. It is located in the Mexicali valley in the Colorado River delta, about 30 km south of the border of Mexico and the United States (Fig. 9). It occupies a relatively flat area of about 30 km^2 and exhibits some surface geothermal manifestations such as mud volcanoes (5 cm to 2 m high), steam and gas vents, hot springs, boiling mud ponds, and a 200-m-high black volcanic cone known as Cerro Prieto, after which the geothermal field is named.

Geologically the Cerro Prieto field is underlain by deltaic sediments that are classified into two units, Unit A and Unit B. Unit A has a thickness of 600 to 2500 m and contains nonconsolidated and semiconsolidated sediments of clay silts, sands, and gravels. Unit B consists of layered consolidated sediment shales and sandstone and is more than 2 km thick. The depths to the producing layers vary over the field between 600 to 900 m and 1300 to 1600 m to the west of the railroad track and between 1800 to 2000 m and 2200 to 2500 m to the east.

The structure of the Cerro Prieto field is controlled by numerous faults related to the San Andreas fault system. The locations of some important faults are shown in Figure 10. Hydrologically, these faults may or may not act as conduits for the influx of fluids from the basement. For example, the Cerro Prieto fault is thought to act as a western hydrologic boundary to the field, whereas the Morelia fault acts as a leaky boundary to

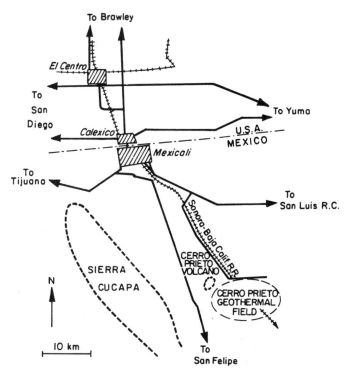

Figure 9. Location of the Cerro Prieto geothermal field, Mexico.

the north. The Delta, the Patzcuaro, and the Hidalgo faults appear to act as conduits to fluid flow. The eastern boundary of the field is not yet well established. On the basis of geophysical data and interference tests, it has been inferred that both Cerro Prieto I and Cerro Prieto II areas, lying on the west and east of the railroad track, respectively, are hydrologically interconnected. According to Mercado (1976), hot water in the eastern part of the field rises up and flows toward the west.

Electric power generation in the Cerro Prieto field began in 1973. In April 1979 the capacity of the plant was doubled to 150 MW, as two new units came into operation (Lippmann and Goyal, 1980). Consequently, fluid production rate has increased from about 2.8×10^6 to 4.2×10^6 kg (2800 to 4200 tonnes per hour. Total heat and mass produced as of November 1980 has been estimated to be 6×10^{13} kcal (2.4×10^{14} BTU) and 1.9×10^{11} kg (1.9×10^8 tonnes), respectively (Goyal et al., 1981). Figure 10 shows the location of over 60 deep wells that have been completed in the field. These wells produce a water-steam mixture, the weight ratio of which varies from well to well from 0.5:1 to 4:1. Under natural conditions, the waters in the producing strata are thought to have existed at or below the boiling point (A. H. Truesdell, 1980, personal commun.). This view is supported by temperature logs which show that the highest temperature of the water has been equal to the saturation temperature corresponding to the hydrostatic pressure of the hot to saline water. Enthalpies of the produced fluid were found to vary from well to well from about 200 kcal/kg to 450 kcal/kg. Temperatures of about 300 to 310 °C and pressure about 10,000 kPa (100 bars) are thought to exist in the field at a depth of about 1300 m (Lippmann and Mañón, 1980). Some approximate locations of isotherms, expected to exist at different depths, are also shown in Figure 10. These profiles are likely to have changed because fluid production has been in progress since 1973. Pressure drops and temperature drops of about 500 to 2000 kPa (5 to 20 bars) and 10 to 15 °C have been observed in some wells.

Subsidence associated with this large-scale fluid extraction was anticipated to occur in the Cerro Prieto field. Therefore, the Dirección General de Estudios del Territorio Nacional (DETENAL) and the U.S. Geological Survey jointly laid out the first network to measure horizontal and vertical deformations in the Mexicali Valley in 1977. The second survey conducted in 1978 and reported by Garcia (1980) showed both uplift (max 33 mm) and subsidence (max 28 mm) over an area extending from the United States–Mexico border to the south of the field. According to these results, subsidence in the producing area was very small. Nevertheless, it must be noted that the interpretation of these results will depend upon the location of the chosen datum of zero subsidence. Horizontal contraction of about 31×10^{-6} strain/yr in a northwest-southeast direction and extension of 0.7×10^{-6} strain/yr in a northeast-southwest direction also observed in the field during the second survey of 1978 (Massey, 1980). It is likely that the Cerro Prieto area might also be undergoing some tectonic deformations similar to those observed in the Imperial Valley. Zelwer and Grannell (1982) provided gravimetric evidence to the effect that between 1977 and 1981 approximately 45 cm of vertical subsidence has occurred east of the power plant, caused by the 6.1-magnitude earthquake of June 8, 1980. Their evidence also suggests that almost all the fluid withdrawn is replenished by recharge.

Increased fluid withdrawal at Cerro Prieto may cause changes in pore pressure, temperature gradients, and volume and stress patterns, which in turn may influence the seismicity of the area (Majer et al., 1979). Seismic studies have been conducted at the Cerro Prieto field since 1971 and have been reported by Albores et al. (1979) and Majer et al. (1979). It was found that microearthquake activity in the production area was lower compared to that in the surrounding region. One explanation for this may be that the effective stress in the production zone is increased due to fluid exploitation. This leads to an increase in the effective strength of the rock against slippage, which in turn reduces the seismic activity in the production zone. The regional seismicity may be due to tectonic stresses rather than the geothermal activity.

As of August 8, 1979, reinjection was underway into well M9 with untreated water separated from well M29. Response of well M9 and that of peripheral wells is being monitored over a period of time. About 40,000 kg/h (40 t/h) of approximately 165 °C fluid is injected into an aquifer located between 721 and 864 m depth. The injection rate had decreased to about 25,000 kg/h (25 t/h) by December 1979 (Alonso et al., 1979). No subsidence-related damages have been reported from the Cerro Prieto geothermal field so far.

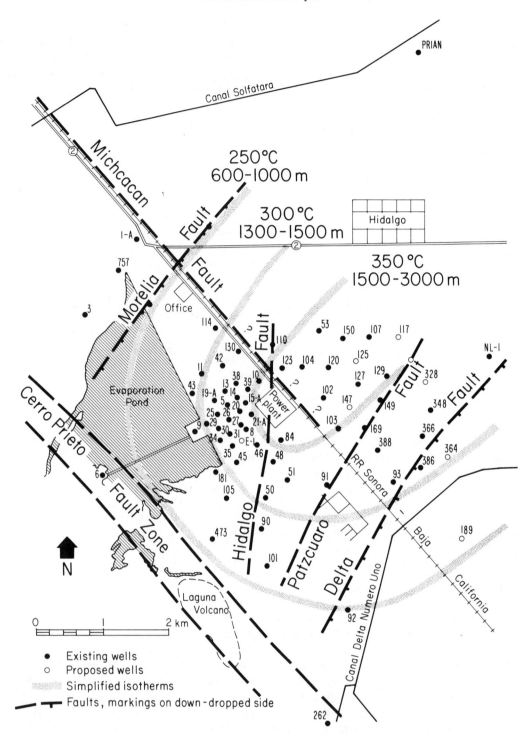

Figure 10. Location of major faults and wells, Cerro Prieto geothermal field, Mexico (from Lippmann and Mañón, 1980).

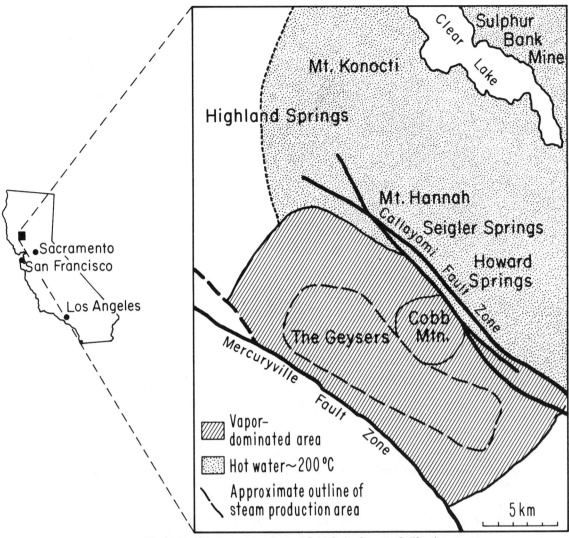

Figure 11. The Geysers geothermal field, Lake County, California.

The Geysers, California, U.S.A.

The Geysers is a vapor-dominated geothermal system and is the largest producer of geothermal electric power in the world. As of early 1982, Pacific Gas and Electric Company was generating approximately 960 MW of electricity from steam supplied by Union Oil of California, Magma Power Company, Aminoil U.S.A., and Thermogenics, Inc. The Geysers field is located about 120 km north of San Francisco in the northern coast ranges of California (Fig. 11). Electric power production at The Geysers began in 1960 when a 12.5-MW generating plant went on line using about 114,000 kg (250,000 lb) per hour of steam supplied by four wells. Since then power production at The Geysers has been steadily increasing through the addition of more wells to the production line. The field is being exploited by private companies, and much of the reservoir performance data is not in the public domain. Good reviews of subsidence-related literature of the Geysers are contained in Grimsrud et al. (1978) and Miller et al. (1980a, 1980b).

The Geysers field, a tectonically active area, can be characterized by a series of generally northwest-trending fault blocks and thrust plates. It is underlain by four major geologic units: the Franciscan assemblage, the ophiolite complex, the Great Valley sequence and the Clear Lake volcanics. The Franciscan graywacke, which has undergone slight to moderate metamorphism, constitutes the reservoir rock. These sandstones are very dense and have low permeabilities ($\simeq 1$ mD) and porosities ($\simeq 10\%$). The steam is thus expected to be confined to open fractures and fault zones, the presence of which have been confirmed by drillers' logs and tested cores. Two reservoirs are thought to exist at The Geysers: a small, shallow reservoir and a deep, extensive one. The depth to the shallow reservoir, which has produced about 50 billion kg (110 billion lb) of steam (Garrison, 1972), is about 640 m (2100 ft). The main deep reservoir is located between 760 to 1520 m (2500 to 5000 ft). These two reservoirs are in communication with each other at some locations and at oth-

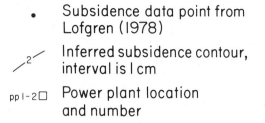

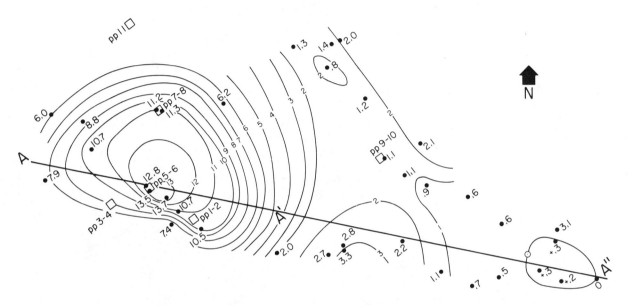

Figure 12. Vertical displacement field, 1973 to 1977, The Geysers, California. Circles denote areas of steam production supporting the power plant at the center (from Grimsrud et al., 1978).

ers they are not. The vertical extent of the reservoir is estimated to be greater than 3050 m (10,040 ft) (Lipman et al, 1977), and the lateral extent is thought to be 4580 m by 4580 m (15,000 ft by 15,000 ft). The top of the reservoir is estimated to be near sea level in elevation.

Prior to 1968 the shallow reservoir was the source of steam for power production. But by the early 1970s most of the fluid was being produced from the deep reservoir that is believed to have an initial temperature and pressure of about 240 °C and 3545 kPa (514 psi), respectively. It is speculated that a deep, boiling-water table may exist at a depth of about 4580 to 6100 m (15,000 to 20,000 ft) which supplies the steam to the producing wells. Noncondensable gases up to 2% by weight are also produced at The Geysers along with the steam. By 1975 there were 110 wells in the field, providing about 3.65 million kg (8 million lb) of steam per hour to generate about 500 MW of electricity. At present, about 900 MW of electricity are being generated. Future plans to increase the capacity are underway. This commercial steam production has reduced reservoir pressures considerably and has caused land deformations. A pressure drop of 1240 kPa (180 psi) was observed in the reservoir between 1969 and 1977. The relative changes in the elevations of the ground surface in The Geysers area are shown in Figure 12. The maximum subsidence to about 13 cm has occurred in the area of maximum fluid withdrawal. It is interesting to note that the vertical changes in the vicinity of the power plants 9 and 10 are minimal, even though there have been large-scale steam productions from these units since 1972. It was also observed that the reservoir pressure dropped and the rates of subsidence were largest soon after the new sources of steam were put on line, and they gradually diminished as recharge gradients reached steady-state conditions (Grimsrud et al., 1978). The vertical displacements observed from 1973 to 1975 and 1975 to 1977 along section AA″ (Fig. 12) are shown in Figure 13. Two types of ground movements may be observed here. (1) a downward local tilt of about 3.5 cm toward west-northwest and (2) a substantial subsidence in the areas of steam production overlapped by circles in Figure 12. A maximum subsidence rate of about 4 cm/yr from 1973 to 1975 decreased to about 2 cm/yr from 1975 to 1977 near power plant 5-6. A nearly uniform uplift from 1975 to 1977 in the east-southeast area of the main production zone can be attributed to the thermal expansions of the overburden in the newly drilled areas or to the assumption of zero subsidence at bench mark R1243. Reservoir pressure drop and subsidence along section

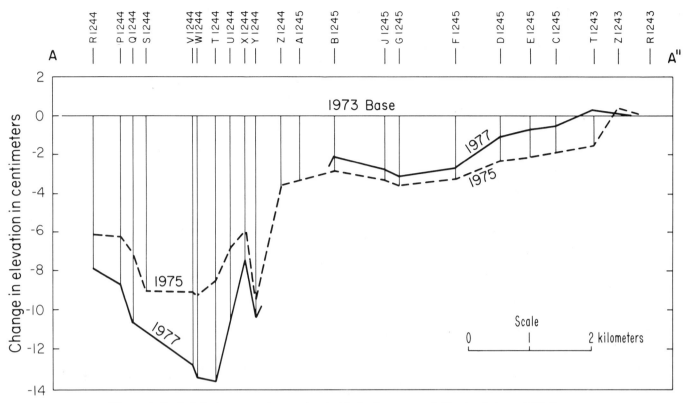

Figure 13. Vertical displacements along section A-A″, the Geyser area, California relative to 1973 (from Lofgren, 1978).

AA″ (Fig. 12) are shown in Figure 14. As can be seen from this figure, the areas of maximum subsidence are the areas of maximum pressure drop. Horizontal displacement rates were found to vary from 1.5 cm/yr in the areas of heaviest fluid withdrawal to 0.4 cm/yr in the peripheral areas (Lofgren, 1978).

In an effort to reduce subsidence and extract maximum thermal energy, reinjection of the steam condensate back into the formation began in 1969 at The Geysers. By 1975, six wells were being used to reinject condensate of about $1.78 \times m^4\ m^3$ (4.7 million gal) per day, about 25% of daily steam output, back into the reservoir between 720 and 2450 m (2364 and 8045 ft) depths. The injected condensate flows down by gravity to depths below adjacent producing wells, in situ steam pressure being less than the hydrostatic pressures. It is thought that the injected fluid can contribute to an increase in the seismicity of the region by inducing slippage along the planes of weaknesses. In addition, microearthquake activity can also be caused by volume changes due to fluid-withdrawal and subsidence (Majer and McEvilly, 1979). A comparison of the seismic activity during preproduction (1962–63) and peak production (1975–77) times showed that the regional seismicity (magnitude ≥2) in the area increased to 47 events per year in the latter period as opposed to 25 events per year in the former (Marks et al., 1979). Also the microearthquakes at The Geysers are strongly clustered around the regions of steam production and fluid injection. It appears likely that much of the present seismicity at The Geysers is induced by one

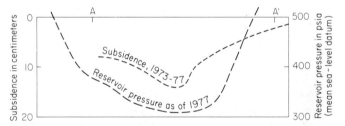

Figure 14. Profiles of reservoir pressure drop and subsidence along section AA″ of Figure 12, The Geyser area, California (from Lofgren, 1978).

or more of the following phenomena: steam withdrawal, injection of condensate, and subsidence. The exact relationship between these phenomena is not firmly established. It is suspected that the slippage of fault blocks past one another due to subsidence movements may in part contribute to the microearthquakes. Or subsidence may give rise to the formation and propagation of microfissures which may enhance microseismic activity. Initiation and propagation of microcracks are also attributed to the thermal stresses produced by circulating geothermal fluids, and this phenomenon is termed thermal stress cracking (Nelson and Hunsbedt, 1979). The mechanisms by which this increase in seismicity has occurred warrants further study. No environmental hazards are reported in The Geysers area due to land subsidence.

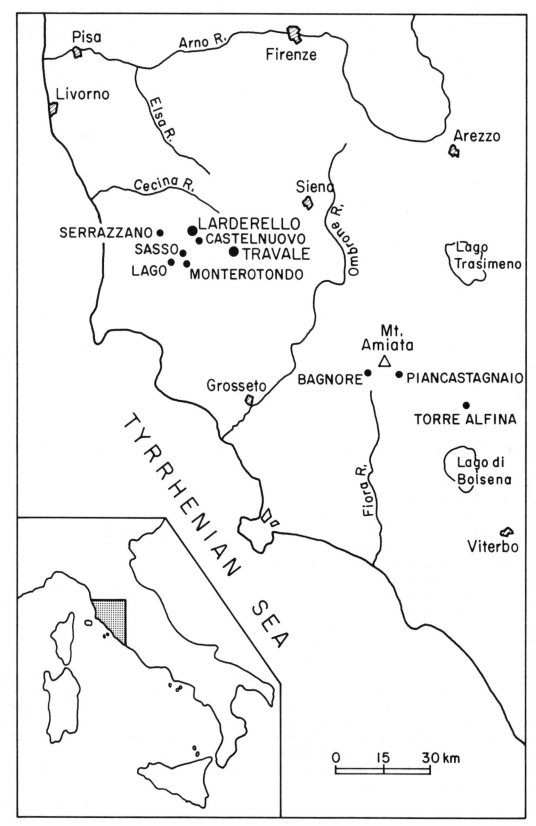

Figure 15. Location map of the Larderello geothermal field, Italy (from Atkinson et al., 1977).

Larderello Geothermal Field, Italy

Electric power generation at the Larderello field began as early as 1913, making it the first geothermal field in the world producing electric power from geothermal steam (DiPippo, 1980). This field is part of an arch of high heat flow extending along the east coast of the Italian peninsula from Tuscany to Sicily (Mongelli and Laddo, 1976). The Larderello system contains many geothermal anomalies. As of March 1975, the producing anomalies included San Ippolito, Gabbro, Larderello, Serrazzano, Castelnuovo V.C., Sasso Pisano, Lagoni Rossi, Lago, Monterotondo, and Molinetto. Figure 15 shows the location of the field and its various anomalies. The Larderello field extends over a distance of about 20 km from Monterotondo in the south to the Gabbro in the north. The area covered by this field is about 170 km^2 (Ceron et al., 1976). As of March 1975, 511 wells were drilled in the region with an average depth of about 650 m. Of these, 194 wells were connected to the production line and 9 were used as observation wells for reservoir engineering studies. Installed capacity at this time was about 380 MW with an average fluid production of about 15,000 kg/h (15 t/h) per productive well at an operating well-head pressure of 100 to 800 kPa abs (Ceron et al., 1976).

The geologic map of the Larderello field along with a cross-section view is shown in Figure 16. From a hydrogeological point of view, the lithology here can be grouped into three main complexes: The first is an impermeable cap-rock complex made of outcrops of "Argille Scagliose" comprising shales, limestones, etc., and "Macigno" and "Polychrome Shales" overlain in places by clay, sand, and conglomeratic sediments. The second is the Tuscan Formation, which constitutes the principal reservoir and forms the circulation region for the endogenous fluid. It comprises radiolarite and evaporite deposits. The third is the basement complex, consisting of phyllitic-quartzitic formations, which is highly impervious where phyllites predominate but may be locally permeable where intercalations of quartzites and crystalline limestones are present. Because the cap rock is not continuous, the Tuscan Formation is exposed at some places and allows the geothermal aquifer to be recharged by rainfall. The steam produced at the Larderello field originates from the meteoric water that may have undergone either a deep regional circulation or a local shallow one (Petracco and Squarci, 1976).

Geophysical studies indicate that the reservoir is characterized by a distinct resistivity high of greater than 100 ohm-meters and is located at depths of less than 1000 m. Thermal gradients of the order of 300 °C/km to a maximum of 1000 °C/km at some places are found in the area. The accepted normal gradient for the area is about 30 °C/km with a heat flow of greater than 3 HFU (heat flow units). At some places heat flows of 5 to 10 HFU exist. The highest reservoir temperatures and pressures encountered in the field are 300 °C and 3000 kPa (440 psi), respectively. The geothermal fluid produced consists of dry, saturated or slightly superheated steam and some noncondensible gases. The amount of noncondensible gases varies over the field from 1% to 20% by weight, the average being about 5%, of which CO_2 is dominant. Temperatures, pressures, and flow rates vary from well to well and from area to area. There has been a significant decrease in the mass flow rates and reservoir pressures over the decades of production. For example, wells 85 and Fabiani of the Larderello anomaly show a significant decrease in the mass flow rate over a 20- to 30-year period (Fig. 17). This figure indicates that flow rates are apparently tending to a steady state. The productive

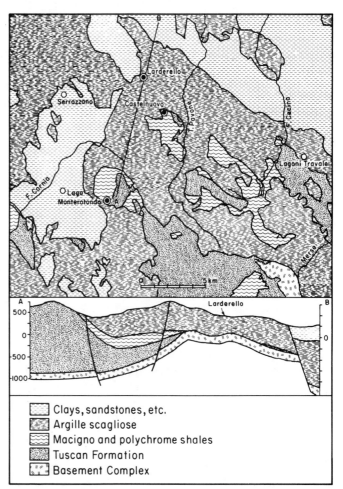

Figure 16. Geologic map and cross section of the Larderello geothermal field (from ENEL, 1976).

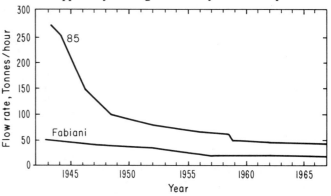

Figure 17. Production history of well 85 and the Fabiani well, Larderello, Italy (from ENEL, 1976).

wells in the Larderello field were first shut-in in 1942, but systematic measurements of relative pressures began only in 1955 (Celati et al., 1977). The water-table data were used by Celati et al. (1976) to determine formation pressures during this period. It was found that the initial pressures in the Larderello region varied from 1960 to 3920 kPa (284 to 568 psi) and that these values were affected by nearby producing zones as a result of the expansion of the explored area. Pressure depth plots indicated that in different parts of the field both water-dominated and vapor-dominated systems had existed before intensive exploitation began. Reservoir pressure distributions in the Serrazzano area for 1970 are shown in Figure 18. The highest downhole pressure observed among all Italian steam fields was measured in the Travale field where a pressure of about 6000 kPa (about 870 psi) existed in new wells drilled fro 1972 onward. Such high pressures indicate that the wells in this field have reached a water-dominated reservoir. Considering that Larderello is in much the same hydrogeological and thermal situation as the Travale field, we may deduce that Larderello, in undisturbed conditions, might also have had similar pressures (Celati et al., 1976). Continued production from the field has led to a considerable drop in the reservoir pressures and water levels. Since 1963–64, water levels in the western part of the field have dropped by about 100 m compared to the other parts where the drop is less than 50 m (Celati et al., 1977). In a related study, Atkinson et al. (1978) calculated an initial pressure of 3920 kPa (570 psi) at the Serrazzano field and steam reserves of about 1.7×10^{11} kg (170 million tonnes).

To monitor the effects of the injection of liquid wastes on surface and ground waters, a reinjection program was started in the Larderello region during the early 1970s. About 20% of the waste liquid was returned to the reservoir by reinjection through the wells at the periphery of the field, and the remaining 80% was discharged directly into the local streams (Defferding and Walter, 1978), despite high concentrations of boron. Generally, reinjection was successful. However, in one case, cold reinjected liquids reached a production well.

No subsidence has been reported at the Larderello field, although production has been in progress for 60 years and on a relatively large scale for 30 years (Kruger and Otte, 1976). Microearthquake studies in the Larderello area do not seem to have been repeated. However, a few earthquakes of magnitude 4 in the Larderello area and one earthquake centered about 15 km east of Larderello have been reported. It seems that microseismicity could be recorded at the Larderello field if sensitive instruments are used.

Geopressured Systems

Several reservoirs of brine at high pressures and moderate temperatures are known along the Gulf Coast of Louisiana and Texas. Wallace (1979) estimated 17.1×10^{21} J of thermal and methane energy in place in the reservoir fluids underlying a 310,000 km^2 area of the northern Gulf of Mexico basin. Temper-

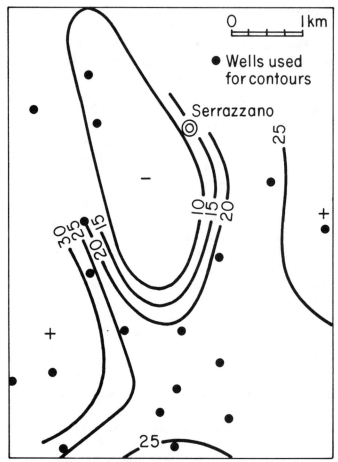

Figure 18. Spatial pressure distribution, Serrazano area, Larderello, Italy during 1970. Contour values are in kg/cm^2 (from Celati et al., 1976).

atures and pressures of most waters in these areas vary between 163 and 204 °C and 69 and 103 MPa (10,000 and 15,000 psi), respectively, in the depth intervals of above 3.5 to 5 km (Wilson et al., 1974). Temperature gradients of about 20 to 40 °C/km exist in the Gulf Coast region in the upper 2 km. Geothermal gradients exceeding 100 °C/km are found within and immediately below the depth interval where maximum pressure-gradient change occurred (Jones, 1970). The depth to the top of the geopressured zone conforms in a general way with a 120 °C isotherm that occurs in the depth range of a 2.5 to 5 km below sea level (Jones, 1970). The loss of load-bearing strength due to thermal diagenesis that takes place between 80 and 120 °C is considered most responsible for creating the top of the geopressured zone. The location and depth of occurrence of the geopressured zones in the Gulf Coast region are shown in Figure 19.

In this paper we shall confine our attention to one geopressued field that is under exploitation. It is the Chocolate Bayou Field in Texas from which oil and gas have been produced since the early 1940s.

Chocolate Bayau, Texas. Chocolate Bayou is an oil and gas field in Brazoria County, Texas, and is located about 30 mi south of Houston (Fig. 20). Oil and gas production from both

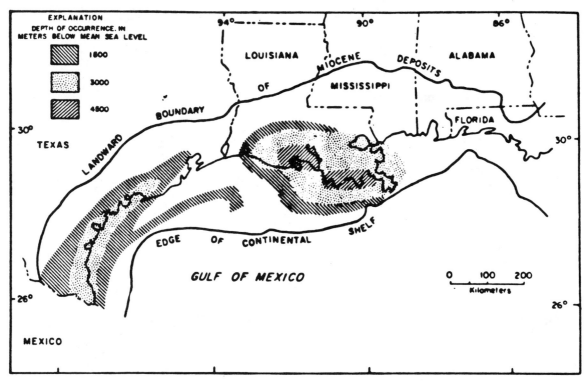

Figure 19. Location and depth of occurrence of the geopressured zone in the northern Gulf of Mexico basin (from Wallace, 1979).

normally pressured and geopressured zones have been responsible for a land subsidence of about 0.6 m (2 ft) in this area. The Austin Bayou prospect, a proposed geothermal exploration site, is located 8 km (5 mi) southwest of Chocolate Bayou and has essentially the same geohydrological conditions. One should expect that the subsidence at the Austin Bayou prospect will not be much different from that at Chocolate Bayou under geothermal fluid production. The Chocolate Bayou field occupies an area of about 69 km^2 (25 mi^2) and has a surface elevation of 3 to 12 m (10 to 40 ft) above sea level with a gentle southeast dip. The formations from surface downward include Pliocene to Holocene sand and clay beds in the upper 760 m (2500 ft), Miocene and Pliocene sands in the next 1200 m (4000 ft), and Oligocene and Miocene shales in the remaining 610 m (2000 ft) down to a depth of about 2650 m (8700 ft). Underlying these formations are the productive geopressured sediments, which might occur down to depths of about 4900 m (16,000 ft). The producing zones are underlain and overlain by thick shales and vary in thickness from less than 3 to more than 60 m (10 to more than 200 ft). The depth to the top of the geopressured zone varies from 2,440 m to 3,050 m (8000–10,000 ft). Numerous faults with no surface expressions exist in the field. Relative fault movement is thought to be responsible for the generations of abnormal pressure zones by bringing shales into contact with sands and thereby preventing the communication between upper and lower aquifers (Miller et al., 1980b). Faults are also thought to act as complete or partial barriers to the fluid flow (Gustavson and Kreitler, 1976).

Since the early 1940s the Chocolate Bayou field has produced more than 4.5×10^{11} m^3 (16×10^{12} ft^3) of natural gas and 5.6×10^6 m^3 (35 million barrels) of oil, respectively. Gas wells have contributed an additional 6.5×10^6 m^3 (41 million barrels) of liquid hydrocarbons. Annual production of the field is shown in Figure 21. The annual production of brine from the field is not shown in this figure because these data are not readily available (Grimsrud et al., 1978). The reinjection of brine since 1965, for the Phillips Petroleum Company wells, is also shown in this figure. This reinjection is thought to be about 10% of the brine produced. As may be noted from this figure, brine is the predominant liquid produced at the field since 1965. Current total production of hydrocarbons is less than 8×10^4 m^3 (0.5 million barrels) per year and that of brine is more than 4.8×10^5 m^3 (3 million barrels) per year (Miller et al., 1980b; Grimsrud et al., 1978). Initial conditions of the reservoir vary within the field from one location to another. The producing zones in west Chocolate Bayou area are all normally pressured; in East Chocolate Bayou, both normally pressured and abnormally pressured zones are present; and in South Chocolate Bayou all zones are abnormally pressured. Pressure versus depth relationships in wells located west of the Chocolate Bayou area are shown in Figure 22. Based on well logs, a temperature gradient of about 3 °C/100 m seems quite reasonable down to about 5000 m. Data presented by Bebout et al. (1978) for a well from the South Chocolate Bayou field indicate that bottom-hole pressures had declined by 55 to 62 MPa (8000 to 9000 psi) during a 10-year period from 1964 to 1974, although bottom-hole temperatures remained fairly stable at about 162 °C (323 °F). The reduction in the

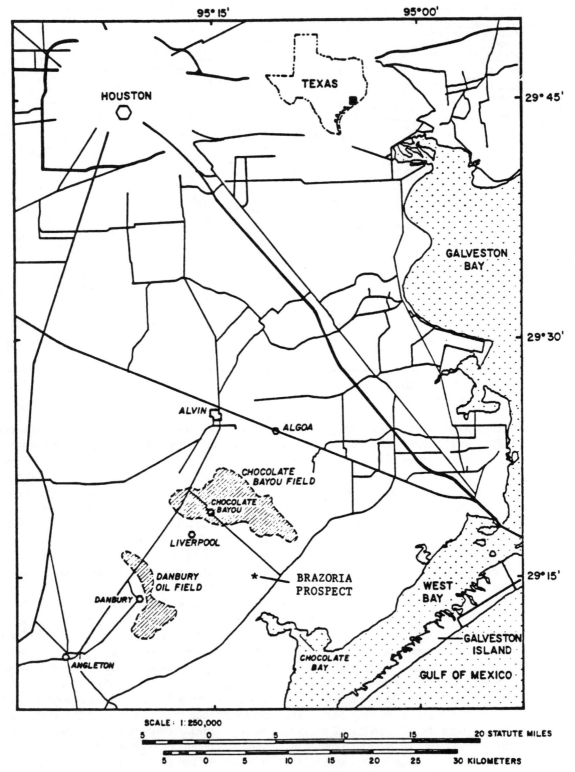

Figure 20. Chocolate Bayou location map (from Grimsrud et al., 1978).

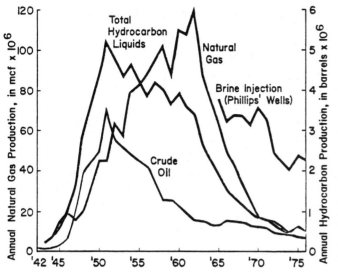

Figure 21. History of fluid production and injection, Chocolate Bayou Area, Texas (from Grimsrud et al., 1978).

formation pressures is expected to cause some land subsidence. In fact, bench mark K691 located in the Chocolate Bayou area has subsided 0.55 m (1.8 ft) since 1943 (Fig. 23). Besides oil and gas production, groundwater withdrawals and tectonic movements are also considered as potential causes of this subsidence (Grimsrud et al., 1978). Groundwater withdrawals between 1943 and 1974 have caused a subsidence of more than 2.13 m (7 ft) in the Houston-Galveston area and are thought to be responsible for some subsidence in the Chocolate Bayou area (Miller et al.,

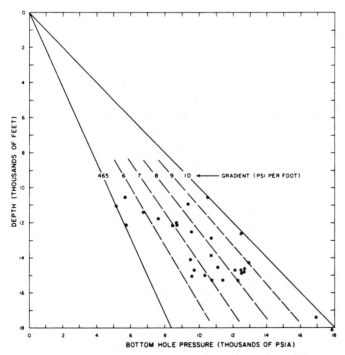

Figure 22. Subsurface fluid pressure gradients from wells in the Austin Bayou area, 8 km southwest of Chocolate Bayou area (from Bebout et al., 1978).

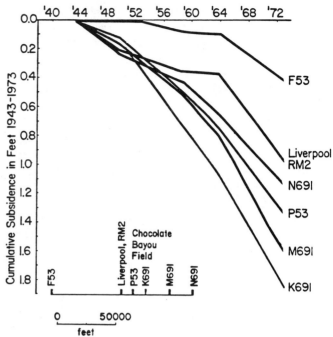

Figure 23. Vertical displacements individual bench marks in the Chocolate Bayou area (from Grimsrud et al., 1978).

1980a). Some estimates have been made on the component of Chocolate Bayou subsidence attributable to groundwater pumpage. It has been suggested that the subsidence of about 0.3 m (1 ft) out of a total of 0.55 m (1.8 ft) can be attributed to the groundwater withdrawal (Sandeen and Wesselman, 1973; Grimsrud et al., 1978). It may be noted from Figure 23 that bench marks P53 and M691 located in the Chocolate Bayou field show a change of slope in the late 1940s and early 1960s. This increased subsidence can be related to increased production of hydrocarbons as shown in Figure 21. Another interesting point may also be noted in Figures 21 and 23 that although the hydrocarbon production at the Chocolate Bayou field has been decreasing since 1964, the average rate of subsidence from 1964 to 1973 was greater than that from 1959 to 1964. This suggests either a lag time of at least several years between extraction of deep fluids and the appearance of subsidence effects at the surface or a transition of the sediments from a state of over consolidation to that of normal consolidation (Holzer, 1981). No surface effects, such as faulting, ground cracking, disruption of well casings, damages to structures, are reported in the Chocolate Bayou area due to ground subsidence.

Summary of Field Observations

There is clear field evidence from different parts of the world to confirm that geothermal fluid extraction can lead to vertical as well as horizontal displacements at the land surface. These deformations, which can cause significant damage to property, often show specific patterns of variations in space and in time, indicative of complex interaction of physical phenomena.

At Wairakei in New Zealand a well-defined subsidence bowl, with an area of about 1.5 km^2, has developed due to fluid production, causing a maximum vertical displacement in excess of 4.5 m. Within this bowl, pronounced horizontal displacements directed toward the center have also been observed, the maximum magnitude being about 0.5 m. Spatially, the subsidence bowl is centered about 1.5 km east-northeast of the center of the main production area (Fig. 2), presumably controlled by peculiarities of local geology. The annual rate of vertical displacement also shows a significant relation to time. An analysis of the subsidence at bench mark A97 in the south-southwest part of the bowl suggests that the rate was about 4 cm/yr between 1953 and 1962. Around 1963 this rate underwent a marked increase, and between 1971 and 1974 the rate attained a magnitude of about 15 cm/yr. An examination of the pressure drop at the same bench mark suggests that the marked change in the rate of subsidence is indicative of a marked increase in the compressibility of the material undergoing subsidence. Although recent studies indicate an increase in the microseismic and macroseismic activity in the Taupo fault belt area (which passes to the west of the field in a northerly direction), it is difficult to decide at present whether this increase is to be attributed to the exploitation activity or to natural tectonism.

The Broadlands field, New Zealand, also exhibits conclusive evidence of subsidence associated with geothermal fluid production. Unlike at Wairakei, the subsidence bowl at Broadlands is centered almost directly over the production area, with a maximum displacement of the order of 0.2 m since the mid-1960s. The field was almost completely shut down during August 1971 and again during September 1974. These events were followed by a marked decrease or even a reversal in the direction of ground displacement. Between 1969 and 1972 the average rate of subsidence was of the order of 5 cm/yr. As at Wairakei, the horizontal movements at Broadlands are also directed toward the center of the bowl, with maximum displacements of the order of 10 cm. Recent investigation indicate no noticeable seismic activities attributable to geothermal exploitations at Broadlands.

The liquid-dominated geothermal field at Cerro Prieto, Mexico, has been producing since 1973. Observations to date indicate that small vertical displacements, of the order of a few centimeters, have occurred over the field since 1977. However, these displacements cannot be conclusively attributed to fluid production. It is probable that the ground deformations and seismicity that have been measured recently in Cerro Prieto could be attributed to tectonic activity in this structurally active part of the earth's crust.

The Geysers field in Lake County, California, constitutes an excellent example of a vapor-dominated system. Relative to a 1973 datum, vertical movements of as much as 13 cm have been observed in The Geysers up until 1977. The maximum subsidence rate was about 4 cm/yr between 1973 and 1975 and declined to about 2 cm/yr near power plant 5-6. At the same time, a nearly uniform uplift of about 2 cm occurred from 1975 to 1977 and was also observed in the east-southeast part of the production area. There is evidence that the areas of maximum subsidence over the field are correlated with areas of maximum pressure drop. Since 1969, spent condensate fluids have been reinjected into the formation and provide pressure support to the reservoir. A comparison of the preproduction seismicity to that of the peak production period, 1975 to 1977, shows that regional seismicity (magnitude $\geqslant 2$) had increased from 25 to 47 events per year. In addition, microearthquake activity is strongly clustered in regions of production and injection.

Another well-known vapor-dominated system is the one in and around Larderello in Italy. Although this field has been under production since the early 1940s, no noticeable subsidence has occurred. A few earthquakes of approximate magnitude 4 have been reported. No data are available on the microearthquake activity in this region.

The geopressured geothermal systems of the Gulf Coast in Texas and Louisiana are currently under exploration. As such, no case histories are now available to evaluate their deformation behavior in response to fluid withdrawal. Some clues to their possible behavior, however, can be obtained by studying the deep oil and gas fields that have been exploited near the exploration sites. In the Chocolate Bayou oil fields of Texas, producing oil and gas horizons are known between depths of 2650 and 4900 m (8900 and 16,000 ft). These zones may be either normally pressured or geopressured. Within the Chocolate Bayou area maximum subsidence of up to 0.55 m subsidence attributable to oil and gas production between 1944 and 1972 has been measured. It is thought that part of this subsidence could be due to shallow groundwater development.

PHYSICAL BASIS

As evidenced by field observations, the important questions requiring consideration in analyzing subsidence due to geothermal fluid withdrawal are as follows: (1) the nature of the size and the shape of the subsidence bowl; distributions of horizontal and vertical displacements; (2) the location of the subsidence bowl in relation to the location of the area of fluid production; (3) the variations in the time-rate of subsidence as a function of time; (4) differential subsidence; and (5) fault movement and induced seismicity, if any.

The fundamental sequence of events leading to land subsidence is as follows: (1) fluid withdrawal causes reduction in fluid pressure; (2) fluid pressure reduction causes an increase in stresses on the rock matrix, accompanied by a reduction in the reservoir bulk volume; (3) the reduction of reservoir volume leads to the generation of a three-dimensional displacement field within the reservoir and some deformation may also be induced by contractions associated with temperature declines; and (4) the reservoir displacements propagate to the land surface to cause horizontal and vertical ground displacements.

For purposes of analysis, it is convenient to distinguish between the reservoir proper where the displacements originate and the overburden through which the reservoir displacements are merely transmitted. We shall define the "reservoir" to include those portions of the system from which geothermal fluids are drained (released from storage) to compensate for the fluids removed at the wells. Specifically, the reservoir includes the highly permeable horizons (the aquifers) as well as the slowly draining formations (aquitards or cap rocks). The overburden, on the other hand, has little hydraulic continuity with the reservoir. Thus, there is no drainage of fluid from the overburden to make up for the geothermal fluids exploited.

The reservoir and the overburden basically differ in the manner in which they are subject to loading and deformation. The reservoir is subjected to loads originating from within the pores (endogeneous loading), whereas the overburden is subject to stresses or displacements imposed on its boundary (exogeneous loading). Stated differently, the reservoir is subject to drained loading, whereas the overburden is subjected to undrained loading.

In practice, it may be hard to define the exact location of the reservoir-overburden interface. This boundary will obviously change its disposition with time, unless there exists a sharp impermeable contact between the two. Nevertheless, there is reason to suspect that results of overall analysis may not be very sensitive to the uncertainties inherent in locating this contact.

Deformation of the Reservoir

In a geothermal reservoir, deformation may occur due to mechanical as well as thermal causes. Although a reasonable theoretical basis is currently available to discuss mechanically induced deformations, much remains to be done to properly explain the complex interactions existing between thermal and mechanical deformations that occur within a geothermal reservoir under exploitation.

In the following discussions on reservoir deformation, we shall restrict ourselves to subsidence caused by fluid withdrawal (that is, endogeneous loading). The duration of reservoir exploitation is considered to be much smaller than that over which tectonic stresses change; therefore, we will treat the total stresses on the system to remain unchanged in time.

Mechanical Deformation. This phenomenon can perhaps be best explained by considering an elemental volume of the reservoir and its response to an imposed change of fluid mass at constant temperature. Such a change in fluid mass is indeed induced when geothermal fluid is mined from the reservoir. The mass of fluid contained in an arbitrary volume element is given by $M_f = V_v \rho_f S_f$, where M_f is mass of fluid, V_v is volume of voids, ρ_f is fluid density, and S_f is fluid saturation. Consequently,

$$\Delta M_f = \rho_f S_f \Delta V_v + V_v S_f \Delta \rho_f + V_v \rho_f \Delta S_f. \tag{1}$$

Of the three quantities on the right-hand side of (1), the first, which denotes the component of ΔM_f arising due to pore volume change, is the phenomenon that directly determines the magnitude of subsidence. The remaining two terms, which govern the dynamics of pore-fluid pressure change, indirectly contribute to subsidence. Therefore, insofar as the mechanism of pore volume change is concerned, we may focus attention on the first term on the right hand side of (1). Thus, only that term needs further consideration. As already stated, ΔV_v is caused by a reduction in the pore fluid pressure following removal of fluid mass. In addition to the change in void volume, one has also to consider the change in the volume of the solid grains in order to evaluate the change in the bulk volume of the element. Note that it is the change in the bulk volume that controls subsidence. The basic ideas of combining ΔV_v and ΔV_s (where V_s is volume of solids) to compute ΔV have been discussed in detail by Skempton (1961). In accordance with Skempton's development, we can show that when the total stress is constant, that is, $\Delta \sigma = 0$,

$$\frac{\Delta V_v}{V} = -c\Delta p, \tag{2}$$

where c is the compressibility under drained flow conditions in which external stress is increased with no change in pore pressure (p), and

$$\frac{\Delta V_s}{V} = c_s \Delta p, \tag{3}$$

where c_s is the bulk compressibility of the solids. As justified experimentally (Skempton, 1961), the assumption inherent in (3) is that a medium subjected to the same magnitude of internal fluid pressure and external stress will behave as if the entire medium was made up of the solids. Based on the theory of elasticity, Nur and Byerlee (1971) provided a theoretical justification for this. In (2) and (3) a reduction in volume is associated with a positive sign. Combining (2) and (3),

$$\frac{\Delta V}{V} = -c\alpha\Delta p, \tag{4}$$

where $\alpha = (1-c_s/c)$. Equation (2) pertains to volume change solely due to grain-grain slippage, and (3) relates to volume change due to elastic compression of the grains. Physically, (4) implies that falling pore pressure will be accompanied by compaction due to grain-grain slippage and a dilation due to the expansion of the solids. Usually $c > c_s$; hence, grain-grain slippage will dominate the deformation process.

Furthermore, (2) implies that for void volume change and bulk volume change, respectively, the following constitutive relations between effective stress, σ', and p are valid.

For void volume change:

$$\Delta\sigma' = -\Delta p. \tag{5}$$

For bulk volume change:

$$\Delta\sigma' = -\left(1 - \frac{c_s}{c}\right)\Delta p = -\alpha\Delta p. \qquad (6)$$

When more than one fluid phase is present in a geothermal system (e.g., steam and water), the relation between change is fluid pressure and change skeletal stresses becomes more complex than (6). No published work, to our knowledge, addresses this relationship for a two-phase, steam-water system. The somewhat analogous problem of deformation of soils partially saturated by water and partially by air has been addressed experimentally by soil mechanisms. The discussions immediately below outline their findings.

Bishop (1955) suggested that (2) would need to be modified. Skempton (1961) generalized Bishop's ideas and proposed, for a porous medium with a water phase and an air phase,

$$\sigma' = \sigma - S_\chi p_w, \qquad (7)$$

where

$$S_\chi = 1 + (1 - \chi)\frac{p_a - p_w}{p_w}, \qquad (8)$$

in which p_w is the pressure in the water phase, p_a is the pressure in the air phase, and χ is a parameter dependent on the capillary pressure $(p_a - p_w)$. If one assumes σ to be constant, (4) implies that (5) should be modified to:

$$\Delta\sigma' = -\left[S_\chi \Delta p_w + p_w \Delta S_\chi\right], \qquad (9)$$

and (6) should be modified to:

$$\Delta\sigma' = -\alpha\left[S_\chi \Delta p_w + p_w \Delta S_\chi\right]. \qquad (10)$$

More recent work by Fredlund and Morgenstern (1976) suggested that χ may be dependent on σ in addition to $p_a - p_w$. It is important to take note of the fact that the change in pore pressure in response to fluid drained will be governed by (2). The Δp controlled by (2) will then govern the solid volume change as in (3) or the bulk volume change as in (4).

In view of the foregoing, we may express the rate of change of void volume with reference to fluid pressure by the relation,

$$\frac{\Delta V_v}{\Delta p_w} = \frac{\Delta V_v}{\Delta \sigma'} \frac{\Delta \sigma'}{\Delta p_w} = -S_\chi' \frac{\Delta V_v}{\Delta \sigma'}, \qquad (11)$$

where

$$S_\chi' = \left[S_\chi + p_w \frac{\Delta S_\chi}{\Delta p_w}\right].$$

Noting that $c = -(\Delta V_v/V)(1/\Delta\sigma')$, we may replace (11) by

$$\frac{\Delta V_v}{\Delta p_w} = S_\chi' c V. \qquad (12)$$

Skempton (1961) presented data indicating that c_s/c is very nearly zero in almost all unconsolidated sediments. But in rocks such as granite and quartzite, it may attain values of 0.7 or more. Also, for fully saturated materials, $S_\chi = S_\chi' = 1.0$.

Very little is known about the nature of the χ and S_χ functions for steam-water systems at elevated temperatures, although one would suspect that a steam-water system would obey an expression similar to (12).

Insofar as the phenomenon of fluid flow is concerned, it is the void volume change, a scalar quantity, that is of critical importance. However, for purposes of subsidence analysis, it is necessary to be able to evaluate the vertical and lateral displacements that accompany bulk volume change. How much of the horizontal displacement seen at the land surface is directly related to horizontal displacements in the reservoir is one of the important questions that needs resolution at the present time.

One of the simplest methods of converting volume change to displacement is to assume that because of the large lateral dimensions of geologic systems, horizontal strains are essentially negligible and that all volume change is caused by vertical displacements. If so, the change in the vertical dimension of a regular prism of cross-sectional area A will be given by, $\Delta h = -(c\alpha\Delta p)h$, where h is the height of the prism. This is the assumption of one-dimensional consolidation theory, which has proven engineering validity in many field situations. In this case, c is usually obtained by testing samples in an oedometer or a device in which lateral strains are prevented.

For those field situations in which lateral strains may not be neglected, the general three dimensional deformation field accompanying volume change has to be considered. The boundary conditions obtaining in such systems cannot generally be duplicated in the laboratory and hence it is not possible to know c a priori. In this case, change in bulk volume is a function of linear displacements in different directions. For simplicity, if we consider a system with elastic, isotropic materials undergoing small strains, the change in bulk volume may be related to directional displacements by the relation,

$$\epsilon_v = \epsilon_x + \epsilon_y + \epsilon_z, \qquad (13)$$

where ϵ_v is the volumetric strain $\Delta V/V$ and ϵ_x, ϵ_y, and ϵ_z are linear strains in the direction of the coordinate axes. The task here is to evaluate ϵ_x, ϵ_y, and ϵ_z based on the linear moduli of the material and the boundary conditions. Distortions of the elemental volume also arise in addition to displacements. These distortions, caused by shear forces, do not contribute to volume change. If we assume the porous medium to be an elastic solid obeying Hooke's Law, then the three-dimensional strain components, ϵ_{ij}, $i = 1, 2, 3$ can be related to the three-dimensional stress components, σ_{ij}', through three material properties, Young's modulus E, shear modulus G, and Poisson's ratio ν (see, for example, Popov, 1968). For purposes of analyzing three-dimensional deformation, the effective stress relations (5) and (6) may be generalized to (Garg and Nur, 1973):

$$\sigma'_{ij} = \sigma_{ij} - S_\chi \delta_{ij} p_w \qquad (14)$$

for void volume change, and

$$\sigma'_{ij} = \sigma_{ij} - \alpha S_\chi \delta_{ij} p_w \qquad (15)$$

for bulk volume change.

It is widely known from experience that the parameters governing volume change, c and E, are often strong functions of effective stress, in addition to being dependent on the direction of the loading path and having a memory of past maximum loads. It is practically most expedient to treat these complexities of behavior by imposing the elasticity assumption over small ranges of stress increment. Accordingly, we will restrict ourselves to the assumption of linear elasticity.

Having considered the phenomenon of deformation of our elemental volume, we now proceed to consider the forcing function that causes volume change. In the present case, the forcing function is the withdrawal of fluid from a geothermal field and the consequent reduction in fluid storage. The task then is to relate (12) to the dynamics of transient fluid flow in a deformable porous medium.

The Single-Equation Approach. The simplest way to couple fluid flow and deformation is to assume that the boundary condition controlling deformation in the field can be reproduced in the laboratory (e.g., oedometer test) and that the compressibility of the porous medium is known. In this case, the entire problem may be represented by a single governing equation,

$$\rho_w G + \nabla \cdot \frac{\rho_w k}{\mu} (\rho_w g \nabla z + \nabla p_w) = m_c^* \frac{\partial p_w}{\partial t}, \qquad (16)$$

where G is the source/sink term, k is absolute permeability, μ is viscosity, ρ_w is fluid density, z is elevation above datum, g is acceleration due to gravity, p_w is fluid pressure, and m_c^* is a generalized storage coefficient,

$$m_c^* = \rho_w \left[nS_w c_w + S_w S'_\chi c + n\, dS_w/dp_w \right], \qquad (17)$$

where n is porosity and c_w is compressibility of water. When two phases are present, an equation similar to (16) has to be set up for the second phase. The solution of the above equation(s) merely leads to the evaluation of the bulk volume change. If one makes an assumption of the pattern of displacement (e.g., zero horizontal displacement), one can easily compute vertical displacements in the reservoir. Essentially such an assumption was used by Helm (1975) in his one-dimensional simulation of land subsidence. Narasimhan and Witherspoon (1978) combined this one-dimensional deformation assumption in conjunction with a three-dimensional fluid flow field. This assumption is perhaps realized within the low-permeability, high-storage aquitards that may exist in geothermal systems. If this assumption is used, the horizontal displacements observed at the land surface will have to be explained solely in terms of the deformation of the overburden. Another limitation of the simple equation approach is that the stress field evaluation is treated in a perfunctory manner; that is, only the change in mean principal stress or the vertical stress is accounted for.

The Coupled-Equation Approach. A more comprehensive solution of the reservoir deformation problem requires that the physics of the problem be described in terms of two coupled equations: one for fluid flow and the other for porous medium deformation. These equations will have to be reinforced by an energy transport equation in geothermal systems, as we shall see later.

The coupled approach was originally propounded by Biot (1941) and later revised by him in 1955. Biot's approach has since been applied extensively in the fields of soil mechanics and rock mechanics (e.g., Sandhu and Wilson (1969). In the general three-dimensional situation, we need to separate out the change in void volume and the expansion of water in (16) and then rewrite equation as,

$$\rho_w G + \nabla \cdot \rho_w \frac{k}{\mu m} (\rho_w g \nabla z + \nabla p_w) = \rho_w \frac{\partial \delta_{ij} \epsilon_{ij}}{\partial t} + \rho_w n \beta_w \frac{\partial p_w}{\partial t}, \qquad (18)$$

where ϵ_{ij} are the elements of the strain tensor and δ_{ij} is Kronecker Delta. Also, $\delta_{ij}\epsilon_{ij} = \epsilon_v$, the volumetric strain. Because the ϵ_{ij}'s are unknown in (18), we need to solve a second equilibrium equation which balances total loads. That is,

$$\frac{\partial \sigma_{ij}}{\partial x_j} + F_i = 0, \qquad (19)$$

where σ_{ij} denotes the total stress tensor and F_i denotes body forces. In order to couple (18) and (19), we may express σ_{ij} in (19) in terms of ϵ_{ij} and p_w using appropriate constitutive, stress-strain laws. For an elastic isotropic material, σ_{ij} may be expressed as (Biot, 1941):

$$\sigma_{ij} = \frac{E}{1+\nu} \epsilon_{ij} + \frac{E\nu}{(1+\nu)(1-2\nu)} \delta_{ij} \delta_{kl} \epsilon_{kl} + \delta_{ij} p_w, \qquad (20)$$

where E is Young's modulus and ν is Poisson's ratio. The physical basis for fluid flow in a deformable porous medium is provided by (18) and (19) modified by (20). These equations are subject to appropriate initial conditions, boundary conditions, and sources.

Thermal Deformation. Although the phenomenon of thermal expansion of solid materials is extremely well known, the role of thermal expansion in relation to fluid flow in deformable media has so far been treated only in a simplisitc manner (see, for example, Golder Associates, 1980). In this approach, the volume change due to temperature change is explicitly added to the volume change due to pore pressure change. In particular, an increase in temperature leads to an increase in the volume of the solid. Intuitively, one would expect that part of the solid volume expansion will tend to decrease the pore volume while part of it will contribute to an increase in the bulk volume. Whether the ensuing decrease in pore volume will generate a pore pressure component or not will have to depend on the relative thermal expansivities of the solid and water. If we neglect these questions, the expansion of the solids due to an increase in temperature will affect bulk volume change in the same sense as a decrease in pore-fluid pressure. Thus, if β_s is the coefficient of thermal expansion of the solids defined as $\beta_s = 1/V_s \, (dV_s/dT)$, then the bulk volume change due to a simultaneous change of Δp_w and ΔT is (under conditions of constant σ):

$$\frac{\Delta V}{V} = -c\left(1 - \frac{c_s}{c}\right)\Delta p_w - (1-n)\,\beta_s \Delta T. \qquad (21)$$

It is obvious that ΔT has to be obtained by solving a separate energy balance equation in addition to the fluid flow equation (19) and the force balance equation (20). In recent years many papers have appeared in the literature on the formulation of the energy balance equation in single-phase and two-phase systems. Notable among these are Coats (1977), Garg et al. (1975), Mercer and Faust (1979), Pinder (1979), and Witherspoon et al. (1977).

In summary, then, a physical description of the subsidence phenomenon accompanying geothermal fluid exploitation involves the simultaneous consideration of three coupled equations: one for the conservation of fluid mass, one for the maintenance of force balance, and one for the maintenance of energy balance. These equations would need to be supplemented by appropriate relations between pore pressure on the one hand and skeletal stresses on the other, as well as information on the compressibilities of the bulk medium and the solids and the thermal expansivity of the solids.

Overburden Deformation

The overall effect of the reservoir deformation is that its interface with the overburden is deflected downward. In addition, points on this interface may also be subjected to some horizontal displacements. In response to these displacements on its bottom boundary, the overburden itself deforms, leading to vertical as well as horizontal displacements at the land surface.

It is clear at the outset that the overburden deforms essentially in an undrained fashion. The deformation of the overburden is therefore governed by a force balance equation such as (19), subject to a prescribed displacement boundary condition at the bottom and a zero stress, free surface boundary condition at the top. Additionally, the overburden may be constrained by other lateral boundaries such as faults and basin margins. The response of the overburden is largely governed by the ratio of its thickness to the radius of the deformed region as well as the properties of the materials constituting it. Where the thickness of the overburden is relatively small compared to the areal extent of deformation, the land subsidence observed will almost be a replica of the deformation pattern at the reservoir overburden interface. However, as the thickness increases, the displacements at the reservoir-overburden interface may be modified and attenuated before reaching the land surface.

One of the intriguing questions still unanswered concerns the nature of the mechanisms that cause horizontal displacements at the land surface; does the horizontal displacement at the surface definitely imply significant horizontal displacement in the reservoir? In principle it is conceivable that horizontal displacements may be caused at the land surface simply because the elastic overburden system responds to the curvature of the underlying subsidence bowl. Such horizontal displacements could be accounted for by means of the force balance equation (19) already mentioned. However, Helm (1982) has been investigating the possible importance of horizontal deformations originating within the reservoir by treating the solid grains as constituting a viscous fluid. The chief difficulty in answering this question is that no field data are presently available on the variation of horizontal displacements with depth within the overburden. Indeed, a recent report by O'Rourke and Ranson (1979) indicated that instruments are nonexistent at present to measure horizontal displacements along a vertical profile, and they may not be available in the foreseeable future. Until sufficient field data are forthcoming, all the hypotheses that attempt to explain horizontal displacements will remain largely untested.

Role of Fractures

Our discussion of physical bases has so far centered exclusively on porous materials. Yet there is reason to believe that many geothermal reservoirs (e.g., Wairakei, New Zealand; The Geysers, United States) are dominated by fractures. Additionally, the overburden may also be traversed by individual faults or a system of faults. Despite these physical realities, incorporating fractures into the physical basis is not easy. Fractures may control subsidence at a microscope level, in terms of microfracturing and development of secondary porosity (Noble and Von der Haar, 1980), or on a macroscopic scale, in terms of differential subsidence across major faults as has been noted at Long Beach, California. In fractured reservoirs, fractures act more as highly permeable channels of fluid flow rather than as spaces for storage. Fractures may indirectly govern deformation by influencing the rate of fluid transmission, and discrete discontinuities may contribute to deformation by acting as failure planes. However, it is doubtful whether the deformation of fractures themselves will contribute greatly to bulk volume changes in the reservoir. Apart from this qualitative reasoning, very little quantitative information is currently available in the literature to evaluate the role of fractures in the subsidence process.

Range of Values of Parameters

Because of the difficulties associated with the collection of undisturbed samples from geothermal reservoirs and the difficulties associated with measuring physical properties of rocks under simulated in situ conditions, reliable data on physical properties relevant to geothermal subsidence are very limited in extent. Within the last decade, the U.S. Department of Energy has funded a few projects aimed at understanding the mechanism of subsidence in hydrothermal systems as well as geopressured geothermal systems. As part of this effort physical properties of core samples from the Wairakei reservoir in New Zealand, the East Mesa reservoir of the Imperial Valley in California, the Cerro Prieto geothermal system in Mexico, and the Pleasant Bayou exploratory geopressured well in Texas have been measured. Even among these, only the East Mesa samples and the Cerro Prieto samples were subjected to elevated temperatures and pressures. The ranges of values, as evidenced by these studies, are as follows.

Wairakei, New Zealand. The producing formation at Wairakei is the Waiora Formation, which is a volcanic tuff. This is overlain by the Huka Falls Formation, a mudstone of lacustrine origin which in turn is overlain by the pumice zone. Both permeability and mechanical properties of cores from these formations were measured at room temperature by Hendrickson (1976). The effective porosities of the aforesaid formations were as follows: Waiora, 35.6% to 41.6%, Huka Falls, 39% to 41%, and Pumice, 48.8%.

The permeability of the Waiora Formation was found to be in the microdarcy range and was found to be distinctly sensitive to effective stress. In the effective stress range of 5 to 15 MPa, absolute permeability of the Waiora was found to decrease from 50 μD (4.93×10^{-17} m^2) to about 10 μD (9.86×10^{-18} m^2). On unloading, the permeability was distinctly lower than it was during loading. A sample of the Huka Falls Formation indicated a permeability of 63 μD (6.22×10^{-17} m^2).

The bulk compressibility of the Waiora formation was found to vary from 3.5×10^{-10} to 2.44×10^{-9}Pa^{-1}, with compressibility decreasing with increasing confining pressure. The compressibility of Huka Falls mudstone varied from 4.5×10^{-10} to 1.2×10^{-9}Pa^{-1}. The Pumice was found to be far more compressible than the other two rock types, compressibility varying from 3.45×10^{-9} to 3.13×10^{-8}Pa^{-1}. The Waiora rock indicated a linear thermal expansion of 8.2×10^{-6} m/m°K and specific heat of about 0.18 cal/g °C.

East Mesa, California, and Cerro Prieto, Mexico. Recently, Schatz (1982) studied the physical properties of cores from East Mesa and from Cerro Prieto under elevated conditions of temperature and pressure and in the presence of fluids similar in chemical composition to the reservoir fluids. In addition to permeability and compressibility, Schatz also studied the creep behavior of the samples under elevated temperatures and pressures.

The observations, in regard to mechanical properties, are summarized in Tables 1 and 2. In all the tests the samples were first carefully subjected to confining pressures and pore fluid pressures expected at the appropriate depth in the reservoir. After this, the confining pressure was maintained constant and the pore pressure was dropped by 6.9 MPa (1000 psi) to simulated pressure drop due to fluid production. The accompanying instantane-

TABLE 1. MECHANICAL PROPERTIES OF RESERVOIR ROCKS FROM THE EAST MESA GEOTHERMAL RESERVOIR, CALIFORNIA

| Rock Type | Depth (m) | Porosity (%) | Type of test | Initial Values | | | | Compressibility Pa^{-1} | | | Temp (°C) |
				Mean effective stress (MPa)	Deviator stress (MPa)	Pore Pressure (MPa)	Δp (MPa)	Normal compaction	Rebound	Uniaxial	
Medium-Grained Gray Siltstone	1000	19.3	Isotropic compression	13.3	0	9.5	6.9	1.6×10^{-10}	1.0×10^{-10}		150
		20.0	Isotropic compression	10.7	3.8	9.5	6.9	2.8×10^{-10}	1.6×10^{-10}		150
		19.7	Isotropic compression	8.2	7.6	9.5	6.9	3.0×10^{-10}	2.0×10^{-10}		150
		17.7	Uniaxial	10.6	3.8	9.5	6.9			4.4×10^{-10}	150
Fine-Grained Sandstone	1676	18.5	Isotropic compression	22.8	0	15.9	6.9	7.3×10^{-11}	5.8×10^{-11}		150
		19.9	Isotropic compression	18.4	6.5	15.9	6.9	2.2×10^{-10}	1.2×10^{-10}		150
		20.3	Isotropic compression	14	13	15.9	6.9	3.4×10^{-10}	2.3×10^{-10}		150
		19.9	Uniaxial	18.4	6.5	15.9	6.9			1.3×10^{-10}	150
Very Fine-Grained Sandstone	2180	16.9	Isotropic compression	29.3	0	20.1	6.9	1.2×10^{-10}	9×10^{-11}		150
		15	Isotropic compression	23.8	8.4	20.1	6.9	1.0×10^{-10}	7.3×10^{-11}		150
		15.2	Isotropic compression	18.2	16.8	20.1	6.9	1.5×10^{-10}	8.7×10^{-11}		150
		15.9	Uniaxial	23.8	8.4	20.1	6.9			1.7×10^{-10}	150

Note: From Schatz (1982). In all the tests external stresses were maintained constant and pore pressure discussed by Δp.

TABLE 2. MECHANICAL PROPERTIES OF ROCKS FROM THE CERRO PRIETO GEOTHERMAL RESERVOIR, MEXICO

Rock Type	Depth (m)	Porosity (%)	Type of test	Initial Values				Compressibility Pa^{-1}			Temp (°C)
				Mean effective stress (MPa)	Deviator stress (MPa)	Pore Pressure (MPa)	Δp (MPa)	Normal compaction	Rebound	Uniaxial	
Coarse-Grained Siltstone	2115	18.9	Isotropic compression	26.5	0	21.4	6.9	8.7×10^{-11}	7.3×10^{-11}		150
		18.96	Isotropic compression	15.9	15.9	21.4	6.9	1.0×10^{-10}	7.3×10^{-11}		150
		18.4	Uniaxial	15.9	15.9	21.4	6.9			1.3×10^{-10}	150
Fine-Grained Sandstone	2175	19.6	Isotropic compression	27.2	0	21.1	6.9	1.2×10^{-10}	6×10^{-11}		150
		19.4	Isotropic compression	16.2	16.4	22.1	6.9	1.3×10^{-10}	1.0×10^{-10}		150
		20.4	Uniaxial	16.2	16.4	22.1	6.9			1×10^{-10}	150
Very Fine-Grained Sandstone	2440	19.6	Isotropic compression	30.4	0	24.8	6.9	8.7×10^{-11}	4.4×10^{-11}		150
		19.3	Isotropic compression	18.1	18.4	24.8	6.9	7.3×10^{-11}	4.4×10^{-11}		150
		19.3	Uniaxial	18.1	18.4	24.8	6.9			1.2×10^{-10}	150

Note: From Schatz (1982). In all the tests external stresses were maintained constant and pore pressure discussed by Δp.

ous strains were then measured. The compressibilities given in Tables 1 and 2 are the ratios of observed strains to the change in pore pressure. Both the East Mesa samples and the Cerro Prieto samples clearly exhibited increased deformation when loaded beyond the preconsolidation stress level. The rebound compressibility varied between 50% and 75% of the virgin compressibility. Porosities of the rocks varied between 15% and 20%, these being functions of lithology as well as depth. The ranges in the compressibilities of rocks from both reservoirs are remarkably similar, varying between 6×10^{-11} aand 3×10^{-10} Pa^{-1}, depending on lithology and conditions of testing. The uniaxial compressibility is generally higher than the corresponding isotropic case.

In order to verify the assumption of the Terzaghi effective stress concept, Schatz also conducted tests increasing the confining pressure by 6.9 MPa (1000 psi) rather than dropping the pore pressure by that amount, after attaining simulated reservoir conditions. He found that within statistical limits the strains observed in either case were approximately equal, suggesting that the concept of effective stress is indeed useful for geothermal reservoirs.

Considering the fact that hydrothermal systems are very active physically and chemically, one should expect that the deformation of the system to any imposed load would require time to equilibrate. Thus, compaction due to creep or, in other words, the dependence of strains on time at constant loads could be very important. Schatz (1982) addressed this issue experimentally as well as theoretically using cores from East Mesa and Cerro Prieto. He also monitored the chemical composition of the waters expelled during the creep tests to decipher the physicochemical mechanisms accompanying creep.

The duration of the creep experiments varied from about 1 day to a maximum of about 9 days. The average long-term creep rate (4 days or more) was of the order of $1 \times 10^{-9} s^{-1}$ for East Mesa and $0.3 \times 10^{-9} s^{-1}$ for Cerro Prieto. Because of experimental considerations, longer term creep tests were not feasible. The results suggest that creep rate tended to decrease with increasing grain size and decrease with decreasing porosity. Under elevated temperatures and pressures, pressure solution effects exert significant influence in creep. As a result, less altered materials which are not yet fully in equilibrium with existing physicochemical conditions are likely to creep more than already hydrothermally altered materials. For the 6.9 MPa (1,000 psi) pore pressure reduction imposed in the experiments, the observed instantaneous compressibility was about $1.5 \times 10^{-10} Pa^{-1}$ ($1 \times 10^{-6} psi^{-1}$) for the East Mesa and Cerro Prieto rocks. Schatz (1982) estimated that assuming a long-term creep rate of $1 \times 10^{-9} s^{-1}$, the bulk strain could increase by a factor of two in 1 day and by a factor of ten in about 3 months over that suggested by the instantaneous value.

During the creep tests, permeability measurements were also made on the Cerro Prieto cores. The permeability at room temperature varied from 0.5 to 14.5 mD (4.9×10^{-16} to $1.4 \times 10^{-14} m^2$) with an average of about 4 mD ($4 \times 10^{-15} m^2$). Measured data indicate that temperature increase from room temperature to 150 °C, as well as creep, causes permeability to change by a statistically significant 40%. However, permeability reduction merely due to pore pressure reduction was not significant. There is a possibility that pressure solution accompanying creep could selectively close throats connecting individual pores, thereby leading to permeability reduction.

In regard to permeability values, it must be emphasized that effective field permeabilities in the field are likely to be larger due to the presence of interconnected macroscopic fractures and lithology changes, too large to be manifest in the small cores tested in the laboratory.

Pleasant Bayou, Texas. The U.S. Department of Energy has drilled two exploratory wells at Pleasant Bayou, Brazoria County, Texas. These wells reached a depth of 4775 m. Core samples from these wells have been studied by Gray et al. (1979) with reference to instantaneous compressibility and permeability and by Thompson et al. (1979) for creep behavior. All these studies have been carried out under room temperatures. Little is known about the possible mechanical behavior of these materials under the observed reservoir temperatures of about 160 °C (320 °F).

Data from 25 different depth intervals between 4478 and 4775 m indicate bulk compressibilities with pore pressure set at 1 atmosphere varied from 4.4×10^{-11} to $5.1 \times 10^{-10} Pa^{-1}$ (0.3×10^{-6} to $3.5 \times 10^{-6} psi^{-1}$). Under conditions of elevated pore pressures, however, bulk compressibilities were somewhat lower, varying from 2.9×10^{-11} to $3.6 \times 10^{-10} Pa^{-1}$ (0.2×10^{-6} to $2.5 \times 10^{-6} psi^{-1}$). Gray et al take this to be definitely indicative that the compressibility of grains cannot be ignored. The uniaxial compressibility for these samples varied from 5.1×10^{-11} to $2.2 \times 10^{-10} Pa^{-1}$ (0.35×10^{-6} to $1.5 \times 10^{-6} psi^{-1}$), and the porosities varied from 2% to 20%. Gray et al. (1979) reported that all samples showed that deformation was stress-path dependent and that the materials stiffened noticeably with increased effective stress. Thompson et al. (1979) studied the creep behavior of some of the Pleasant Bayou core samples by holding the stresses constant for up to 8 hours and observing the dependence of volumetric and distortional strains as a function of time. They found that the volumetric behavior could be treated as that of a modified Kelvin body and that distortional behavior could be treated as that of a Maxwell material. The implication in relation to subsidence is that the effective long-term compressibilities will be much larger than the instantaneous, elastic values, leading to slower pressure declines and larger subsidence potential. The permeabilities of the sandstones from the geopressured horizons varied from 2 to 100 mD (2×10^{-15} to $1 \times 10^{-13} m^2$).

In summary, the compressibilities of geothermal rocks are of the order of $1.5 \times 10^{-6} psi^{-1}$, which is about one-third the compressibility of water. Nonelastic and time-dependent deformation (creep) is a rule rather than the exception. There is reason to believe that pressure solution and reprecipitation effects may be significant in controlling creep behavior of geothermal rocks. The intergranular permeabilities of these rocks are in 1 to 100-mD range and are significantly modified by temperature changes and porosity changes. The effect of temperature on absolute permeability of unconsolidated Ottawa silica sand was investigated by Sageev et al. (1980). They found that up to 300 °F, absolute permeability did not depend on temperature.

DATA SYNTHESIS AND PREDICTION

The ultimate objective of geothermal subsidence analysis is to foresee the pattern and magnitude of subsidence for a given production strategy and then to modify production strategies suitably or prepare for alternate ameliorative measures to minimize adverse consequences of subsidence. Predictive models used for this purpose vary widely in sophistication. Most of these models have been developed in the fields of petroleum reservoir engineering and hydrogeology. The simplest of these models is motivated by a need for engineering solutions in the absence of even a minimum amount of required field data. At the other extreme, highly sophisticated models have been developed within the past decade based on detailed theoretical considerations. Presumably, these models can make detailed predictions, but their important data requirements far exceed our ability to collect appropriate field data.

As already discussed under Physical Basis, the task of prediction entails two aspects: the deformation of the reservoir and the deformation of the overburden. Predictive algorithms could therefore be developed either separately for each aspect, or a single generalized algorithm could be developed to handle both in a single framework. Some of the algorithms that have appeared in the literature under each of these categories are briefly discussed below.

Reservoir Deformation Models

The simplest reservoir deformation models involve the direct application of Terzaghi's one-dimensional consideration theory, neglecting the effects of thermal contraction. The key assumption here is that the reservoir is compressed vertically over a wide area so that lateral strains are negligible. Such an assumption is likely to be realistic in those situations (1) where pore pressures decline over a large area in a well field involving several production wells or (2) when water drains vertically from a low-permeability, high-storage aquitard to the aquifer characterized by high permeability and low storage over a large area. The assumption is likely to be unrealistic where pressure drawdowns are highly localized and strong spatial gradients in pressure drawdown exist. For the one-dimensional approximation, the vertical deformation of a prism of the reservoir material can be computed by

$$\delta H = - c_m H \delta_p, \qquad (22)$$

where δH is the change in the prism height, c_m is the uniaxial compressibility, H is the prism height, and δp is the change in pore-fluid pressure. The simplicity of the expression enables it to be used either for computing the ultimate deformation where δp is the ultimate pressure change or for computing, in a more general fashion, the time-dependent variation of δH in a transient system. In situations where pore pressures may rise and fall, one could account for nonrecoverable or nonelastic compaction by using either the normal consolidation value or the rebound value for c_m. Helm (1975) used this model with success to simulate the observed subsidence history near Pixley in the central valley of

California over an 11-year period. His model has subsequently been used by the U.S. Geological Survey to analyze land subsidence due to groundwater withdrawal in other parts of the United States. Helm's approach consisted in modeling only the aquitard material as a one-dimensional, doubly draining column, subject to prescribed time-dependent variations of fluid potentials at the boundaries.

The one-dimensional approximation of deformation has been used in a more general context by other workers. Following Narasimhan (1975), Lippmann et al. (1977) developed an algorithm in which the one-dimensional deformation approximation is used in conjunction with a general three-dimensional field of nonisothermal fluid flow. For computing fluid pressure changes, this algorithm accounts for the temperature dependencies of fluid density and viscosity and allows for variations of permeability in space or due to stress changes. To the extent that the assumptions are appropriate, this algorithm has the advantage of avoiding the need to solve the force balance equation.

Many numerical models have been proposed in the literature to simulate geothermal reservoirs, neglecting a detailed consideration of pore volume change in response to fluid withdrawal. The primary goal of these models is to follow the evolution in time of the fluid pressure and temperature fields (or equivalently, fluid density and internal energy fields) over the reservoir. Among these simulators one should include the following: the two-dimensional, areal, finite element model of Mercer et al. (1975); the three-dimensional finite difference model of Pritchett et al. (1975); the vertically integrated, two-dimensional, finite difference model of Faust and Mercer (1979); and the three-dimensional, integrated finite difference model of Coats (1977) and Pruess and Schroeder (1980).

Although these models do not in themselves handle matrix deformation, they could be readily coupled with an algorithm designed to solve the stress-strains equation. A good example of such a coupling is the work of Garg et al. (1976) in which they coupled their finite difference heat-mass transfer model with a two-dimensional finite element model for static stress-strain analysis.

Of the algorithms mentioned above, those of Mercer et al. (1975) and Lippmann et al. 1977) are for single-phase liquid-dominated systems, and the rest are for two-phase, water-steam systems.

Overburden Deformation Models

The goal of overburden deformation models is to ignore the presence of fluid in the system and to solve the stress equilibrium equation over the system. The solution itself may be carried out with analytical techniques or through the use of more general numerical models.

Among the analytical techniques, two deserve special mention. One of these is the Nucleus of Strain Method, variants of which have been developed by Gambolati (1972), Geertsma (1973), and others. Essentially, these models involve the superposition of the fundamental exact solution of a uniform pressure drop within a spherical region in an isotropic, homogeneous, elastic half-space. The superposition enables the handling of irregularly shaped regions. The second analytical approach involves the use of the more recent Boundary Integral Element Method approach. In this approach, the required solution for the steady-state problem is found by integrating the product of the boundary values and the normal derivative of the appropriate Green's function over the surface bounding the domain of interest, using numerical techniques.

Since the mid-1960s numerical methods, notably the finite element method, have been successfully used to solve the problem of static equilibrium in a loaded linear elastic continuum (Desai and Abel, 1972). Nonlinear material properties can be included in such models by an iterative process using effective elastic modules. Pritchett et al. (1975) adapted such a model to simulate reservoir deformation and overburden response in a geothermal reservoir. Although not developed specifically for geothermal reservoirs, several two- and three-dimensional finite element deformation models are known from the soil mechanics and the rock mechanics literatures (e.g., Sinha, 1979). These could be easily applied to simulate overburden deformation in geothermal systems, provided that sufficient data are available to characterize the subsidence.

Coupled Reservoir-Overburden Models

The most sophisticated of all the approaches to modeling is, undoubtedly, the fully coupled approach in which the fluid-flow equation, the energy-transport equation, and the force-balance equation are all simultaneously solved over the entire region extending from the land surface down to the base of the reservoir. Although no such model is available for geothermal systems (nor is one apparently warranted due to lack of data), algorithms do exist for isothermal systems in which the fluid-flow equation and the force-balance equation are simultaneously solved. Perhaps the earliest such model is that of Sandhu and Wilson (1969), which considers an elastic, fluid-saturated medium in the light of Biot's (1941, 1955) theory. A more recent example of a fully coupled model is that of Lewis and Schrefler (1978).

Comparison of Geothermal Subsidence Models

Recently, Miller et al. (1980b) carried out an in-depth comparison of a set of typical models available for simulating geothermal subsidence. Their study included reservoir models, deformation models as well as coupled models. In addition to comparing the conceptual contents of the models, they also solved typical problems with different models.

One of their major conclusions is that the lack of suitable data precludes the need for highly sophisticated, fully coupled models. Indeed, in many cases full coupling may increase cost more than it does accuracy. Depending on the stage of activity, exploration, drilling, testing, and development, model sophistication should be commensurate with quality of field data. Currently available reservoir and deformation models are conceptually adequate to handle field problems in relation to the inaccuracies introduced by lack of data. Miller et al. (1980a) found that there is a greater need to have the algorithms in readily usable forms than to develop newer and more sophisticated models.

Some Simulation Results

In their model calculation studies, Miller et al. (1980b) applied models of varying sophistication to simulate the observed subsidence at The Geysers in California and at Wairakei in New Zealand.

For The Geyser simulation, they assumed the following parameters: α, coefficient of linear thermal expansion = 10^{-5} ft/°C; K, bulk modulus of the reservoir rock = 1.44×10^{-8} psf and a temperature versus pressure relation, $\Delta T = 9.31 \times 10^{-4} \Delta p$ where T is in °C and p is in psf. At the outset, they found that the thermal contraction at The Geysers may be over four times as large as contraction due to increasing pore pressure. A comparative study of the Boundary Integral Element Method, the Nucleus of Strain Method, and simple, back-of-the envelope type of calculations indicated that a good match with field observation could be obtained with any of the methods using appropriate assumptions. However, if one wanted to increase model certainty, a major program of field investigations would be essential.

The Miller et al. (1980a) simulation of Wairakei subsidence was restricted purely to deformation modeling; fluid flow was not considered. The required pressure-change profiles were obtained from Pritchett et al. (1978). The methods used for simulation included one-dimensional hand calculations, two-dimensional finite element calculations, and three-dimensional nucleus of strain calculations. Although the simulations yielded overall similarities with several simulations, they could not match the pronounced localization of the subsidence bowl at Wairakei. Miller et al. (1980a) concluded that Wairakei subsidence was dominated by inhomogeneity of pressure drops, strong variabilities in the thickness of beds or pronounced variations in material compressibility, and that data were grossly inadequate to model these phenomena accurately.

It is pertinent here to cite two recent attempts at simulating Wairakei subsidence. Pritchett et al. (1980) have recently carried out detailed one- and two-dimensional simulations of the Wairakei geothermal field, using a two-phase nonisothermal numerical model. In their approach, the authors start with the premise that 90% of the total reservoir compaction occurs within the permeable Waiora Formation. In addition, they also assumed that (1) the Waiora Formation thickens toward the region of maximum subsidence and (2) the late-time subsidence of the Waiora's is about 15 times larger than that at early times, apparently due to preconsolidation effects. On the basis of their simulations they concluded that the subsidence bowl lies close to the margin of the geothermal field and that local phenomena such as a seismic slippage along preexisting faults control the offset location of the subsidence bowl from the main production area. On the basis of available data and parametric studies, Pritchett et al. (1980) thought that pore collapse cannot adequately explain the peculiarities of Wairakei subsidence.

Narasimhan and Goyal (1979) carried out a preliminary three-dimensional analysis of a Wairakei-type idealized system to investigate whether the offset of the subsidence bowl and the plastic deformation noticed in Wairakei could be explained in terms of a leaky aquifer–type situation, the Huka Falls mudstone acting as an aquitard. By assuming variable aquitard thickness and suitable preconsolidation stresses, they were able to show that the observed patterns could indeed by simulated in space and time. As pointed out by Pritchett et al., the compressibility values used by Narasimhan and Goyal were effectively 6 to 12 times higher than the actual measurements on a few core samples from Wairakei. Although Narasimhan and Goyal did not carry out a detailed analysis of the field data, their results did indicate that pore collapse, in the context of heterogeneities and variation of material compressibility, does have a reasonable chance of explaining a major portion of Wairakei subsidence. As pointed out by Pritchett et al. (1980), very little subsurface data is available from the area of the subsidence bowl to resolve this question satisfactorily.

CONCLUDING REMARKS

It is well documented that deformations of the land surface (vertical displacements, horizontal displacements, differential subsidence) may accompany geothermal fluid production under favorable hydrogeological and exploitation conditions. Such deformations are induced by volume changes in the geothermal reservoir caused by depletion of fluid storage as well as thermal contraction. Conceptual models do exist at the present time to explain the phenomena that are involved in the subsidence mechanisms. Compared to our ability to collect field data to characterize the subsidence history as well as the geologic system itself with adequate resolution, our conceptual models are exceedingly sophisticated. Even with the present technological revolution, the cost of collecting input data to do justice to the resolution of sophisticated mathematical models appears to be excessive. It is also doubtful whether certain kinds of data, such as the changes of stresses and horizontal displacements with depth within the geologic system will ever become available in sufficient detail in the near future.

Under the circumstances, the best course of action appears to be to establish an adequate surface and subsurface, deformation monitoring system, so that measurements can be made continuously as the system evolves from the exploration through the exploitation phase. There is a need to develop improved, economic devices to measure deformations, especially as a function of depth. Without this valuable data, most of our sophisticated mathematical models will be practically useless, since they can never be validated in a credible fashion.

Mathematical, predictive models have an important role to play. During the early stages of development, when data are scarce, simple models can be used to predict a range of consequences and to help decide whether a particular field could be developed in environmentally and economically acceptable fashion. Models should grow with a field as the field evolves in time and more and more data are accumulated. The status of modeling at present is such that even with adequate data base, only short-term predictions (over a period of a few years) can be attempted.

ACKNOWLEDGMENTS

We are grateful to Donald R. Coates, D. C. Helm, and M. J. Lippmann for their critical review of the manuscript and constructive criticisms. Their suggestions have greatly helped in improving the presentation. This work was supported by the Assistant Secretary for Conservation and Renewable Energy, Office of Renewable Technology, Division of Hydrothermal and Hydropower Technologies, the U.S. Department of Energy, under Contract No. DE-AC03-76SF0098.

REFERENCES CITED

Albores, L. A., Reyes, C. A., Brune, J. N., Gonzalez, J., Garcilazo, G. L., and Suarez, F., 1979, Seismic studies in the region of the Cerro Prieto field, *in* Proceedings, First Symposium on the Cerro Prieto Geothermal Field, Baja California, Mexico, September 1978: Lawrence Berkeley Laboratory, Berkeley, California, Report no. LBL-7098, p. 227-238.

Alonso E., Dominquez, H. B., Lippmann, M. J., Molinar, R., Schroeder, R. E., and Witherspoon, P. A., 1979, Update of reservoir engineering activities at Cerro Prieto, *in* Proceedings, Fifth Workshop, Geothermal Reservoir Engineering, December 12-14, 1979: Stanford Interdisciplinary Research in Engineering and Earth Sciences, Stanford University SGP-TR-4p, p. 247-256.

Atkinson, P., Barelli, A., Brigham, W., Celati, R., Manetti, G., Miller, F., Neri, G., and Ramey, H. Jr., 1977, Well testing in travale-radiocondoli field, *in* Proceedings, Larderello Workshop on Geothermal Resource Assessment and Reservoir Engineering, Larderello, Italy: Rome, Ente Nazionale per l'Energia Eletrica, September 12-16, 1977, p. 1-75.

Atkinson, P., Miller, F. G., Marcocini, R., Neri, G., and Celati, R., 1978, Analysis of reservoir pressure and decline curves in Serrazzano, Larderello geothermal field: Geothermics, v. 7, p. 133-144.

Bebout, D. G., Loucks, R. G., and Gregory, A. R., 1978, Geopressured geothermal fairway evaluation and test well site location, Frio Formation, Texas Gulf Coast: Bureau of Economic Geology, University of Texas, Austin, Report no. ORO/4891-4.

Biot, M. A., 1941, General theory of three dimensional consolidation: Journal of Applied Physics, v. 12, p. 155-164.

—— 1955, Theory of elasticity and consolidation for a porous anisotropic solid: Journal of Applied Physics, v. 26, p. 182-185.

Bishop, A. W., 1955, The principle of effective stress: Oslo, Norway, Terratek, Ukleblad, no. 39.

Bixley, P. F., and Grant, M. A., 1979, Reinjection testing at Broadlands, *in* Proceedings, Fifth Workshop, Geothermal Reservoir Engineering, December 12-14, 1979: Stanford Interdisciplinary Research in Engineering and Earth Sciences, Stanford University SGP-TR-40, p. 41-47.

Boldizsar, T., 1970, Geothermal energy production from porous sediments in Hungary, *in* Proceedings, United Nations Symposium on the Development and Use of Geothermal Resources, Pisa, 1970: Geothermics Special Issue no. 2, v. 2, part 1, p. 99-109.

Browne, P.R.L., 1970, Hydrothermal alteration as an aid in investigating geothermal fields, *in* Proceedings, United Nations Symposium on the Development and Use of Geothermal Resources, Pisa, 1970: Geothermics Special Issue no. 2, v. 2, part 1, p. 564-570.

Celati, R., Squarci, P., Stefani, G. C., and Taffi, L., 1976, Analysis of water levels and reservoir pressure measurements in geothermal wells, *in* Proceedings, Second United Nations Symposium on the Development and Use of Geothermal Resources, San Francisco, 1975: Washington, D.C., U.S. Government Printing Office, v. 3, p. 1583-1590.

—— 1977, Study of water levels in Larderello region geothermal wells for reconstruction of reservoir pressure trend: Geothermics, v. 6, p. 183-198.

Ceron, P., DiMario, P., and Leardini, T., 1976, Progress report on geothermal development in Italy from 1969 to 1974 and future prospects, *in* Proceedings, Second United Nations Symposium on the Development and Use of Geothermal Resources, San Francisco, 1975: Washington, D.C., U.S. Government Printing Office, v. 1, p. 59-66.

Coats, K. H., 1977 Geothermal reservoir modeling: 52nd Annual Meeting, Denver, Colorado: Society of Petroleum Engineers, AIME, Paper SPE 6892.

Defferding, L. J., and Walter, R. A., 1978, Disposal of ligand effluents from geothermal installations: Geothermal Resources Council, Transactions, v. 2, sec. 1, p. 141-143.

Desai, C. S., and Abel, J. F., 1972, Introduction to the finite element method: New York, Von Nostrand Reinhold Company, 477 p.

DiPippo, R., 1980, Geothermal energy as a source of electricity: Washington, D.C., U.S. Government Printing Office, Stock no. 061-000-00390-8, 370 p.

ENEL, 1976, Geothermoelectric power plants of Larderello and Monte Amiata, electric power from endogenous steam: Publishing and Public Relations Office of ENEL, Italy, Serie Grandi Impianti ENEL 3.

Evison, F. F., Robinson, R., and Arabosz, W. J., 1976, Micro-earthquakes, geothermal activity and structure, Central North Island, New Zealand: New Zealand Journal of Geology and Geophysics, v. 19, p. 625-637.

Faust, C. R., and Mercer, J. W., 1979, Geothermal reservoir simulation II: Numerical solution techniques for liquid- and vapor-dominated hydrothermal systems: Water Resources Research, v. 15, p. 31-46.

Fredlund, D., and Morgenstern, N. R., 1976, Constitutive relations for volume change in unsaturated soils: Canadian Geotechnical Journal, v. 13, p. 261-276.

Gambolati, G., 1972, A three dimensional model to compute land subsidence: International Association of Scientific Hydrology Bulletin, v. 17, p. 219-227.

Garcia, G., 1980, Geodetic control, first order leveling in the Cerro Prieto geothermal zone, Mexicali, Baja California, *in* Proceedings, Second Symposium on Cerro Prieto Geothermal Field, Baja California, Mexico: Comision Federal de Electricidad, Mexicali, Mexico, p. 274-293.

Garg, S. K., and Nur, A., 1973, Effective stress laws for fluid saturated porous rocks: Journal of Geophysical Research, v. 78, p. 5911-5921.

Garg, S. K., Pritchett, J. W., and Brownell, D. H., 1975, Transport of mass and energy in porous media, *in* Proceedings, Second United Nations Symposium on the Development and Use of Geothermal Resources, San Francisco, 1975: Washington, D.C., U.S. Government Printing Office, v. 3, p. 1651-1656.

Garg, S. K., Pritchett, J. W., Rice, L. F., and Brownell, D. H., Jr., 1976, Study of the geothermal production and subsidence history of the Wairakei Field, *in* Proceedings, 17th U.S. Symposium on Rock Mechanics: Salt Lake City, Utah Engineering Experiment Station, University of Utah, p. 3831-3835.

Garrison, L. W., 1972, Geothermal steam in the Geysers in the Clear Lake Region, California: Geological Society of America Bulletin, v. 83, no. 5, p. 1449-1468.

Geertsma, J., 1973, Land subsidence above compacting oil and gas reservoirs: Journal of Petroleum Technology, v. 25, p. 734–744.

Golder Associated, 1980, Simulating geothermal subsidence, comparison report No. 1, Physical processes of compaction: Lawrence Berkeley Laboratory, Berkeley, California, Report no. LBL-10794.

Goyal, K. P., Miller, C. W., Lippmann, M. J., and Von der Haar, S. P., 1981, Analysis of Cerro Prieto production data, *in* Proceedings, Third Symposium on the Cerro Prieto Geothermal Field, Baja California, Mexico: Lawrence Berkeley Laboratory, Berkeley, California, Report no. LBL-11967, p. 496–500.

Grange, L. I., 1937, The geology of the Rotorua-Taupa Subdivision, Rotorua and Kaimanawa Divisions: New Zealand Department of Scientific and Industrial Research, Bulletin 37, 138 p.

Grant, M. A., 1977, Broadlands—a gas-dominated geothermal field: Geothermics, v. 6, p. 9–29.

Gray, K. E., Jogi, P. N., Morita, N., and Thompson, T. W., 1979, The deformation behavior of rocks from the Pleasant Bayou wells, *in* Proceedings, Fourth U.S. Gulf Coast Geopressured-Geothermal Energy Conference, Dorfman, M. H., and Fisher, W. L., eds.: Center for Energy Studies, The University of Texas, Austin, p. 1031–1059.

Grimsrud, G. P., Turner, B. L., and Frame, P. H., 1978, Areas of ground subsidence due to geofluid withdrawal: Lawrence Berkeley Laboratory, California, Report no. LBL-8618, 330 p.

Grindley, G. W., 1965, The geology, structure, exploitation of the Wairakei Field, Taupo, New Zealand: New Zealand Geological Survey Bulletin, N.S. 75, 131 p.

—— 1970, Subsurface structures and relations to steam productions in the Broadlands Geothermal Field, New Zealand, *in* Proceedings, United Nations Symposium on the Development and Use of Geothermal Resources, Pisa, 1970: Geothermics Special Issue no. 2, v. 2, part 1, p. 248–261.

Grindley, G. W., and Browne, P.R.L., 1976, Structural and hydrological factors controlling the permeabilities of some hot-water geothermal fields, *in* Proceedings, Second United Nations Symposium on the Development and Use of Geothermal Resources, San Francisco, 1975: Washington, D.C., U.S. Government Printing Office, v. 1, p. 377–386.

Gustavson, T. C., and Kreitler, C. W., 1976, Geothermal resources of the Texas Gulf Coast: Environmental concerns arising from the production and disposal of geothermal waters: Bureau of Economic Geology, University of Texas, Circular o. 76-7.

Hatton, J. W., 1970, Ground subsidence of a geothermal field during exploitation, *in* Proceedings, United Nations Symposium on the Development and Use of Geothermal Resources, Pisa, 1970: Geothermics Special Issue no. 2, v. 2, part 2, p. 1294–1296.

Healy, J., 1956, Geology of the Wairakei Geothermal Field, *in* Proceedings, Eighth Commonwealth Mining and Metallurgical Congress, Australia and New Zealand: New Zealand Section, v. 7, paper 218, p. 1–13; Melbourne, Australian Institute of Mining and Metallurgy.

Helm, D. C., 1975, One dimensional simulation of aquifer system compaction near Pixley, California, 1. Constant parameters: Water Resources Research, v. 11, p. 465–478.

—— 1982, Conceptual aspects of subsidence due to fluid withdrawal, *in* Narasimham, T. N., ed., Recent trends in hydrogeology: Geological Society of America Special Paper 189, p. 103–142.

Hendrickson, R. R., 1976, Tests on cores from the Wairakei Geothermal Project, Wairakei, New Zealand: Salt Lake City, Utah, Terratek, Report no. TR75-63, 58 p.

Hitchcock, G. W., and Bixley, P. F., 1976, Observations of the effect of the three-year shutdown at Broadlands Geothermal Field, New Zealand, *in* Proceedings, Second United Nations Symposium on the Development and Use of Geothermal Resources, San Francisco, 1975: Washington, D.C., U.S. Government Printing Office, v. 3, p. 1657–1661.

Hochstein, M. P., and Hunt, T. M., 1970, Seismic, gravity and magnetic studies, Broadlands Geothermal Field, New Zealand, *in* Proceedings, United Nations Symposium on the Development and Use of Geothermal Resources, Pisa, 1970: Geothermics Special Issue no. 2, v. 2, part 1, p. 333–346.

Holzer, T. L., 1981, Preconsolidation stress of aquifer systems in areas of induced land subsidence: Water Resources Research, v. 17, p. 693–704.

Jones, P. H., 1969, Hydrodynamics of geopressure in the Northern Gulf and Mexico Basin: Journal of Petroleum Technology, v. 21, p. 803–810.

—— 1970, Geothermal resources of the Northern Gulf of Mexico basin, *in* Proceedings, United Nations Symposium on the Development and Use of Geothermal Resources, Pisa, 1970: Geothermics Special Issue no. 2, v. 2, part 1, p. 14–26.

—— 1976, Geothermal and hydrodynamic regimes in the Northern Gulf of Mexico basin, *in* Proceedings, Second United Nations Symposium on the Development and Use of Geothermal Resources, San Francisco, 1975: Washington, D.C., U.S. Government Printing Office, v. 1, p. 429–440.

Kruger, P., and Otte, C., 1976, Geothermal energy-resources production, stimulation: Stanford, California, Stanford University Press, 360 p.

Lewis, R. W., and Schrefler, B. A., 1978, A fully coupled consolidation model of the subsidence of Venice: Water Resources Research, v. 14, p. 223–230.

Lipman, S. C., Strobel, C. J., and Gulati, M. S., 1977, Reservoir performance of the Geysers Field, *in* Proceedings, Larderello Workshop on Geothermal Resource Assessment and Reservoir Engineering, Larderello, Italy: Rome, Ente Nazionale per l'Energia Elettrica, p. 233–255.

Lippmann, M. J., and Goyal, K. P., 1980, Numerical modeling studies of the Cerro Prieto Reservoir, *in* Proceedings, Second Symposium on the Cerro Prieto Field, Baja California, Mexico, October 1979: Comision Federal de Electricidad, Mexicali, Mexico, p. 497–507.

Lippmann, M. J., and Alfredo Mañón, M., 1980, Minutes of the Cerro Prieto Internal DOE/CFE Workshop, Vallombrosa Center, Menlo Park, California: Lawrence Berkeley Laboratory Internal Document, Berkeley, California.

Lippmann, M. J., Narasimhan, T. N., and Witherspoon, P. A., 1977, Numerical simulation of reservoir compaction in liquid-dominated geothermal systems: *in* Proceedings, Land Subsidence Symposium, Anaheim, California: International Association of Hydrological Sciences Publication 121, p. 179–189.

Lofgren, B. E., 1978, Monitoring crustal deformation in The Geysers-Clear Lake Geothermal Area, California: U.S. Geological Survey, Open-File Report 78-597.

Macdonald, W.J.P., 1976, The useful heat contained in the Broadlands Geothermal Field, *in* Proceedings, Second United Nations Symposium on the Development and Use of Geothermal Resources, San Francisco, 1975: Washington, D.C., U.S. Government Printing Office, v. 2, p. 1113–1119.

Majer, E. L., and McEvilly, T. V., 1979, Seismological investigations at the Geysers Geothermal Field: Geophysics, v. 44, p. 246–269.

Majer, E. L., McEvilly, T. V., Albores, L. A., and Diaz, S., 1979, Seismological studies at Cerro Prieto, *in* Proceedings, First Symposium on the Cerro Prieto Geothermal Field, Baja California, Mexico, September 1978: Lawrence Berkeley Laboratory, Berkeley, California, Report no. 7098, p. 239–249.

Marks, S. M., Ludwin, R. S., Louie, K. B., and Bute, C. G., 1979, Seismic monitoring at the Geysers Geothermal Field, California: U.S. Geological Survey Open-File Report 78-798.

Massey, B. L., 1980, Measured crustal strain, Cerro Prieto Geothermal Field, Baja California, Mexico, *in* Proceedings, Second Symposium on the Cerro Prieto Geothermal Field, Baja California, Mexico, October 1979: Comision Federal de Electricidad, Mexicali, Mexico, p. 294–298.

Mercado, G. S., 1976, Movement of geothermal fluids and temperature distributions in the Cerro Prieto Geothermal Field, Baja California, Mexico, *in* Proceedings, Second United Nations Symposium on the Development and Use of Geothermal Resources, San Francisco, May 1975: Washington, D.C., U.S. Government Printing Office, v. 1, p. 487–494.

Mercer, J. W., and Faust, C. R., 1979, Geothermal resource simulation: Application of liquid and vapor dominated hydrothermal modeling techniques to Wairakei, New Zealand: Water Resources Research, v. 15(3), p. 653–671.

Mercer, J. W., Pinder, G. F., and Donaldson, I. G., 1975, A Galerkin-Finite Element analysis of the hydrothermal system at Wairakei, New Zealand: Journal of Geophysical Research, v. 80, p. 2608–2621.

Miller, I., Dershowitz, W., Jones, K., Meyer, L., Roman, K., and Schauer, M., 1980a, Simulation of geothermal subsidence: Lawrence Berkeley Laboratory, Berkeley, California, Report no. LBL-10794, 160 p.

Miller, I., Dershowitz, W., Jones, K., Myer, L., Roman, K., and Schauer, M., 1980b, Case study data base: Companion Report no. 3 to simulation of geothermal subsidence: Lawrence Berkeley Laboratory, Berkeley, California, Report no. LBL-10839, 66 p.

Ministry of Works and Development, 1977, Broadlands Geothermal Field investigation report: Compiled by the Office of the Chief Power Engineer.

Mongelli, F., and Laddo, M., 1976, Regional heat flow and geothermal fields in Italy, in Proceedings, Second United Nations Symposium on the Development and Use of Geothermal Resources, San Francisco, 1975: Washington, D.C., U.S. Government Printing Office, v. 1, p. 495–498.

Muffler, L.J.P., and White, D. E., 1972, Geothermal Energy: The Science Teacher, v. 39, p. 40–43.

Narasimhan, T. N., 1975, A unified numerical model for saturated-unsaturated ground water flow [Ph.D. dissertation]: Department of Civil Engineering, University of California, Berkeley.

Narasimhan, T. N., and Goyal, K. P., 1979, A preliminary simulation of land subsidence at the Wairakei Geothermal Field in New Zealand: Lawrence Berkeley Laboratory, Berkeley, California, Report no. LBL-10299, 7 p.

Narasimhan, T. N., and Witherspoon, P. A., 1978, Numerical model for saturated-unsaturated flow in deformable porous media, Part 3, Applications: Water Resources Research, v. 14, p. 1017–1034.

Nelson, D. V., and Hunsbedt, A., 1979, Progress in studies of energy extraction from geothermal reservoir, in Proceedings, Fifth Workshop, Geothermal Reservoir Engineering, December 12–14, 1979: Stanford Interdisciplinary Research in Engineering and Earth Sciences, Stanford University, SGP-TR-40, p. 317–325.

Noble, J. E., and Von der Haar, S. P., 1980, Evaluation of secondary dissolution porosity in the Cerro Prieto geothermal reservoir [abs.]: 55th Technical Conference, September 21–24, Dallas, Texas, Society of Petroleum Engineers, AIME.

Nur, A., and Byerlee, J. D., 1971, An exact effective stress law for elastic deformation of rock with fluids: Journal of Geophysical Research, v. 76, p. 6414–6420.

O'Rourke, J. E., and Ranson, B. B., 1979, Instruments for subsurface monitoring of geothermal subsidence: Lawrence Berkeley Laboratory, Berkeley, California, Report no. LBL-8616, 116 p.

Petracco, C., and Squarci, P., 1976, Hydrological balance of Landerello geothermal region, in Proceedings, Second United Nations Symposium on the Development and Use of Geothermal Resources, San Francisco, 1975: Washington, D.C., U.S. Government Printing Office, v. 1, p. 521–530.

Pinder, G. F., 1979, State of the art review of geothermal reservoir modeling: Lawrence Berkeley Laboratory, Berkeley, California, Report no. LBL-9093, 144 p.

Popov, E. P., 1968, Introduction to mechanics of solids: Prentice Hall, Inc., 571 p.

Pritchett, J. W., Garg, S. K., Brownell, D. H., Jr., and Levina, H. B., 1975, Geohydrological environmental effects of geothermal power production: Report no. 555-R-75-2733, Systems, Science and Software, La Jolla, California.

Pritchett, J. W., Garg, S. K., and Brownell, D. H., Jr., 1976, Numerical simulation of production and subsidence at Wairakei, New Zealand: Stanford Interdisciplinary Research in Engineering and Earth Sciences, Stanford University, SGP-TR-20, p. 310–323.

Pritchett, J. W., Rice, L. F., and Garg, S. K., 1978, Reservoir engineering data: Wairakei Geothermal Field, New Zealand, Volume 1: Report no. SSS-R-78-3597-1, Systems, Science and Software, La Jolla, California.

Pritchett, J. W., Rice, L. F., and Garg, S. K., 1980, Reservoir simulation studies: Wairakei Geothermal Field, New Zealand: Lawrence Berkeley Laboratory, Berkeley, California, Report no. LBL-11497, 103 p.

Pruess, K., and Schroeder, R. C., 1980, SHAFT79 users' manual, Lawrence Berkeley Laboratory, Berkeley, California, Report no. 10861, 47 p.

Risk, G. F., 1976, Monitoring the boundary of the Broadlands Geothermal Field, New Zealand, in Proceedings, Second United Nations Symposium on the Development and Use of Geothermal Resources, San Francisco, 1975: Washington, D.C., U.S. Government Printing Office, v. 2, p. 1185–1189.

Sageev, A., Gobran, B. D., Brighamn, W. E., and Ramey, H.J.J., 1980, The effect of temperature on the absolute permeability to distilled water of unconsolidated sand cores, in Proceedings, Sixth Workshop, Geothermal Reservoir Engineering, December, 1980: Stanford Interdisciplinary Research in Engineering and Earth Sciences, Stanford University, SAP-TR-50, p. 297–302.

Sandeen, W. N., and Wesselman, J. B., 1973, Ground water resources of Brazoria County, Texas: Texas Water Development Board Report no. 163, 199 p.

Sandhu, R. S., and Wilson, E. L., 1969, Finite element analysis of land subsidence: Proceedings, International Association of Hydrologic Sciences Publication 88, p. 383–400.

Schatz, J. F., 1982, Physical processes of subsidence in geothermal reservoirs: Salt Lake City, Utah, Terratek, Report no. TR82-39, 136 p.

Sinha, K. P., 1979, Displacement-discontinuity technique for analyzing stress and displacements in steam deposits [Ph.D. dissertation]: University of Minnesota, Minneapolis.

Skempton, A. W., 1961, Effective stress in soils concrete and rocks, in Pore pressure and suction in soils: London, Butterworths, p. 4–16.

Stilwell, W. B., Hall, W. K., and Tawhai, J., 1975, Ground movement in New Zealand Geothermal fields, in Proceedings, Second United Nations Symposium on the Development and Use of Geothermal Resources, San Francisco, 1975: Washington, D.C., U.S. Government Printing Office, v. 2, p. 1427–1434.

Thompson, T. W., Gray, K. E., MacDonald, R. C., and Jogi, P. N., 1979, A preliminary creep analysis of rocks from the Pleasant Bayou Wells, in Proceedings, Fourth U.S. Gulf Coast Geopressured-Geothermal Energy Conference, Dorfman, M. H., and Fisher, W. L., eds.: Center for Energy Studies, The University of Texas, Austin, p. 1060–1080.

Viets, V. F., Vaughan, C. K., and Harding, R. C., 1979, Environmental and economic effects of subsidence: Lawrence Berkeley Laboratory, Berkeley, California, Report no. 8615, 232 p.

Wallace, R. H., Jr., 1979, Distribution of geopressured-geothermal energy in reservoir fluids of the Northern Gulf of Mexico Basin, in Proceedings, Fourth United States Gulf Coast Geopressured-Geothermal Energy Conference, Dorfman, M. H., and Fisher, W. L., eds.: Center for Energy Studies, University of Texas, Austin, p. 1097–1136.

Wilson, J. S., Shepherd, B. P., and Kaufman, S., 1974, An analysis of the potential use of geothermal energy for power generation along the Texas Gulf Coast: Dow Chemical U.S.A., Texas Division, 63 p.

Witherspoon, P. A., Neuman, S. P., Sorey, M. L., and Lippman, M. J., 1977, Modeling geothermal systems, in Proceedings, Il Fenomeno Geotermico e sue Applicazioni, Rome-Pisa, March 3–5, 1975: Atti dei Convegni Lincei d'Accademia Nazionale dei Lincei, v. 30, p. 173–221.

Zelwer, R., and Grannell, R. B., 1982, Correlation between precision gravity and subsidence measurements at Cerro Prieto, in Proceedings, Fourth Symposium on the Cerro Prieto Geothermal Field, Baja California, Mexico, August 10–12, 1982: Comision Federal de Electricidad, Mexicali, Mexico.

MANUSCRIPT ACCEPTED BY THE SOCIETY APRIL 18, 1984

Ground failure induced by ground-water withdrawal from unconsolidated sediment

Thomas L. Holzer*
U.S. Geological Survey
345 Middlefield Road
Menlo Park, California 94025

ABSTRACT

Ground failures, ranging from long tension cracks or fissures to surface faults, are caused by man-induced water-level declines in more than 14 areas in the contiguous United States. These failures are associated with land subsidence caused by compaction of underlying unconsolidated sediment. Fissures, which range in length from dekameters to kilometers, typically open only a few centimeters by displacement but are eroded by surface runoff into gullies 1 to 2 m wide and 2 to 3 m deep. Surface faults commonly attain scarp heights of 0.5 m and lengths of 1 km; the highest and longest scarps are 1 m and 16.7 km, respectively. Scarps grow by aseismic creep at rates approximately ranging from 4 to 60 mm/yr; modern fault movement is high angle and normal. Fault movement commonly correlates with seasonal water-level fluctuations, and examples of seasonal water-level recoveries halting fault movement have been reported. The greatest economic impact from ground failure is in the Houston-Galveston, Texas, metropolitan region where more than 86 surface faults have caused millions of dollars of damage and losses of property value.

Most ground failures probably are caused by localized differential compaction, although this mechanism has not been demonstrated everywhere. Earth fissures formed by this mechanism are caused by stretching related to bending of the overburden that overlies the differentially compacting zone. Surface faults form when differential compaction is discrete across preexisting faults. Fissures that form complex polygonal patterns probably are caused by tension induced by capillary stresses in the zone above a declining water table. Ground failures can be predicted either by determining potential areas of differential compaction or by monitoring surface deformation in areas of ongoing water-level decline. Potential ground-failure sites can be resolved by either technique to within a few dekameters.

INTRODUCTION

Surface deformation related to ground-water withdrawal from unconsolidated aquifer systems has affected an aggregate area of more than 22,000 km² in the United States. The best documented types of deformation are land subsidence and ground failure. Subsidence (loss of elevation of the land surface) has been historically more significant in both areal extent and economic impact; it has influenced the design and operation of canals and has caused inundation and permanently increased the flood hazard in some coastal and low-lying areas. Subsidence, which results from compaction within the aquifer system, has exceeded 1 m in many areas; the maximum subsidence was 8.5 m in the San Joaquin Valley, California. Commonly associated with this regional land subsidence is aseismic ground failure, that is, localized rupture of surficial materials. Two types of ground failure—earth fissures and surface faults—are recognized, although these types commonly are indistinguishable in the field.

---

*Present address: U.S. Geological Survey, 104 National Center, Reston, Virginia 22092.

Figure 1. Approximately 1-year-old earth fissure in south-central Arizona that has been enlarged by erosion. Fissure is approximately 1 m across.

Earth fissures are tensile failures; that is, opposing sides move perpendicular to the plane of failure. Surface faults are shear failures; that is, opposing sides move parallel to the plane. Horizontal separations across earth fissures typically are less than a few centimeters, whereas fault offsets of as much as 50 cm are common. Because deformation associated with ground failure is highly localized, the principal hazard is to engineered structures, including buildings, utilities, and transportation facilities. In addition, earth fissures commonly are enlarged by erosion and may pose a hazard to people or animals that fall into them.

In addition to failures related to ground-water withdrawal (the subject of this report), a few surface faults have been attributed to petroleum withdrawal (Pratt and Johnson, 1926; Yerkes and Castle, 1969; Castle and Yerkes, 1976; Verbeek and Clanton, 1981; Castle and others, 1983). These faults are associated with land subsidence caused by the petroleum withdrawal. Although most of these faults grow by aseismic creep, seismicity has occurred in a few fields. Pratt and Johnson (1926) described earthquakes associated with subsidence and surface faulting in the Goose Creek, Texas, oil field. Seismicity associated with the Wilmington, California, oil field was attributed to slippage along approximately horizontal bedding plane surfaces that was induced by flexing related to differential subsidence (Kovach, 1974). The ruptures did not extend to the surface. Yerkes and Castle (1976) described 6 additional cases of seismicity possibly related to petroleum withdrawal. Yerkes and Castle (1969) attributed surface faulting associated with oil fields to reductions in horizontal stress induced by the differential subsidence. Evidence for this mechanism consists of both theoretical considerations (Castle and Youd, 1972) and the concentric position of many of these faults in the inferred zone of extension around the margin of the subsidence bowl (Yerkes and Castle, 1969).

In general, mechanisms of formation of earth fissures and surface faults induced by ground-water withdrawal are poorly documented, although a few detailed investigations have been reported. Two mechanisms appear to apply. Surface faults and straight to arcuate earth fissures probably are caused by localized differential compaction related in part to subsurface conditions. Surface faults form when differential compaction is discrete as may occur across preexisting faults behaving as partial ground-water barriers. Earth fissures form when differential compaction is distributed over a narrow zone and causes the overburden to bend. Fissures that form in polygonal patterns probably are caused by contractions induced by capillary stresses in the dewatered zone above a declining water table. Field evidence suggests that horizontal strains sufficient to cause tensile failure can be induced by these capillary stresses.

Both earth fissures and surface faults may form by natural as well as by man-induced processes; in fact, the cause of failure at some localities is ambiguous because of the similarity between natural and man-induced failures. Accordingly, a review of the field occurrence of ground failure and of the evidence for its relation to ground-water withdrawal is desirable. Because one of the principal arguments advanced heretofore for a relation between ground failure and water-level declines has been the temporal and areal association of failures with declines, particular emphasis here is placed on reviewing this evidence. Such a review also serves to emphasize the regional scope of the problem as well as to demonstrate the common association of ground failure with subsidence in the western United States. In addition to the field occurrence of man-induced ground failure, evidence bearing on the mechanisms of ground failure is reviewed. An understanding of these mechanisms is necessary to predict ground failure confidently.

CHARACTERISTICS OF GROUND FAILURE

Earth Fissures

Earth fissures are the most spectacular type of surface deformation associated with ground-water withdrawal (Fig. 1); their visual impact derives from their length and enlargement by erosion. The longest fissure zone that has been described was 3.5 km long (Holzer, 1980a), but lengths of hundreds of meters are common. Fissures typically are enlarged by erosion into wide deep gullies. Gully widths of 1 to 2 m are commonplace. The greatest open depth measured in a fissure is 25 m (Johnson, 1980), but depths of 2 to 3 m are more common. The precise sequence of events in fissure formation has not been fully documented because sites of fissures have been examined only after the fissures began to form. Fissures commonly are first noticed after erosion along the fissure has commenced in association with large rainstorms. Fissures in their early stage typically consist of a series of collapse features aligned along the trend of the fissure; commonly, these collapse features are connected by a throughgoing hairline crack. Small tunnels generally occur near the base of these collapse features, and the rapid drainage of surface runoff suggests that the features are in hydraulic connection with each other. At some localities, however, collapse features do not appear to be connected by a throughgoing crack at the land surface. Although it cannot be completely dismissed that eolian or fluvial processes have obscured a throughgoing crack, it is more likely that cracking has occurred at depth and not propagated to the surface; the surface appearance of these fissures probably is due to piping rather than rupture. Usually the collapse features along fissures eventually coalesce into a single gully once sufficient runoff has entered the fissure. Figure 2 shows a newly formed fissure along which collapse has just begun; presumably, after

Figure 2. New earth fissure in south-central Arizona that is being enlarged by erosion. Part of fissure in foreground probably will undergo collapse with continued erosion

additional runoff, collapse will occur along the narrow crack in the foreground, and a gully similar to that shown in Figure 1 will form.

Earth fissures initially have closed drainage and are sinks for sediment and surface runoff. The near-surface void space in large fissures immediately after their formation commonly exceeds 5 m^3 per lineal meter of fissure; the largest volume measured is 16.4 m^3/m (Johnson, 1980). Because field evidence indicates that this void space is generated by transport of shallow material farther down into the crack, the surficial void space is a measure of the total void space created during fissure formation. The volume of void space in the original fissure as estimated by this approach, however, is conservative because significant volumes of material eroded from natural gullies intercepted by the fissure also are commonly deposited in fissures. Johnson (1980) indicated that approximately half of the sediment fill in two fissures in south-central Arizona was derived from headward erosion of gullies intercepted by the fissures.

The volume of void space generated during formation of an earth fissure equals the product of fissure length, depth, and average horizontal separation. Because the length can be measured and the volume of void space can be inferred approximately, the depth to which cracking extends can be estimated if information on the separation across the fissure is available. Fissure separations associated with virgin ground breakage, however, can only be inferred from limited field data. Measurements of openings in caliche, trees, and engineered structures indicate that few fissure separations exceed a few centimeters; the maximum measured separation was 6.4 cm, measured on a split mesquite tree in south-central Arizona (Holzer, 1977). Such small separations imply that some fissures may extend to dekameter depths. For example, if a separation of 6.4 cm is assumed for a fissure with a void volume of 5 m^3 per lineal meter of crack, then a depth of 78 m is implied; if a smaller separation or larger void volume is assumed, the computed depth is even greater. Such estimates of depth by this approach are only approximate. Measured open depths in fissures confirm that at least some fissures are deep; depths ranging from 5 to 10 m measured with weighted lines lowered into fissures are commonly reported. The only reported effort to follow a crack downward to its base was done with a bucket auger boring on a fissure in the San Joaquin Valley near Pixley, California. The fissure, which was partly filled with sediment, was logged to a depth of 16.8 m, where logging was terminated at the water table (Guacci, 1979).

Although the uncemented condition of most of the alluvial deposits affected by earth fissures evidently is a major factor contributing to the erosion associated with fissures, the presence of dispersive or erodible clays in these deposits may also be significant. Dispersive clays are structurally unstable when brought into contact with fresh water (Mitchell, 1976). The instability results from deflocculation of clay minerals in the soil upon wetting. Two tests, designed to evaluate soils for this property (Kinney, 1979), were used on alluvial materials collected from 5 different fissure areas in Arizona, California, and Nevada (Table 1). One of these tests, the double-hydrometer test, measures for two identical samples, the ratio of the percentage of particles finer than 5 μm in suspensions formed with and without dispersing agents and mechanical mixing (Kinney, 1979). Table 1 lists the ratio (percentage of dispersion) for the 5 areas. According to Mitchell (1976), dispersion values greater than 20% to 25% indicate a potential dispersive soil. Values greater than 50% nearly always indicate severe erodibility. Table 1 also lists the results of the Emerson crumb test, the other test that was used to determine dispersion potential. In this test, a small parcel of soil is set gently into distilled water and the tendency of the colloidal-size particles to hydrate and maintain a suspended cloud is observed (Kinney, 1979). Although the results of these two tests are not completely consistent, they suggest that many of the tested soils in fissure areas are dispersive and therefore particularly suspectible to erosion.

Surface Faults

Scarps formed by faulting related to ground-water withdrawal generally resemble fault scarps of natural origin and can be confused with them. Arguments that have been used for a relation between individual faults and ground-water withdrawal include: (1) seasonal fault movement or correlations of fault movement with ground-water level fluctuations (Reid, 1973; Nason and others, 1974; Holzer and others, 1979), (2) temporal and areal association of modern faulting with ground-water withdrawal (Lockwood, 1954; Clark and others, 1978; Holzer and others, 1979), and (3) restrictions of modern subsurface fault offset to within the aquifer system (Holzer and others, 1979). Faults that, on one or more of these bases, are suspected to be related to ground-water withdrawal commonly have scarps more than 1 km long and more than 0.5 m high. The longest scarp measured to date was 16.7 km long. Scarps range from discrete shear failures to narrow, visually detectable flexures. This range of fault-scarp morphology commonly is exhibited along individual faults, as illustrated in selected topographic profiles across the Picacho fault in south-central Arizona (Fig. 3). In addition to fault offset, fault-associated deformation occurs in the footwall and hanging-wall blocks (Holzer and Thatcher, 1979; Elsbury and others, 1980). This deformation has been measured at the surface by precise geodetic surveys to extend to more than 200 m from some scarps (T. L. Holzer, 1982, unpub.).

Scarps attributed to ground-water withdrawal generally grow in height by creep at rates that range from a few millimeters to a few centimeters per year. Neither abrupt movement nor

TABLE 1. RESULTS OF DISPERSIVE SOIL TESTS

Sample locality	Depth (cm)	Percentage of dispersion*	Emerson crumb test†
Fremont Valley, CA			
1	90	Nonplastic	--
2	60	Nonplastic	--
3	30	67	--
4	150	8	--
5	90	10	--
6	90	5	--
7	90	62	--
8	120	2	--
9	100	2	--
Las Vegas Valley, NV			
1	120	0	4
2	30	30	4
3	90	34	1
4	90	5	1
5	90	63	2
Lucerne Lake Playa, CA			
1	30	72	--
Picacho basin, AZ			
1	75	Nonplastic	2
2	150	35	1
3	75	17	1
4	30	Nonplastic	2
5	90	Nonplastic	3
6	180	9	1
7	45	7	1
8	60	100	4
Willcox area, AZ			
1	90	42	2
2	180	19	1
3	60	45	4
4	90	44	4
4	180	30	1
5	30	72	4
6	60	58	1

*Ratio of percentage of soil smaller than 5 m in diameter in suspension in modified hydrometer test to that in standard hydrometer test, times 100.
†1, Not dispersive; 2, slightly dispersive; 3, moderately dispersive; 4, strongly dispersive.

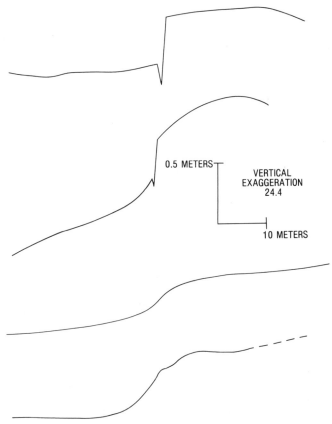

Figure 3. Selected topographic profiles across Picacho fault in south-central Arizona.

seismicity has been reported to be associated with these faults. Detailed monitoring of differential vertical displacements across a few faults, however, has revealed that creep rates vary. Both seasonal movement that correlates with water-level fluctuations (Reid, 1973; Holzer, 1978b, 1980b) and short-term episodic movement has been reported (Reid, 1973; Gabrysch and Holzer, 1978). In addition, long-term changes in creep rate have been described for a few faults (Van Siclen, 1967; Holzer and others, 1979). Only dip-slip displacements have been reported; such field evidence as offset roads and engineered structures suggests that strike-slip displacements are small to negligible. Measured ratios of horizontal to vertical displacement and field evidence suggest that these faults are high angle and normal.

IMPACT ON LAND USE

Earth fissures associated with man-induced land subsidence have been reported in at least 18 alluvial basins in 12 areas in the western United States (Fig. 4). The density of fissures varies greatly between areas. In some areas only a few isolated fissures have formed, whereas in others, many fissures occur. Although some of this variation evidently is related to differences between areas in the amount of ground-water development, part of the variation is related to subsurface conditions conducive to the formation of fissures. Two distinct hazards are posed by fissures: (1) displacements associated with their formation, and (2) deep, steep-walled gullies formed by their erosional modification. Although horizontal displacements across fissures during their formation are small, such displacements are sufficient to damage rigid engineered structures. In addition, differential vertical displacements in narrow zones near fissures may affect structures whose operation is sensitive to small tilts. Gullies associated with fissures commonly are large enough to trap and injure livestock and other animals, as well as to pose a potential hazard to humans.

Even in areas where the density of fissures is high, the economic impact from fissures has been small, primarily because of the land-use practices. Most of the areas affected by fissures are agricultural, and it has proved feasible to fill in most fissures occurring in fields. In a few fields, the impracticality of such filling has forced abandonment of part of the cultivated acreage (Winikka and Wold, 1977, Fig. 3). In addition, fissures in some uncultivated, undeveloped lands have had a modest affect on property value. Earth fissures are known in only two metropolitan areas—Las Vegas, Nevada, and Phoenix, Arizona. The only documented damage to a building by a fissure was a single-family, masonry residence in North Las Vegas. The house was subsequently demolished during construction of a freeway. An earth fissure in the Phoenix suburb of Mesa has affected the use of land near the fissure, causing the city to refuse to award building permits for this undeveloped land.

Modern surface faults are associated with land subsidence in at least 5 areas in the United States (Fig. 4). The density of faults varies greatly from area to area; in some areas, isolated faulting occurs, whereas in other areas, faulting is widespread. Fault density is greatest in the Houston-Galveston, Texas, and Fremont Valley, California, subsidence areas. In the Houston-Galveston metropolitan region, alone, the cumulative scarp length of modern surface faults is estimated to exceed 240 km (E. R. Verbeek, 1982, written commun.). Although variations from area to area in the intensity of ground-water development contribute to the variation in the density of faults, subsurface conditions probably are dominant.

The primary potential for damage from surface faults is from offset across the scarp itself. Because these faults are normal, offset consists of both horizontal and vertical components, although the high angle of faulting generally causes the vertical component to be greater than the horizontal component. In addition to fault offset, tilting of the land surface near the scarp in the footwall and hanging-wall blocks can locally affect structures sensitive to tilts. Most of the damage caused by faults has been in urbanized parts of the Houston-Galveston region. Although the cost of fault damage there has not been systematically surveyed, millions of dollars of damage and losses of property value probably have been incurred (Clanton and Amsbury, 1975); these costs derive principally from damage to houses and other buildings, utilities, and roads. Costs of damage from faults in other areas have been small because the faults occur in agricultural areas; minor and reparable damage to roads and utilities has been the principal impact. Be-

Figure 4. Ground-failure areas in the western United States.

cause fault offset is by creep, continuous maintenance is required, as is releveling of the affected fields.

GROUND-FAILURE AREAS

South-Central Arizona

More than 50 areas of earth fissures and 4 surface faults occur in the south-central Arizona region (Fig. 5). The affected region lies within the southern part of the Basin and Range province, which is characterized by fault-bounded alluviul basins and mountains. Faulting and major uplift in south-central Arizona occurred relatively early in the late Cenozoic (Eaton, 1972), and the area is now seismically inactive or nearly so (Sturgal and Irwin, 1971; Scott and Moore, 1976). Regional tilting of paired river terraces in the Phoenix area, however, suggests that slow tectonic movements have occurred at least into the Pliocene and, possibly, the Pleistocene (Péwé, 1978). Pleistocene deformation in the Tucson area was inferred by Davidson (1973). The basinal surfaces consist primarily of large bajadas underlain by unconsolidated alluvium that thickens basinward. These surfaces grade gently into each other and appear to be slowly burying the intervening mountains. Inselbergs of crystalline and volcanic bedrock also are common.

The unconsolidated alluvial deposits, which compose the principal aquifer, range from fine- to coarse-grained materials (Lee, 1905; Stulik and Twenter, 1964; Hardt and Cattany, 1965). The thickness of these deposits range from a featheredge around basin margins to more than 700 m in depocenters (Hardt and Cattany, 1965). The underlying materials range from evaporites (Eaton and others, 1972; Peirce, 1976) to volcanic rocks, crystalline bedrock, and indurated conglomerate (Lee, 1905; Hardt and Cattany, 1965).

Withdrawal of ground water from the unconsolidated alluvium has induced local declines in water level of more than 100 m (Ross, 1978; Laney and others, 1978b; Konieczki and English, 1979). The withdrawn water is used principally for crop irrigation, and most of it is lost from the region by evapotranspiration.

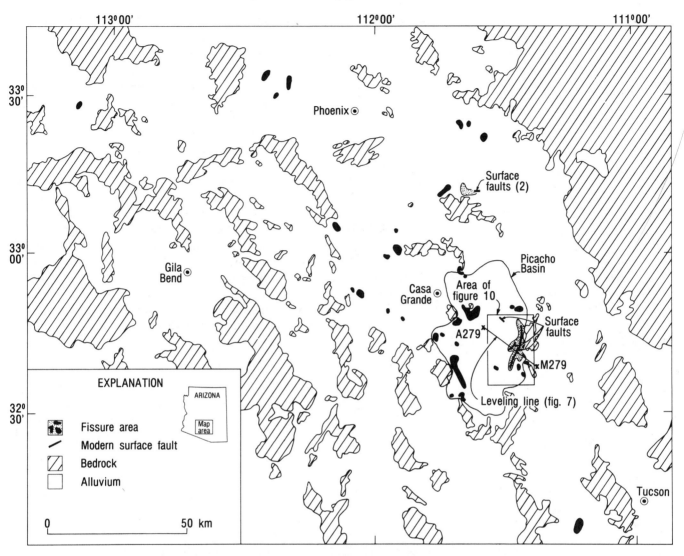

Figure 5. Modern surface faults and areas with earth fissures (stippled) in south-central Arizona. Based on Laney and others (1978a), Raymond and others (1978), Holzer and others (1979), and T. L. Holzer (1981, unpub.). Some of the more than 50 areas with fissures are grouped together because of scale of map.

Cones of depression are areally extensive and extend to the margins of the basins. The aquifer is unconfined. Ground-water pumping for agriculture began by the early 1900s (Lee, 1905), but ground-water development was modest until 1919 (Arizona Water Commission, 1975). The rate of pumpage increased significantly in the 1920s and 1930s in response to changing economic and technical factors. By the early 1940s, all the basins had undergone major ground-water development. Some of this history is reflected in a hydrograph (Fig. 6) based on wells near the center of Picacho basin, one of the later basins to be heavily developed.

Extensive land subsidence within the alluvial basins has been induced by water-level declines (Schumann and Poland, 1969; Schumann, 1974; Laney and others, 1978a). Although the margin of each basinal subsidence area is poorly defined by leveling data, the total area affected may be more than 3000 km^2; the maximum documented subsidence is 3.8 m (Laney and others, 1978a). The history of such land subsidence is best documented with leveling data (Fig. 7) in the southern part of the south-central Arizona in Picacho basin (Fig. 5). Subsidence in this basin began sometime between 1934 and 1948 (Holzer and others, 1979). Holzer (1981) has proposed that sediments within the aquifer system in Picacho basin were overconsolidated before ground-water withdrawal began and that major subsidence did not begin until water levels declined more than approximately 30 m. The vertical distribution of the compaction causing the subsidence is poorly documented. Comparison of compaction, recorded by a vertical extensometer, with measured subsidence in the center of Picacho basin, indicates that 82% of the compaction from 1964 to 1977 occurred in the upper 252 m of unconsolidated alluvium (T. L. Holzer, 1981, unpub.).

The first earth fissure reported in south-central Arizona

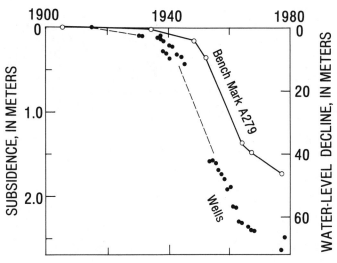

Figure 6. Subsidence of bench mark A279 and water-level declines measured in nearby wells near center of Picacho basin, Arizona (Holzer, 1981). See Figure 5 for location of A279; wells are within 2 km of A279.

formed in 1927 in Picacho basin (Leonard, 1929). The next earth fissure formed in 1949 (Feth, 1951). It was situated 0.87 km southeast of the 1927 fissure (Holzer and others, 1979). Since then, both the number of fissures and the areas affected by them have increased steadily (for example, Peterson, 1964; Schumann, 1974; Laney and others, 1978a). Although delineation of separate fissure areas is somewhat arbitrary, more than 50 fissure areas were mapped in 1976–77 by Laney and others (1978a). Most areas are near the margins of the basins and near exposed bedrock; a few fissures, however, are more centrally situated within the basins. The complexity of fissure patterns varies greatly from area to area; at some localities, only single straight or arcuate cracks occur, whereas at others, complex networks of intersecting fissure zones occur (Fig. 8). In addition to the complex patterns formed by some fissures, a few fissures show complex histories of growth and activity. For example, Jachens and Holzer (1982, Fig. 6) documented fissure zones that had grown in length for more than 26 years. They also described fissures within these zones that were still opening and receiving sediment more than 24 years after their initial formation.

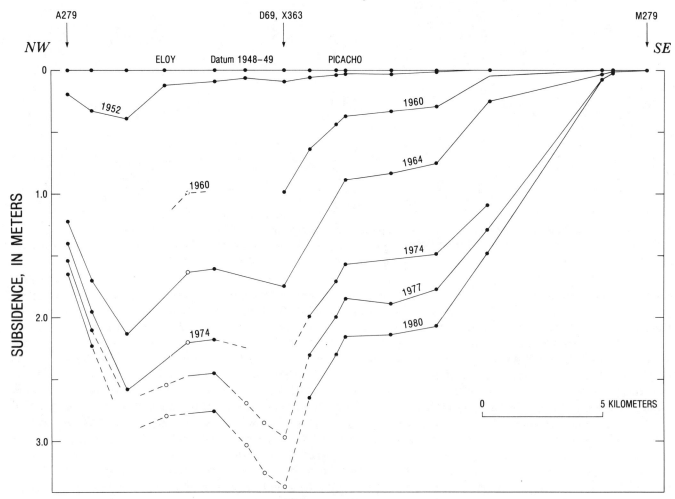

Figure 7. Selected subsidence profiles extending partly across Picacho basin, Arizona (updated from Holzer and others, 1979); dashed where inferred. Subsidence shown by open circles is based on bench marks set after 1948–49. See Figure 5 for location.

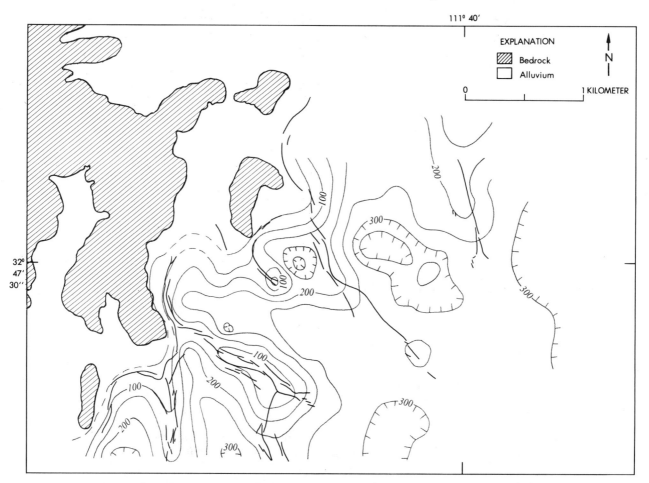

Figure 8. Isopach map of unconsolidated alluvium beneath fissure area near west margin of Picacho basin, Arizona. Contour interval is 50 m. Fissures are shown by heavy dark lines. Hachures on contours point toward increasing sediment thickness. From Jachens and Holzer (1982).

The temporal and areal correlation of earth fissures with ground-water withdrawal indicates that the earth fissures are man-induced. The relation of the 1927 Picacho fissure to ground-water withdrawal, however, is problematic. Although ground-water development began in 1914 in the Picacho basin, water-level declines in 1927 at the fissure probably were very small (Smith, 1940). The possibility of some, but very modest, decline is suggested by extrapolation of a hydrograph (Fig. 9) from a well near the fissure. This hydrograph indicates that major water-level decline related to migration of the cone of depression did not occur until about 1935 (subsidence in Picacho basin began during the period 1934–1948). Therefore, it may be that the 1927 fissure was of natural origin; if it was man-induced, then only modest water-level declines were required to induce it.

Subsurface conditions beneath many earth fissures have been inferred principally from gravity surveys. This technique is particularly useful in south-central Arizona because of the large density contrast between unconsolidated alluvium constituting the aquifer system and the underlying material. The surveys (Anderson, 1973; Jennings, 1977; Pankratz and others, 1978b; R. Beruff, 1979, written commun.; Jachens and Holzer, 1979, 1982) indicate that most earth fissures occur above ridges or "steps" in the buried bedrock surface. Detailed and precise surveys by Jachens and Holzer (1979, 1982) indicate that fissures controlled by bedrock occur above the loci of points of maximum convex-upward curvature on the bedrock surface. Even highly complex patterns formed by fissures have been related to bedrock control (for example, Fig. 8). A few fissures, however, have been investigated whose locations do not appear to be controlled by relief on the bedrock surface (Jennings, 1977; Jachens and Holzer, 1979). Detailed subsurface data, other than gravity data, beneath these fissures generally are absent, although Holzer (1978b), on the basis of data from closely spaced boreholes, concluded that the location of a fissure not controlled by buried bedrock relief was determined by a preexisting fault that behaved as a partial ground-water barrier. In what is probably an unusual condition, the locations of fissures in an area west of Phoenix appear to be related to the upper surface of a buried salt body (Eaton and others, 1972).

Despite the widespread occurrence of earth fissures, research on specific mechanisms of their formation has been primarily qualitative and necessarily speculative until recently. Several

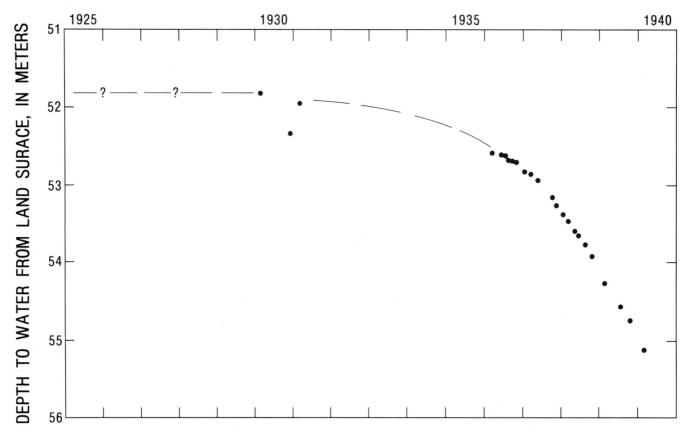

Figure 9. Hydrograph of well approximately 600 m southeast of 1927 fissure in Picacho basin, Arizona. Depth to water in 1915, before major pumping, was approximately 51.8 m (Smith, 1938, pl. 3). Hydrograph data from Smith (1940).

mechanisms have been hypothesized to explain their formation: localized differential compaction (Feth, 1951), hydrocompaction (Pashley, 1961), horizontal seepage forces (Lofgren, 1971), regional differential compaction (Bouwer, 1977), and desiccation caused by water-table declines (Holzer and Davis, 1976). The association of most earth fissures with variations in aquifer-system thickness suggests that stretching caused by localized differential compaction probably is the dominant mechanism. Detailed gravity surveys by Jachens and Holzer (1979, 1982), which illustrate the highly localized bedrock control of even complex fissure patterns, particularly support this mechanism. Localized differential subsidence and horizontal strain across two fissures have been measured directly by Holzer and Pampeyan (1981). In addition, Jachens and Holzer (1982) infer such differential subsidence from topographic profiles across fissures.

Although earth fissures are the dominant mode of ground failure in south-central Arizona, prominent fault scarps have also developed at 4 locations (Fig. 5). Because initial ground failure along parts of at least some of these scarps consisted of tensile breaks and because collapse features have continued to form along segments of the scarps, distinction between these scarps and earth fissures may appear somewhat arbitrary. Both field (Raymond and others, 1978, Fig. 5) and survey (Holzer and Thatcher, 1979) data suggest, however, that the relative displacements across these scarps are dominantly vertical and parallel to the plane of failure in contrast to the relative displacements across fissures, which are dominantly horizontal and perpendicular to the plane of failure. In addition, the absence of collapse features along many parts of these scarps, some of which are 0.6 m high, demonstrates that the relative displacements across the scarps, at least at these localities, parallel the plane of failure. Such observations suggest that these failures are shear failures rather than tensile failures, and so on this basis, they are here referred to as surface faults.

Peterson (1962, 1964) was the first investigator to report vertical offset across a ground failure in south-central Arizona. The offset, which was noted in May 1961 on a highway, formed across a fissure that had first opened in 1949 (Feth, 1951). This scarp, which has been named the Picacho fault, is now 15 km long and has grown to a height that ranges from 0.2 to 0.6 m along its length (Holzer and others, 1979). It occurs along the east flank of Picacho basin and the subsidence bowl that has formed within the basin (Fig. 10). Since Peterson's (1962) recognition of vertical offset, two additional scarps, with relief of approximately 0.15 m, have been identified in Picacho basin (Fig. 10). The positions of these three faults within the subsidence bowl are not systematic in that they occupy positions that range from near the center to the flanks of the subsidence bowl. On the basis of field

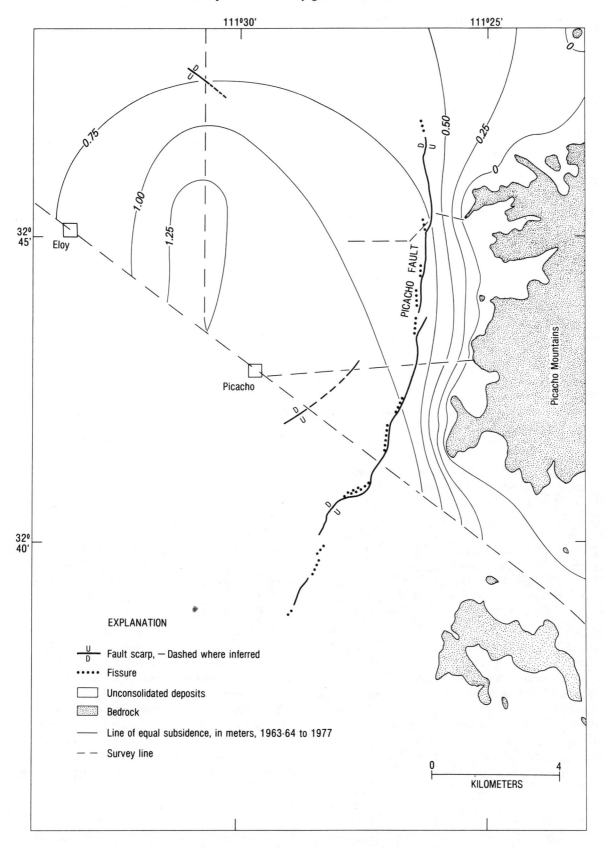

Figure 10. Surface faults and land subsidence from 1963–64 to 1977 in Picacho basin, Arizona. Modified from Holzer and others (1979). See Figure 5 for location of map.

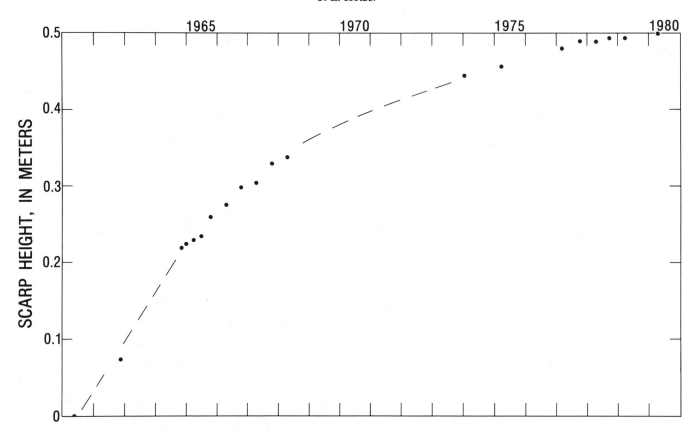

Figure 11. Height of Picacho fault scarp over time. Updated from Holzer and others (1979).

evidence, the faults are high angle and normal. In addition to the faults in Picacho basin, Raymond and others (1978, Fig. 5) described a small graben that had local relief of 0.46 m (northernmost fault in Fig. 5). All of these faults probably have grown by aseismic creep, although some controversial seismicity has occurred (Peirce, 1975; Yerkes and Castle, 1976).

The Picacho fault is the most thoroughly studied of these faults (Holzer and others, 1979). The fault scarp, which began to form in 1961, initially increased in height at a rate of approximately 60 mm/yr but slowed to an annual rate of growth of approximately 9 mm/yr from 1975 to 1980 (Fig. 11). The creep rate also varies seasonally and correlates with water-level fluctuations (Holzer, 1978b; Holzer and others, 1979). Analysis of both near-fault surface deformation and a 12-year record of the relative displacements of bench marks adjacent to the fault indicates faulting is high angle and normal (Holzer and Thatcher, 1979). Surface deformation near the fault indicates that formation of the fault scarp was preceded by man-induced differential subsidence which was distributed over an approximately 300-m-wide zone (Holzer and others, 1979).

Several lines of evidence suggest that modern surface faulting in south-central Arizona is related to ground-water withdrawal. First, faulting is restricted to basins undergoing water-level declines. Second, the region appears to be tectonically quiet, on the basis of both regional geodetic data (Holzer, 1979a) and the geomorphology of the basins. And third, scarp growth by aseismic creep is unusual in the Basin and Range province (Holzer and others, 1979). In addition to these general arguments, Holzer and others (1979) advanced two specific arguments for relating movement of the Picacho fault to ground-water withdrawal. First, seasonal fault movement correlates with seasonal water-level fluctuations and compaction. And second, measured vertical displacements associated with faulting are compatible with the results from an analytical model of subsurface faulting in which modern rupturing does not extend beneath the zone stressed by water-level decline.

Seismic refraction surveys (Pankratz and others, 1978a) and closely spaced exploratory borings (Holzer, 1978b) indicate that the Picacho fault is associated with a preexisting fault which is a partial ground-water barrier in the alluvial aquifer: this preexisting fault is high angle and normal. Monitoring of piezometers in the boreholes established that the water-level difference across the fault is seasonal and virtually disappears during annual periods of water-level recovery. The magnitude of this seasonal difference in water level appears sufficient to account for the modern fault offset by localized differential compaction (Holzer, 1978b).

Houston-Galveston, Texas, Region

Within the Houston-Galveston, Texas, subsidence region (Fig. 4, area 13) more than 160 faults with an aggregate length of more than 500 km have been identified at the land surface (Verbeek and others, 1979; Verbeek and Clanton, 1981). Of these 160 faults, at least 86, with an aggregate length of 240 km, have been historically active (E. R. Verbeek, 1982, written commun.). The affected region, part of the Texas Gulf Coast, is underlain by a thick sequence of late Mesozoic and Cenozoic miogeosynclinal sedimentary deposits that record a long history of geologic subsidence (Murray, 1961). According to Lehner (1969), currently the most structurally active section of the Gulf Coast is along the continental slope where deformation occurs in conjunction with seaward progradation of the sedimentary prism. Although Lehner (1969) suggested that the rate of offset for a given fault diminishes to insignificant amounts as the continental slope progrades seaward beyond it, Pleistocene faulting far landward of the present continental slope, such as in the Houston area, indicates that this structural model is oversimplified. Although the Pleistocene sedimentary record has been variously interpreted (Bernard and others, 1962; Proctor, 1973), the increasing coastal dip of the sediment with increasing age suggests that basinal subsidence and accompanying tilting are ongoing. In addition to the regional deformation, normal faulting and rising salt domes have locally deformed Pleistocene sediment. Geologically youthful natural deformation is thus indicated (Verbeek and Clanton, 1981). Therefore, the relative contribution of natural processes to historical faulting in the Houston-Galveston region is an important consideration.

Determination of the cause of historical faulting in the Houston-Galveston region is further complicated by the observation that the faulting probably is occurring along preexisting faults. For at least 10 historical scarps that have formed on preexisting scarps, the relation is unambiguous (Van Siclen, 1967; Elsbury and others, 1980; Verbeek and Clanton, 1981). In addition, exploratory drilling has documented connections between historical scarps and preexisting faults at many other localities (Verbeek and Clanton, 1981). The best documented of these investigations is for the Ethyl fault (Fig. 13). Shallow seismic-reflection profiles and data from closely spaced boreholes indicate that the historical scarp connects to a preexisting fault with a long history of at least Pleistocene displacements (Woodward-Lundgren and Associates, 1974, unpub. report for Brown and Root, Inc.). Finally, Van Siclen (1967) recognized that the gross patterns formed by the historical scarps suggest that they can be separated into two structural provinces that overlie structural provinces defined on the basis of the preexisting fault offset of subsurface horizons. In the southeastern province, faults are short, generally 1 to 3 km, and orientations are diverse; these faults are associated with salt domes. In the northwestern province, which lacks salt domes, faults are as long as 16.7 km and trend subparallel to the coast. Although Van Siclen (1967) suggested that the historical surface faults connect to the underlying preexisting faults and that none are virgin ground failures, documentation of this relation is modest and varies in detail from fault to fault (for example, Van Siclen, 1967; Kreitler, 1977a; Verbeek and Clanton, 1981).

Ground water in the Houston-Galveston region is with-

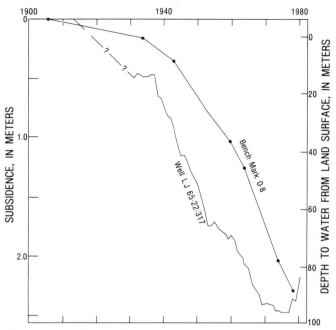

Figure 12. Subsidence of bench mark 0-8 and depth to water from land surface in well LJ 65-22-317 in Evangeline aquifer near the center of the Houston-Galveston, Texas, subsidence bowl (see Fig. 13 for locations). Post-1976 water-level recovery caused by reduced pumpage in general area near well.

drawn from a thick sequence of fluvial and marine unconsolidated sediment. The maximum depth to the base of the aquifer system is approximately 800 m. Inland the base is defined as the top of the Burkeville confining bed, whereas coastward its base is the interface between fresh and salt water (Turcan and others, 1966). The aquifer system, consisting primarily of discontinuous layers of sand and clay, has been subdivided into two confined aquifers (Turcan and others, 1966), both of which are heavily pumped (Jorgensen, 1975). The Chicot, the upper aquifer, and the Evangeline, the lower aquifer, attain approximate maximum thicknesses of 400 and 600 m, respectively. The aquifers are distinguished from each other primarily on the basis of permeability; the Chicot is more permeable.

Significant development of ground water began toward the end of the nineteenth century with the drilling of free-flowing wells (Taylor, 1907). Pumpage increased gradually until the 1930s, when the rate of ground-water withdrawal accelerated (Jorgensen, 1975). In response to this long-term ground-water development, water levels have declined since the early 1900s (Fig. 12). Maximum declines in both aquifers exceeded 100 m. The regional water-level declines have created one of the largest subsidence bowls in the United States (Fig. 13); approximately 12,200 km^2 of land subsided more than 0.15 m from 1943 to 1973 (Gabrysch and Bonnet, 1975).

The early history of this man-induced subsidence is poorly documented. Releveling in 1933 of bench marks set in 1905–1906 and 1918 provided the first indication of subsidence (for example, Fig. 12). The historical relation between subsidence

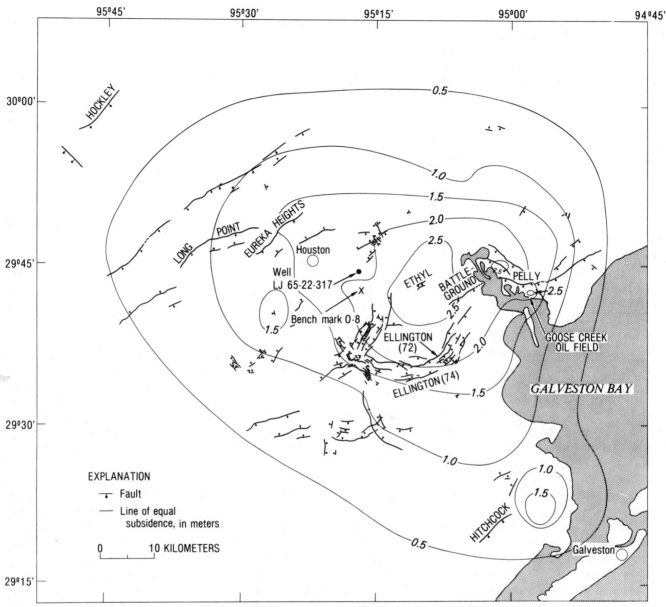

Figure 13. Land subsidence from 1906 to 1978 (adapted from Gabrysch, 1980) and surface faults (Elsbury and others, 1980; Verbeek and Clanton, 1981; E. R. Verbeek, 1981, unpub. data) in the Houston-Galveston, Texas, region. Ball-and-bar symbols on faults indicate downthrown side. Named faults are referred to in text or figures.

and water-level decline suggests that subsidence probably began with the earliest water-level declines, although it was modest until the preconsolidation stress of the aquifer system was exceeded (Holzer, 1981). Both aquifers have compacted. Gabrysch (1980) estimated maximum subsidence from 1906 to 1978 exceeded 2.7 m.

In addition to regional ground-water withdrawals, oil and gas production is regionally extensive. The contribution of petroleum withdrawal to regional subsidence appears to be minor, however. Subsidence profiles through or adjacent to 29 oil and gas fields indicate a general absence of local increases of subsidence centered over the fields (Holzer and Bluntzer, 1984). A few exceptions have been recognized (Yerkes and Castle, 1969; Kreitler, 1977a). The most notable is the Goose Creek oil field (Fig. 13), which subsided more than 1 m from 1917 to 1925 in response to rapid exploitation of shallow petroleum reservoirs (Pratt and Johnson, 1926). Subsidence within a few fields, however, actually has been less than that in the surrounding area; these fields generally are associated with shallow-seated salt domes contained within the aquifer system, which cause the system to thin (Holzer and Bluntzer, 1984).

The earliest historical surface faulting in the Houston-Galveston region occurred at the Goose Creek oil field, where it was attributed to withdrawal of oil, water, gas, and sand from shallow petroleum reservoirs (Pratt and Johnson, 1926). Three faults, with an average scarp length of 0.4 km, formed concentric to the subsidence depression over the field. According to Elsbury and others (1980), the earliest recognition of surface faulting possibly related to ground-water withdrawal was between about 1936 and 1940 when 3 active scarps were recognized. The earliest published report was by Lockwood (1954), who mapped known pavement offsets and noted that most of the displacements had occurred since 1942. Lockwood (1954) also was the first to suggest that the faulting might be related to ground-water withdrawal. Since these early reports, many investigators, including Gray (1958), Weaver and Sheets (1962), Van Siclen (1967), Reid (1973), Kreitler (1977b), Verbeek and Clanton (1978, 1981), Verbeek and others (1979), and Elsbury and others (1980), have mapped surface faults at various scales. An areal compilation from several sources, but principally Verbeek and Clanton (1981, Fig. 1), is shown in Figure 13. Some of the faults mapped by Verbeek and Clanton (1981) were mapped only from aerial photographs. Field evidence suggests that all surface faults are high angle and normal.

As can be seen in Figure 13, faults are neither situated systematically within the subsidence bowl, nor are they oriented systematically with respect to the subsidence contours. Faults occur near both the center and flanks of the subsidence bowl and variously trend parallel, perpendicular, and oblique to the contours of equal subsidence. Finally, the areal density of faulting varies greatly within the subsiding area. As first recognized by Van Siclen (1967), the historical scarps can be grouped into two structural provinces that coincide with the provinces defined by deeper preexisting faults. These two observations—the absence of a relation to the subsidence strain field and the correlation with subsurface structure—suggest that the locations of historical scarps are determined by preexisting faults and not directly by land subsidence.

Heights of historical fault scarps vary greatly. Most reported heights are based on surveyed offsets of pavement and so are relative to a construction datum. The highest modern scarp reported is 1.12 m high, which was measured in 1972 on a road surface originally paved in 1928 (Reid, 1973); heights of approximately 0.4 m are more typical (for example, Van Siclen, 1966; Reid, 1973). With the exception of the early faulting in the Goose Creek oil field, scarps have grown by aseismic creep. Pratt and Johnson (1926, p. 581) reported that the fault movements at Goose Creek "were accompanied by slight earthquakes which shook the houses, displaced dishes, spilled water, and disturbed the inhabitants generally." Rates of creep computed by Van Siclen (1966) on the basis of surveys in 1966 of 24 pavement offsets approximately range from 4 to 27 mm/yr and average about 10 mm/yr; the periods for which these rates were computed range from 12 to 39 years. Repeated surveys of particular road surfaces confirm the creeping style of scarp growth (Fig. 14). Although some surveys suggest long-term variations of creep rates (Van Siclen, 1966; Reid, 1973), the precision and frequency of the surveys, in addition to road repairs, do not permit rigorous

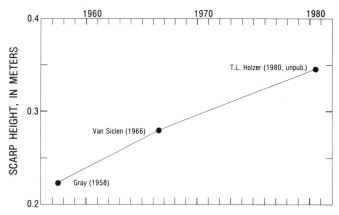

Figure 14. Scarp height of Pelly fault (see Fig. 13) inferred from profiles of offset pavement of Edna Street in Baytown, Texas.

evaluation of these variations. At least at one location, the Battleground fault (Fig. 13) at State Highway 225, faulting ceased by 1959 after a long period of historical activity (Van Siclen, 1967).

Short-term variations in creep rate were indicated by 5 tilt beams and 4 horizontal extensometers established across 3 faults in northwestern Houston (Reid, 1973) and by resurveys of closely spaced bench-marks that were established more recently across 12 faults. (Eleven of the bench-mark sites were established beginning in 1978 by the U.S. Geological Survey and the twelfth by McClelland Engineers in 1973). The tilt beams and extensometers indicated that faulting occurred during episodes of from 1 hour to 4 days and that these episodes were separated by inactive periods lasting from 4 to 60 days (Reid, 1973); vertical displacements during movement episodes ranged from 0.09 to 3.33 mm. Fault movement appeared to correlate with seasonal water-level fluctuations in the aquifer system; scarp height increased during periods of decline and decreased during periods of rapid recovery. After periods of rapid water-level recovery, but still during the overall period of recovery, fault rebound slowed and eventually reversed, so that the scarp height increased again despite the rising water level (Reid, 1973). At 2 of the tilt-beam sites, Kreitler (1977b) computed a significant statistical correlation between faulting and water-level fluctuations for the first 2 years of record; this correlation, however, did not persist for the remaining 4-year period of record (Gabrysch and Holzer, 1978). Figure 15 plots the results from precise leveling surveys of the closely spaced bench marks on opposite sides of 7 historical surface faults; 5 sites with records of less than 1-year duration are omitted. These bench marks were specially constructed to isolate the mark from expansive-soil phenomena. The records demonstrate that short-term creep rates vary, although these records are too short and the resurveys too infrequent to infer systematic variations. Particularly striking is the temporary cessation of movement of the Eureka Heights fault after approximately 20 mm of scarp growth. Periods of inactivity are indicated on all records except for the Long Point fault, which was continuously active.

Fault-associated vertical deformation, as reflected by bench-

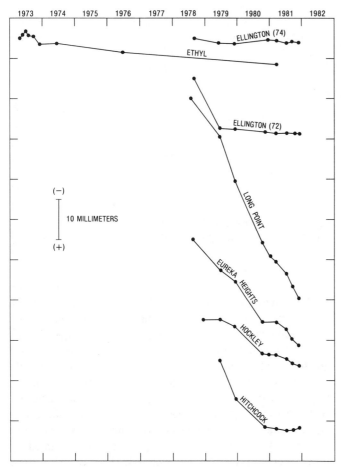

Figure 15. Change of fault scarp height of 7 faults in Houston-Galveston, Texas, region (see Fig. 13). Positive displacement indicates increase in scarp height. Heights are based on elevations of bench-mark pairs spanning scarp.

mark data, generally is concentrated within a few hundred meters of the scarp itself, although fault-associated deformation of a few faults may be more widespread (T. L. Holzer, 1981, unpub.). This is illustrated by the profiles of changes of elevation across the Eureka Heights and Long Point faults (Figs. 16 and 17). The profiles in Figure 16 cross both faults and are based on bench marks that have a typical spacing of 1 km or more, whereas the profiles in Figure 17 cross one fault each and are based on closely spaced bench marks. Consider first the Eureka Heights fault which was recognized to be active by Lockwood (1954) and which had grown to a height of 560 mm by 1982 (T. L. Holzer, 1982, unpub.). Note that the profiles (Fig. 16A), which are based on widely spaced bench marks, across the Eureka Heights fault are not offset in the same sense as the faulting. Figure 17A, based on closely spaced bench marks established in 1978, illustrates that the fault-associated deformation is accommodated within an approximately 400-m-wide zone near the fault. By contrast, differential subsidence of more widespread significance may coincide with the Long Point fault on the basis of profiles with a 1943 datum and the widely spaced bench marks (Fig. 16A). The historical scarp was approximately 80 and 390 mm high in 1966 and 1982, respectively (Van Siclen, 1966; T. L. Holzer, 1982, unpub.). Note, however, that the fault has no expression in the profile with a 1973 datum (Fig. 16B). As with the Eureka Heights fault, closely spaced bench marks established near the Long Point fault in 1978 indicate, at least, that the post-1978 fault-associated deformation is primarily accommodated in a narrow zone (Fig. 17B).

The relation between historical surface faulting and ground-water withdrawal, though suspected for many years, is speculative. The principal uncertainties derive from the unknown contributions of natural processes and oil and gas withdrawal. Despite these uncertainties, the striking temporal and areal association of historical faulting with water-level decline and associated land subsidence suggests that they are casually related. With only a few exceptions, all of which occur in or near oil fields, all the faults reported to have moved during historical time along the Gulf Coast are in the Houston-Galveston subsidence bowl (Verbeek and Clanton, 1981). In addition to the temporal and areal association, the correlation observed by Reid (1973) and quantified by Kreitler (1977b) between some fault movement and seasonal water-level fluctuation at several sites also suggests that modern movement may be at least partly man-induced (Reid, 1973).

Support for the concept of historical natural faulting consists principally of (1) Quaternary fault scarps that antedate fluid withdrawals (Van Siclen, 1967; Elsbury and others, 1980; Verbeek and Clanton, 1981) and (2) ongoing warping of the Gulf Coast geosyncline (Bernard and others, 1962). Topographic maps with a 1-ft contour interval were published by the U.S. Geological Survey early in the twentieth century. Scarps of at least 10 faults show clearly on these maps, and anomalous drainage patterns are associated with the trace of other historical surface faults (Verbeek and Clanton, 1981). Because these scarps and drainage anomalies have formed in Pleistocene sediment, youthful geologic movement is indicated. Average historical rates of fault creep, however, appear to be anomalously high, in that if they had been sustained over geologic time, higher and many more preexisting scarps should be present. This argument does not preclude the possibility that natural faulting is episodic and that many faults within a given region move simultaneously. Analysis of regional geodetic data, however, does not yield evidence for anomalously high rates of historical regional tilting in adjacent areas unaffected by fluid withdrawal. Holdahl and Morrison (1974), on the basis of regional geodetic data from regions unaffected by fluid withdrawal, reported that modern coastward tilting in the Texas Gulf coast occurs at a yearly rate of approximately 0.029 mm/km, although they noted that their result is "not highly regarded." Reevaluation of the same data suggests that natural rates of tilting are too small to be detected with existing data (T. L. Holzer and D. S. Wright, 1981, unpub.), a result compatible with rates of tilting implied by Bernard and others (1962) on the basis of geologic data. Historical natural faulting related to ongoing gulfward tilting of the sedimentary

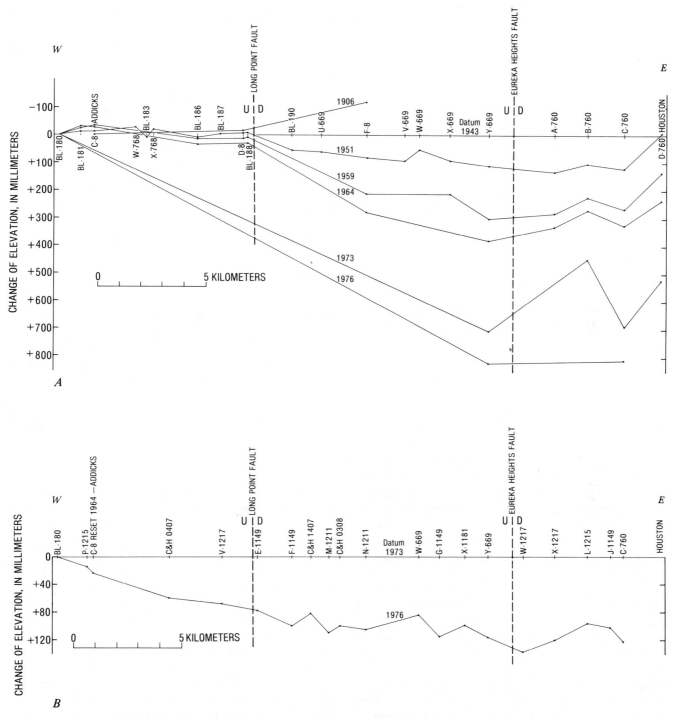

Figure 16. Changes of elevation across Long Point and Eureka Heights faults based on U.S. National Geodetic Survey bench marks in northwest part of the Houston-Galveston, Texas, region. Changes are relative to bench mark BL-180 (west end of profile), which subsided 425 mm from 1943 to 1976. Loss of relative elevation is positive.

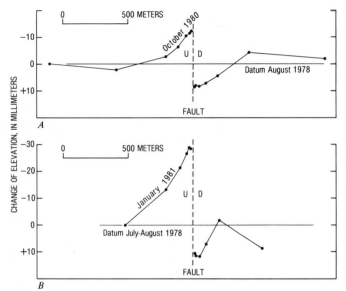

Figure 17. Changes of elevation across (A) Eureka Heights and (B) Long Point faults, based on closely spaced bench marks. Changes are relative to westernmost bench mark in each profile. Closely spaced marks were established along same line shown as in Figure 16. Survey line intersects Long Point fault at 20°, and profile has been projected perpendicular to the fault. Loss of relative elevation is positive.

prism is thus considered to be unlikely. Other plausible causes of natural faulting (for example, rising salt domes) cannot, however, be dismissed on this basis.

Historical surface faulting in the Goose Creek oil field before significant water-level declines had affected the oil-field area indicates that petroleum withdrawal has at least partly contributed to fauting in the Houston-Galveston region (Pratt and Johnson, 1926; Holzer and Verbeek, 1980). The principal question involves the uniqueness of this occurrence in the Houston-Galveston region. Because subsidence caused by petroleum withdrawal in the region generally appears to be minor, Holzer and Bluntzer (1984) have concluded that petroleum withdrawal is not a major contributor to the historical surface faulting, at least by a differential compaction mechanism.

As previously noted, a relation between historical and preexisting faults has not been demonstrated for all the historical faults. Nevertheless, the natural grouping of historical faults into structural provinces that coincide with the provinces defined by preexisting faults mapped on deeply buried stratigraphic horizons suggests that most, if not all, surface faults probably connect to preexisting faults (Van Siclen, 1967). Hydrogeologic conditions beneath the surface faults are even less well known. Reid (1973) and Kreitler (1977b) suggested that the preexisting faults are partial ground-water barriers. Kreitler (1977b), in particular, concluded that water-level differences across faults are sufficient by themselves to cause the surface scarps by localized differential compaction. Gabrysch and Holzer (1978), however, challenged this conclusion on the basis of the ratios of subsidence to water-level decline; they noted that the magnitude of water-level differ-

ence required to cause the observed scarps was incompatible with the differences permitted by water-level data. A potentially significant aspect of preexisting faults in the Gulf Coast is fault activity during deposition of sediments offset by the faults (Murray, 1961). Gabrysch (1969) analyzed the sensitivity to percentage of clay of the ratio of subsidence to water-level decline. If preexisting faults were active during deposition of the sediments within the aquifer system and if this activity caused differences in the amount of clay on opposite sides of a fault, differential subsidence would result even in the absence of differential water-level declines. For example, on the basis of the relation by Gabrysch (199, Fig. 9), a unit increase in percentage of clay would cause the ratio of subsidence to water-level decline to increase from 2% to 9%, depending on the percentage of clay in the aquifer system.

Las Vegas Valley, Nevada

Las Vegas Valley, an alluvial basin in southern Nevada (Fig. 4, area 12), is underlain by unconsolidated Pleistocene(?) alluvial and lacustrine deposits that rest unconformably on the Miocene-Pliocene(?) Muddy Creek Formation (Maxey and Jameson, 1948). This formation is characteristically fine grained with a few thin beds of sand and fine gravel. It is well consolidated where exposed in outcrops. The Muddy Creek Formation, in turn, was deposited in a broader more extensive valley cut into bedrock, the lithology of which is poorly known (Domenico and others, 1964). The maximum thickness of the unconsolidated alluvium and the Muddy Creek Formation encountered in wells is 300 and 1200 m, respectively (Harrill, 1976). Several prominent fault scarps, downthrown to the east, with a topographic relief locally as great as 30 m, occur within the valley (Fig. 18). Although the histories of fault offset are poorly known, displacement across the northernmost fault in Figure 18, the Eglington fault, occurred at least as recently as 11,000 yr B.P. (Haynes, 1967).

The principal ground-water development has been in the unconsolidated alluvium, which was divided by Harrill (1976) into two aquifers—an upper, unconfined to semiconfined aquifer and a lower, confined aquifer. Most ground water is withdrawn from the confined aquifer. According to Harrill (1976), the top of the confined aquifer is more than 60 m below the land surface beneath much of the valley. Significant lateral lithologic variations in the confined aquifer were documented by Domenico and others (1964) and Harrill (1976). The aquifer coarsens westward. Domenico and others (1964) also noted one example where an abrupt westward increase in the percentage of sand and gravel coincides with a prominent fault scarp.

Ground-water development in Las Vegas Valley began with completion of the first free-flowing well in 1907. Although initial development was in part agricultural, most of the subsequent pumping has been for municipal purposes. A basinwide overdraft has caused water levels to decline locally more than 55 m (Malmberg, 1965; Harrill, 1976). Some of the history of these declines is illustrated by the hydrograph of a well near downtown

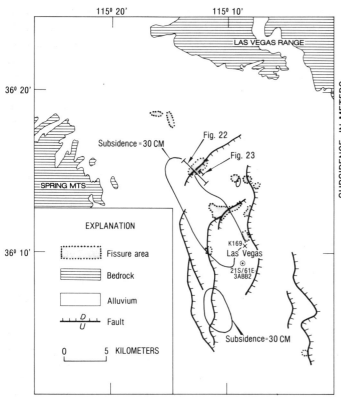

Figure 18. Fissure areas (Patt and Maxey, 1978), preexisting fault scarps, and 30-cm-subsidence contour (1963–73) (Harrill, 1976) in Las Vegas Valley, Nevada.

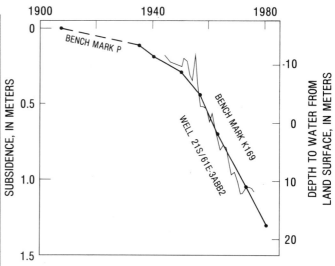

Figure 19. Subsidence of bench mark K169 and depth to water from land surface in well 21S/61E-3ABB2 (water levels above land surface are negative) near center of Las Vegas Valley, Nevada, subsidence bowl (see Fig. 18). Updated from Holzer (1981).

Las Vegas (Fig. 19), which indicates that the rate of decline was low until the 1950s, when it increased abruptly.

More than 125 km^2 of land has been affected by land subsidence induced by ground-water withdrawal (see maps by Malmberg, 1964; Mindling, 1971; and Harrill, 1976). Figures 19 and 20 illustrate some of the subsidence history. Subsidence began before 1935 and continued at a low rate until the 1950s, when it accelerated; the maximum documented subsidence is 1.3 m (Fig. 19). Subsidence has varied areally in response to both localized water-level declines (Harrill, 1976) and variations of aquifer-system compressibility (Domenico and others, 1966). These localized water-level declines are associated with both private and municipal well fields. Regional variations in compressibility are suggested by geotechnical analysis and by the absence of correlation between areas of maximum subsidence and of maximum water-level declines during the pre-1957 period of ground-water development (Domenico and others, 1966). Differential subsidence across preexisting faults was described by Holzer (1978a). Such fault control of land subsidence previously had been hypothesized by Domenico and others (1966) on the basis that at least some of the faults were partial ground-water barriers and that changes in compressibility occurred across faults.

The earliest earth fissures in Las Vegas Valley presumably related to water-level declines formed in 1957 and 1961 (Domenico and others, 1964; Mindling, 1971). No vertical offsets were reported. The longest fissure was approximately 400 m long. Damage to at least one engineered structure was reported (Fig. 21). Most of these early fissures were characterized by their proximity to fault scarps and well fields (Domenico and others, 1964). Since these early reports, the number of areas in Las Vegas Valley affected by fissures has slowly increased (Mindling and John A. Blume and Associates, 1974; Patt and Maxey, 1978). As of 1976, 11 areas with earth fissures had been described (Fig. 18), most of which are on or near preexisting fault scarps (Fig. 18). Trends of fissures near but not on fault scarps typically are parallel to the scarps. Patterns formed by individual fissures in a few areas are complex and do not show systematic trends (for example, Patt and Maxey, 1978, Fig. 4). Dimensions of fissures are modest; most fissures are open to depths of only 1 to 2 m and are less than 1 m wide and 100 m long.

Most earth fissures are areally and temporally correlated with declines of ground-water level and therefore appear to be related to these declines. Patt and Maxey (1978), however, described two earth fissures that formed in 1925 and 1930 in the northern part of the valley. Documentation of the dates of occurrence of these fissures was based on eyewitness accounts of property owners and was unsubstantiated. Patt and Maxey (1978) suggested that the 1930 fissure was associated with a major earthquake in central Nevada. The southernmost area of fissuring shown in Figure 18 is outside the regional cone of depression; these fissures appear to be related to compaction of man-placed fill (G. Lindsey, 1980, oral commun.).

Subsurface conditions beneath earth-fissure areas are poorly known. The areal association of most fissure areas with fault scarps suggest that structural control is important, but subsurface data are insufficient to evaluate this control precisely. As previously noted, Domenico and others (1966) suggested that at

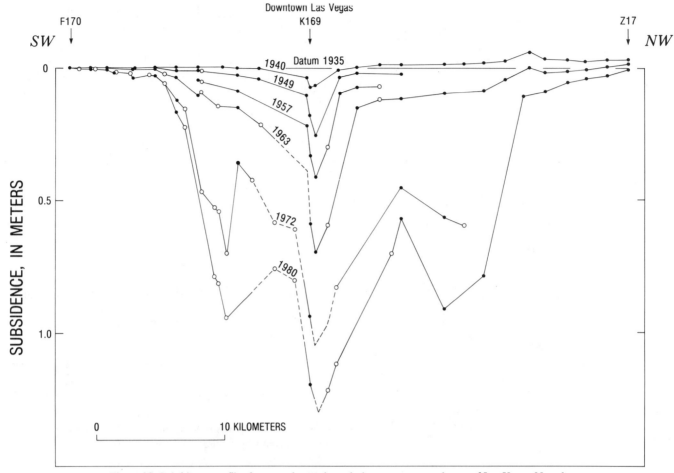

Figure 20. Subsidence profiles from southwest through downtown to northwest of Las Vegas, Nevada; dashed where inferred. Profiles are based on unadjusted data and are relative to bench mark F170 (southwest end of profile), which may have subsided 2 cm from 1935 to 1980. Subsidence shown by open circles is based on bench marks set after 1935.

A

B

Figure 21. House at Owens Avenue and A Street, Las Vegas, Nevada damaged by earth fissure in December 1961. (A) View from adjacent field. (B) Closeup from driveway. Photographs by Fred Houghton (Courtesy of J. F. Poland).

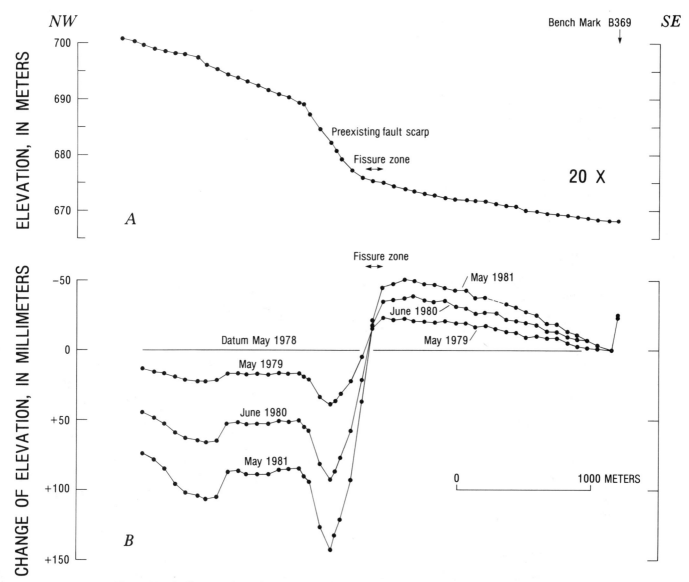

Figure 22. (A) Topography and (B) changes of elevation across preexisting fault scarp in Las Vegas Valley, Nevada (see Fig. 18). Elevation changes are relative to a bench mark 67.4 m northwest of bench mark B369 on downthrown side of scarp. Bench mark B369 subsided 0.284 m from 1963 to 1973. Data from Nevada Department of Transportation.

least some faults in Las Vegas Valley are partial ground-water barriers as well as sites of lithologic changes. Differential subsidence recorded along lines of closely spaced bench marks across several fault scarps provides at least suggestive evidence of detailed structural control; the lines were established in 1978 by the Nevada Highway Department. One of these lines crosses a zone of recently formed fissures (Bell, 1980, Fig. 2), whereas two others cross the trend of fissures (Bell, 1980, Figs. 3, 4). Results from releveling in 1979, 1980, and 1981 indicate that differential subsidence occurs over narrow zones associated with the fault scarps (Holzer, 1979b; Bell, 1980, 1981). Results from the line that crosses the Eglington fault and a zone of recently formed fissures are particularly significant in that they show a localized subsidence depression associated with the fault (Fig. 22). The position of this local subsidence depressional presumably is determined by subsurface conditions rather than local water-level declines, because no pumping wells are located within the area of the depression. Leveling data from lines north and south of the line shown in Figure 22 suggest that the depression has a longitudinal dimension greater than 2 km.

Coincidence between fissures and points of maximum convex-upward curvature in subsidence profiles based on closely spaced bench marks (Figs. 22 and 23) suggests that the fissures are forming in response to stretching related to bending of the alluvium. The bending presumably is caused by differential compaction. This bending is particularly well illustrated in Figure 23, where the resolution of the point of maximum curvature, the point of maximum horizontal tension by a bending mechanism, is

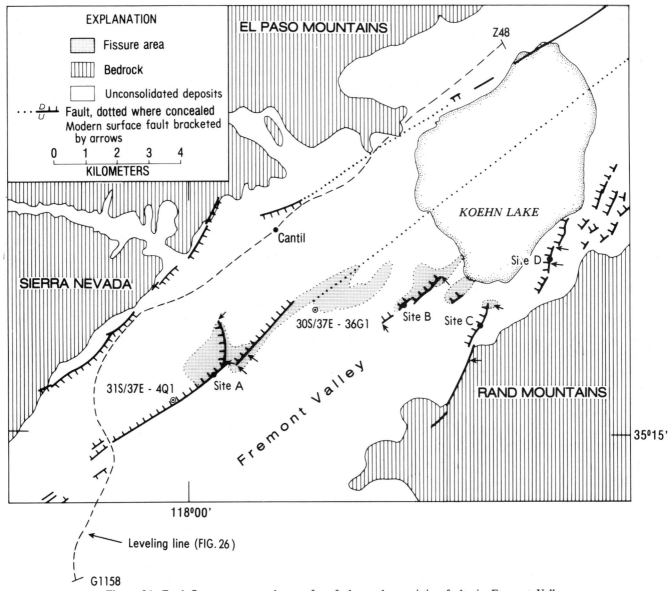

Figure 24. Earth-fissure areas, modern surface faults, and preexisting faults in Fremont Valley, California.

approximately 30 m and encompasses the fissure. It is moot whether this mechanism applies to all the fissures. The unsystematic orientation of fissures in some areas, such as the Nellis offbase well field (Patt and Maxey, 1978, Fig. 4) suggests that these fissures are forming in response to an isotropic, horizontal, tensile stress field. Such a field is difficult to attribute to differential compaction. Desiccation related to water-table declines (Holzer and Davis, 1976) might be one mechanism by which such strains could be generated.

Fremont Valley, California

Fremont Valley is a closed alluvial basin, approximately 40 km long by 10 km wide, on the north margin of the Mojave Desert in southern California (Fig. 4, area 5). Mabey (1960) interpreted the basin (Fig. 24) to be a deep graben within the Garlock fault zone; he estimated that the Cenozoic deposits, consisting of sedimentary and volcanic units, are 3 km thick. The basin is complexly faulted (Clark, 1973), and many of the faults

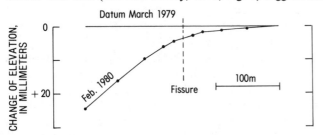

Figure 23. Changes of elevation across earth fissure in Las Vegas Valley, Nevada (Holzer and Pampeyan, 1981). Loss of elevation is positive. See Figure 18 for location.

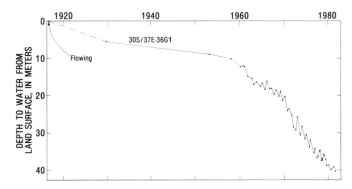

Figure 25. Hydrograph of well 30S/37E-36G1 in Fremont Valley, California (see Fig. 24).

have sharply defined scarps. The Garlock fault zone is a major active fault. Burke (1979) documented evidence for recurrent fault offset since 14,700 B.P. on a fault strand within the fault zone in Fremont Valley. The faults with well-defined scarps presumably are active as well.

Ground water in Fremont Valley has been developed principally for crop irrigation. The aquifer system consists of unconsolidated alluvial deposits that range from coarse-grained and poorly sorted sediment near the mountains at the margins of the valley to fine-grained and better sorted sediment near the center of the valley (Koehler, 1977). These deposits are more than 275 m thick in the middle of the valley and thin toward the valley margins (Koehler, 1977). In addition, variations in thickness across faults have been reported. The degree of confinement of the aquifer system is poorly documented. Although withdrawals have induced water-table declines throughout the valley, the largest water-level declines are in the potentiometeric surface of the lower, confined part of the unconsolidated aquifer. Declines of this surface by 74 m have been measured in the central part of the valley (Koehler, 1977). In some areas, faults offsetting the unconsolidated deposits are partial ground-water barriers (Koehler, 1977). The history of ground-water development is illustrated by the hydrograph of a well from the center of the valley (Fig. 25). Development was modest until the late 1950s, when the amount of water pumped for irrigation increased.

The modern history of land subsidence can be documented only along a leveling line that passes through the northwest margin of the valley (Figs. 24, 26). Data elsewhere are too sparse to map the subsidence. This leveling line intersects the margin of the regional cone of depression and therefore probably does not measure the maximum subsidence that has occurred in the valley. The maximum subsidence from 1962 to 1978 along the line is 0.48 m; maximum subsidence of 0.48 m also was measured from September 1977 to April 1981 along a leveling line near the southeast margin of the valley (E. H. Pampeyan, 1981, unpub.). The tectonic component of modern subsidence probably is small, on several bases. First, the large increase in the rate of subsidence after 1962 is consistent with the increase in the rate of water-level decline in the late 1950s. Second, the temporal migration of subsidence to the southwest, indicated by the difference between the 1973 and 1978 profiles, coincides with increased ground-water development and large water-level declines in the southwestern end of the valley. Third, subsidence is restricted to the area of ground-water withdrawal. And fourth, well protrusion has been observed at a few localities.

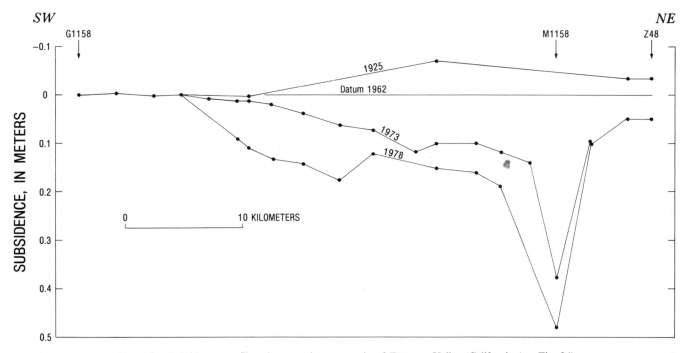

Figure 26. Subsidence profiles along northwest margin of Fremont Valley, California (see Fig. 24). Based on unadjusted data and relative to bench mark G1158 at southwest end of profiles.

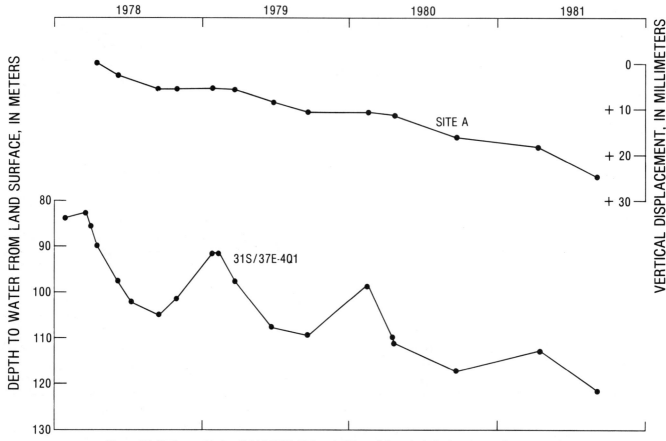

Figure 27. Hydrograph of well 31S/37E-4Q1 and differential vertical displacements of bench marks 20 m apart and on opposite sides of earth fissure at site A (see Fig. 24). Subsidence of mark on downthrown side of preexisting scarp relative to mark an upthrown side is positive.

Modern ground failure consists of both earth fissures and surface faults. At least 12 fissure zones and 5 modern scarps have been recognized (Fig. 24). In Figure 24, some of the fissure zones are lumped together to define a fissure area. Most of the fissure zones consist of isolated, straight fissures, although parallel fissures occur within a few zones. The longest fissures are approximately 1 km long; open depths of 6.7 m have been measured. At several locations the two types of failure are associated; for example, three modern scarps merge along strike into fissures, and collapse is common along the modern scarps. The early history of modern ground failure in Fremont Valley is incomplete. M. M. Clark (1978, oral commun.), while mapping the Garlock fault zone, first observed in 1971 that preexisting fault scarps were increasing in height by creep. He estimated on the basis of observed rates of displacement that the modern episode of creep began after 1961. As of 1981, scarps as high as 0.8 m and as long as 1.5 km had formed. Fissures were first observed after fault creep was recognized.

Demonstration of a relation between ground failure and water-level decline is complicated by occurrence of the failure within an active fault zone. All but one of the modern fault scarps have formed along preexisting scarps. In addition, fissures along faults and caused by tectonic fault creep have been described elsewhere (for example, Clark, 1972). Restriction of modern failures to the area affected by water-level declines, however, suggests a relation to the decline, as does the absence of reports of modern surface deformation elsewhere within the Garlock fault zone. In addition, ground failure appears to postdate the beginning of major withdrawals.

Geodetic and water-level monitoring data (Fig. 27) near an earth fissure on the preexisting Garlock fault scarp provide the most direct evidence for relating earth fissures to water-level declines. The monitoring site (Fig. 24, site A) is located near the center of the regional cone of depression. Seasonal differential vertical displacements across the fault, based on releveling a pair of bench marks 20 m apart, correlated with seasonal water-level fluctuations from 1978 to 1981. This bench mark pair is part of a line of 10 marks established perpendicular to the fault. Releveling of the entire line revealed that during summer periods of seasonal water-level decline, land on the downthrown side of the Garlock fault subsided relative to the upthrown side, and during winter periods of water-level recovery, the land surface partly rebounded (Fig. 28). Subsidence was greater on the side of the fault with greater thickness of alluvium and greater water-level decline and fluctuation. As was shown by Riley (1969), deformation of compressible aquifer systems in response to water-level declines con-

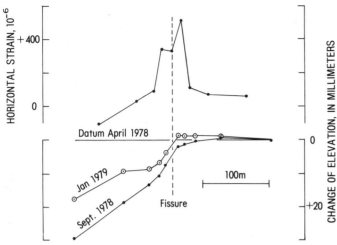

Figure 28. Profiles of changes of elevation and horizontal strain across earth fissure at site A in Fremont Valley, California (see Fig. 24). Horizontal strain measured from April 1978 to September 1978 (Holzer and Pampeyan, 1981).

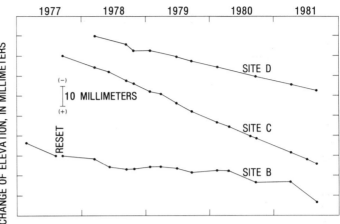

Figure 29. Change of scarp height of faults at sites B, C, and D in Fremont Valley, California, based on bench-mark pairs spanning scarp (see Fig. 24). Positive displacement indicates increase in scarp height.

sists of both recoverable and nonrecoverable components. Accordingly, in aquifer systems subjected to large seasonal water-level fluctuations, a significant part of the gross annual compaction may be recovered each year by elastic expansion during periods of seasonal water-level recovery. The parameters used to measure these properties of an aquifer system are the recoverable and nonrecoverable specific storages, defined, respectively, as the ratio of elastic and permanent compaction per unit thickness per unit water-level decline. By estimating these parameters at site A on the Garlock fault and comparing them with the values for other aquifer systems, the possibility that the deformation is caused solely by water-level changes can be evaluated. On the basis of (1) the 100-m difference in aquifer-system thickness across the fault (Koehler, 1977), (2) the 1978–79 water-level fluctuation at well 31S/37E-4Q1 (Fig. 27), and (3) the 1978–79 differential subsidence of the bench marks on the ends of the monitoring line (Fig. 28), values of nonrecoverable and recoverable specific storage of 4×10^{-6} m^{-1} and 9×10^{-6} m^{-1}, respectively, are indicated. These values compare favorably, for the purpose of this report, with those reported for other aquifer systems in the western United States (for example, see Helm, 1978). Hence, changes of ground-water level appear to be sufficient to cause the observed surface deformation across the Garlock fault.

Results from geodetic monitoring beginning in 1977–78 of three modern scarps formed on preexisting faults also are compatible with ground-water withdrawal as a related cause of modern offset (Fig. 29). Although the three sites (see sites B, C, and D in Fig. 24) are at varying distances from the area of ground-water withdrawal, all are within the regional cone of depression. The site closest to the pumping, site B, is in the region affected by seasonal water-level fluctuations, and the fault there exhibits a well-defined pattern of seasonal creep, whereas the fault at D, the site most distant from the pumping and in a region of constant-rate water-level decline, moves at a constant rate. It is not known if water levels fluctuate seasonally at site C.

Formation of modern scarps on preexisting fault scarps in Fremont Valley demonstrates that the locations of modern scarps are structurally controlled. Subsurface conditions beneath most of the earth fissures are unknown. A few of the fissures appear to be structurally controlled, occurring either directly on preexisting fault scarps or near and parallel to them; other fissures cannot be related to preexisting faults on the basis of available data. The best documented subsurface conditions are for the fissure that is on the Garlock fault (Fig. 24, site A). Koehler (1977) identified differences in both water level and thickness of unconsolidated deposits across the fault.

The differential subsidence measured at site A suggests that bending of the alluvium in response to differential compaction is the mechanism of fissure formation (Fig. 28). Holzer and Pampeyan (1981) showed that horizontal strains were a maximum at the fissure (Fig. 28).

San Jacinto Valley, California

San Jacinto Valley is a deep structural valley 130 km east of Los Angeles, California (Fig. 4, area 8). It is bounded on its northeast and southwest margins by the San Jacinto and Casa Loma faults, respectively (Fig. 30A). Both faults are tectonically active, having distinct scarps and offsetting Holocene sediment (Sharp, 1967). In addition, Proctor (1974) reported 35 mm of right-lateral strike-slip displacement from 1958 to 1973(?) across the Casa Loma fault. Downwarping between these faults has depressed bedrock approximately 3000 m from the land surface (Fett and others, 1967). West of the Casa Loma fault, bedrock is less than approximately 150 m deep (Fett and others, 1967). Relative subsidence of the structural trough has been occurring at an approximately uniform rate of 3 to 5.6 mm/yr for at least the past 15,000 to 40,000 years (Morton, 1977). Unconsolidated

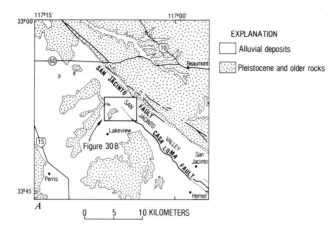

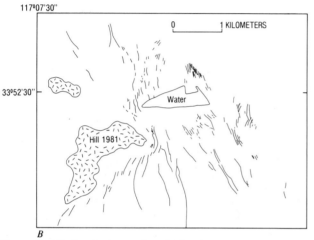

Figure 30. (A) San Jacinto Valley, California. (B) Earth fissures in San Jacinto Valley in 1974. From Morton (1977).

alluvial deposits constitute the upper 610 m of sedimentary fill within the graben (Lofgren, 1976).

Ground-water development has been principally restricted to the unconsolidated deposits. Large withdrawals, which began about 1945, have caused both the water-table and potentiometric surfaces to decline (Fett and others, 1967). By the early 1970s, maximum declines were more than 60 m (Lofgren, 1976). Large vertical variations of head occur because the aquifer system contains multiple fine-grained confining beds (Fett and others, 1967). The Casa Loma fault is a significant partial ground-water barrier where it offsets alluvial units. Marked differences across the fault are noted not only in water level but also in the magnitude of seasonal water-level fluctuations (Lofgren, 1976).

Extensive historical land subsidence has occurred in San Jacinto Valley; the relative contributions of man-induced compaction and tectonic downwarping, however, are poorly documented. Subsidence has been documented along two cross-valley aqueducts, at an extensometer site (Lofgren, 1976), and at a few releveled bench marks in the valley (Proctor, 1962; Fett and others, 1967); the maximum measured subsidence was 0.71 m from 1939 to 1959 (Proctor, 1962). Because the average rate of tectonic basinal subsidence for the past 15,270 years is high, 5.6 mm/yr (Morton, 1977), a significant percentage of the observed subsidence may be tectonic. For example, Lofgren (1976) speculated that approximately 50% and 16% of the subsidence reported by Fett and others (1967) and Proctor (1962), respectively, may have been due to tectonic downwarping. Comparison of compaction and water-level records from the only extensometer in the valley indicates that a significant part of the subsidence, at least at the extensometer site, was caused by man-induced compaction within the upper 377 m of the unconsolidated deposits (Lofgren, 1976). Preliminary analysis of 4 years of records suggests that between 70% and 80% of the subsidence was man-induced.

Historical ground failure contemporaneous with ground-water withdrawal was first reported by Fett and others (1967), who described vertical offset and sinkhole formation along separate 9-km-long traces of the Casa Loma and San Jacinto faults. Because of the high rates of modern movement, they suggested that the movement might be wholly or partly related to ground-water withdrawal. Fett and others (1967) also described long open cracks in an area west of the Casa Loma fault (Fig. 30B). Morton (1977) mapped these fissures on the basis of aerial photographs taken in 1953, 1962, 1966, and 1974 and showed that the area affected by fissures slowly increased from 1 to 12 km^2 during this period. Fissures were not evident on pre-1953 aerial photographs (D. M. Morton, 1981, oral commun.). The easternmost fissures were straight and trended parallel to the Casa Loma fault which they were near. Fissures in the western part of the area were more arcuate and conformed to the contact between alluvium and an adjacent crystalline bedrock hill; the longest fissure was 850 m long.

The possible high rate of tectonic subsidence in San Jacinto Valley precludes a demonstration that the earth fissures mapped by Morton (1977) were induced solely by ground-water withdrawal. The temporal correlation between fissure formation and withdrawal is strongly suggestive of such a relation, however, because no evidence has been recognized for major accelerations of historical tectonic rates of deformation. The arcuate fissures that conform to the alluvium-bedrock contact are comparable to fissures in south-central Arizona that form adjacent to bedrock outcrop. The proximity of these fissures to bedrock outcrop and their shape suggest that locations of these fissures may be controlled by relief on the buried bedrock surface. The location of fissures near the Casa Loma fault may be determined by either bedrock or structural control or both because this fault is documented elsewhere to be a partial ground-water barrier.

Southeastern Arizona

Many earth fissures occur in parts of Sulphur Springs and San Simon Valleys in southeastern Arizona (Fig. 4, area 2, and Fig. 31). Both valleys are structural depressions that are partly filled with lacustrine and alluvial deposits (White, 1963; Brown and Schumann, 1969). The region is currently aseismic and is

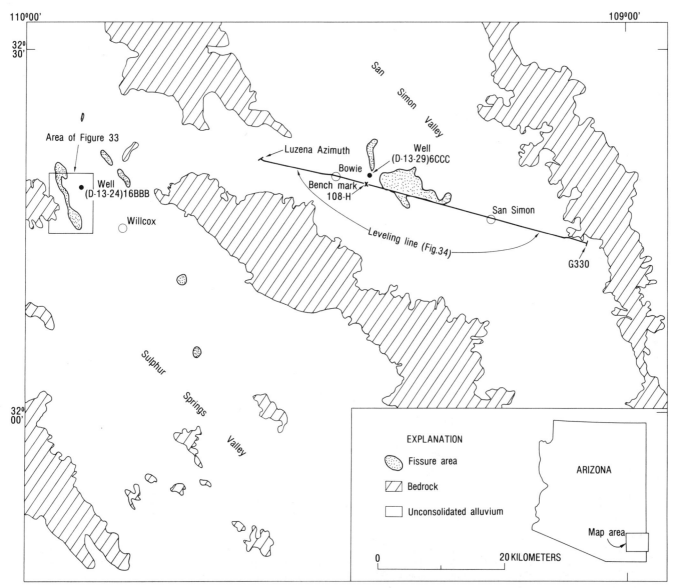

Figure 31. Earth fissure areas in southeastern Arizona.

thought to be atectonic, although Eaton (1972) inferred gentle arching of Quaternary sediment in San Simon Valley. Brown and Schumann (1969) subdivided the sedimentary fill in Sulphur Springs Valley into a lower unit of moderately to poorly consolidated alluvium and an upper unit of unconsolidated alluvium. Where exposed, the unconsolidated alluvium unconformably overlies the poorly consolidated alluvium; well logs also indicate an abrupt transition suggestive of an unconformity between the upper and lower units. The upper alluvial unit is the principal aquifer, although many wells partially penetrate the lower unit (Brown and Schumann, 1969). The unconsolidated alluvium consists of both coarse-grained sediment and areally extensive fine-grained lacustrine deposits. San Simon Valley is underlain by a layer of dense blue clay interbedded between coarse-grained alluvial deposits (White, 1963). The clay attains a maximum thickness of 180 m in the center of the basin; the depth from the land surface to its top ranges from 20 to 60 m. The clay layer forms a confining bed for the lower sand and gravel unit from which most of the ground water is withdrawn.

Minor ground-water development occurred in both Sulphur Springs and San Simon Valleys before 1920. Development was slightly greater in the San Simon Valley because many wells were initially free flowing (White, 1963). The principal ground-water development began in Sulphur Springs and San Simon Valleys in the mid-1940s and early 1950s, respectively (Fig. 32). Regional cones of depression, with maximum declines of more than 30 and 80 m, have formed in Sulphur Springs and San Simon Valleys, respectively.

Land subsidence is associated with water-level declines in both valleys (Holzer, 1980a). The maximum subsidence measured in Sulphur Springs Valley was 1.63 m from 1937 to 1974. Although data are insufficient to document the history of subsidence

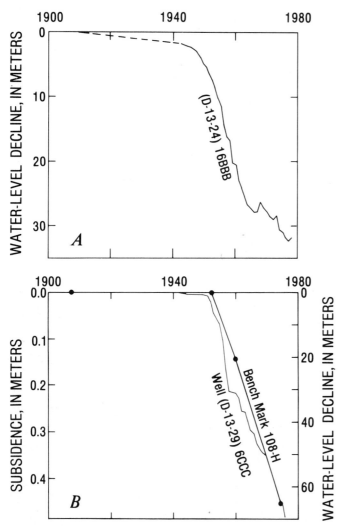

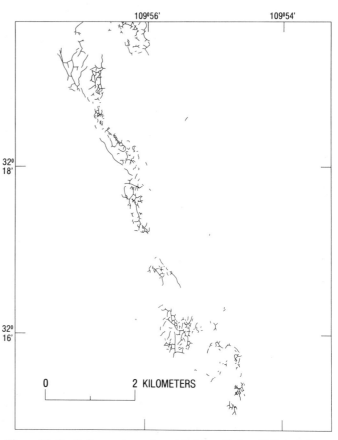

Figure 32. (A) Water-level decline in well (D-13-24) 16BBB near center of cone of depression in northern Sulphur Springs Valley, Arizona. (B) Subsidence of bench mark 108-H and water-level decline in well (D-13-29) 6CCC in San Simon Valley, Arizona (Holzer, 1981). See Figure 31 for locations.

Figure 33. Earth fissures in northwest Sulphur Springs Valley, Arizona (see Fig. 31), on March 8, 1978 (Holzer, 1980a).

in this valley, the areal correlation of subsidence with water-level decline suggests that any tectonic contribution to the measured subsidence is negligible. Shallow-seated aquifer compaction also is suggested by many protruding well casings. Maximum documented subsidence in the San Simon Valley was 1.79 m which was measured at Bowie from 1952 to 1980. Sufficient releveling data are available to document that subsidence in San Simon Valley began between 1941 and 1952, contemporaneous with the beginning of water-level declines. Figure 32B illustrates this temporal correlation of subsidence and water-level decline. Thus, modern tectonic deformation in the San Simon Valley is probably negligible.

The history of fissure formation in both valleys is readily documented by aerial photographs. On this basis, earth fissures began to form in Sulphur Springs Valley sometime between 1935 and 1958. In San Simon Valley, many earth fissures already had formed by 1935, the date of the earliest aerial photographs. As in Sulphur Springs Valley, the number of fissures has slowly increased. Mapping in 1978 of fissures in parts of both valleys by Holzer (1980a) revealed that at least 12.4 and 22.8 km^2 of land in the Sulphur Springs and San Simon Valleys, respectively, are affected by earth fissures. In the part of Sulphur Springs Valley mapped by Holzer (1980a), earth fissures occur in narrow areas near the margin of the valley, whereas in San Simon Valley, the fissures occur near the center of the valley. In most of the areas the fissures form complex quasi-polygonal patterns (Fig. 33). Diameters of closed polygons approximately range from 15 to 150 m. Straight to arcuate fissures, however, occur within the polygonal systems; one of these fissures in the San Simon Valley, southeast of Bowie, is 3.5 km long. In Sulphur Springs Valley, the straight fissures generally trend parallel to the valley margin.

The temporal and areal correlation of earth fissures in Sulphur Springs Valley with water-level declines (Holzer, 1980a) suggests that the fissures there have been induced by the declines. By contrast, the presence of earth fissures that antedate water-level declines (Holzer, 1980a) makes such a relation in San Simon Valley ambiguous. The high modern rate of fissure formation inferred from sequential aerial photographs of San Simon Valley, however, suggests that man-induced declines of water level probably have at least contributd to fissure formation.

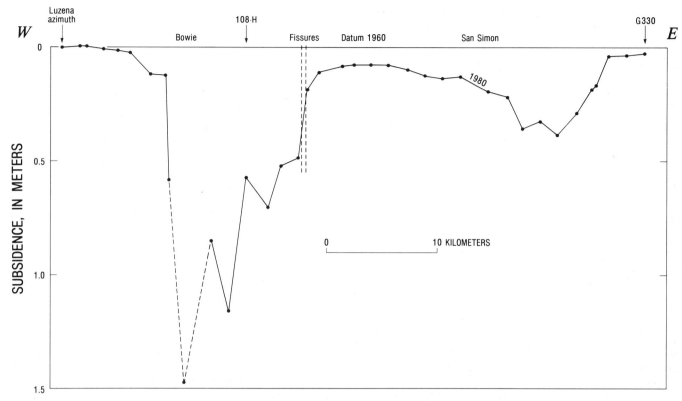

Figure 34. Subsidence from 1960 to 1980 across San Simon Valley, Arizona (see Fig. 31), based on unadjusted data and relative to bench mark Luzena Azimuth at western end of profile. Dashed segment prorated from 1952–80 subsidence.

Knowledge of subsurface conditions beneath the fissure areas is modest. The occurrence of these areas along the distal margins of alluvial fans suggests the possibility of shallow lithologic changes. The areas in Sulphur Springs Valley probably also coincide with the outermost margin of Pleistocene Lake Cochise (Anderson, 1978). Three detailed gravity traverses across fissure areas on the west margin of the north end of Sulphur Springs Valley did not detect evidence for local density contrasts associated with the fissure areas (R. C. Jachens, 1978, written commun.). Eaton (1972) inferred from geophysical data that the location of the 3.5-km-long fissure in San Simon Valley coincides with an underlying bedrock horst.

Earth fissures in southeastern Arizona may be caused by several mechanisms. The quasi-polygonal patterns formed by most of the fissures suggests a horizontally isotropic tensile stress field. Such stress fields have been hypothesized in playas, where naturally occurring fissures form polygonal patterns (Neal and others, 1968). Such a horizontal stress field might be caused by capillary stresses in the drained zone above the water table after man-induced declines. The locations of fissures formed by this mechanism presumably would be determined by the ability of the soils to mobilize the capillary stresses as effective stresses (for example, Bishop and Blight, 1963). The positions of fissure areas around the margins of subsidence bowls in southeastern Arizona and the preferred orientation of long straight earth fissures parallel to the subsidence-bowl margins, however, also suggest that stretching induced by differential subsidence may be significant. Robinson and Peterson (1962) previously attributed the 3.5-km-long fissure in San Simon Valley to this mechanism (Fig. 34).

San Joaquin Valley, California

Three fissures and one surface fault (Fig. 4, area 9) in the 13,000 km^2 San Joaquin Valley, California, subsidence area, one of the two largest in the United States, have been attributed to ground-water withdrawal. The valley comprises the southern part of the Great Valley of California, a nearly flat alluvial plain approximately 720 km long and averaging 80 km in width. The plain is underlain by an elongate, asymmetric, structural trough that has undergone structural warping and infilling with continental and marine sediment from the Jurassic to the Holocene (Hackel, 1966). Modern ground failure is restricted to the southeast part of the San Joaquin Valley, the Tulare-Wasco area, where ground-water levels have declined as much as 60 m (Lofgren and Klausing, 1969); the maximum subsidence measured from 1926 to 1970 was 3.90 m (Poland and others, 1975). The aquifer system in the Tulare-Wasco area consists of unconsolidated flood-plain and lacustrine deposits overlain by alluvial-fan deposits (Lofgren and Klausing, 1969). The base of the aquifer system is defined by the interface between fresh and salt

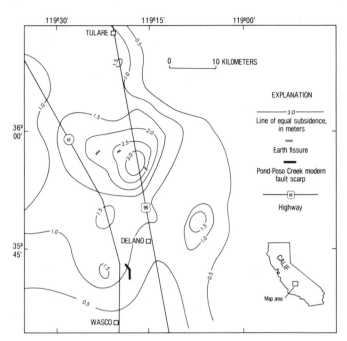

Figure 35. Earth fissures (Guacci, 1979), modern fault scarp (Holzer, 1980b), and land subsidence from 1948 to 1970 (adapted from Poland and others, 1975), in Tulare-Wasco area, California.

water, which is more than 400 m deep (Page, 1973). The aquifer system consists of an upper, unconfined to semiconfined aquifer and a lower, confined aquifer; water levels have declined in both aquifers.

The three fissures occur within 12 km of each other (Fig. 35). Only the easternmost of these fissures has been described in detail (Guacci, 1979); it was first noticed on February 26, 1969, when it appeared in cultivated fields after floods. The fissure was reported to be an approximately 0.8-km-long straight crack. The partly filled fissure was logged to a depth of 16.8 m from a bucket-auger boring, but according to M. Galloway (1976, written commun.), the fissure was actually observed to a depth of approximately 19.8 m. Formation of the fissure was attributed to extensional strains related to regional differential subsidence (Los Angeles Department of Water and Power, 1974). This crack, however, does not trend parallel to the subsidence contours based on regional geodetic data (Fig. 35).

Holzer (1980b) attributed modern movement across an approximately 3.4-km-long segment of the Pond-Poso Creek fault to ground-water withdrawal (Fig. 35). Modern vertical offset is approximately 0.23 m. The active segment is part of a preexisting fault zone that trends northwesterly for approximately 60 km from the east margin of the southern San Joaquin Valley to near the center of the valley. Closely spaced exploratory borings and seismic reflection surveys indicate that the preexisting fault zone consists of a series of subparallel, high-angle, normal faults that offset strata within the aquifer system at least as shallow as 76 m from the land surface (Los Angeles Department of Water and Power, 1974). Cumulative offset across the fault zone of the confining layer between the unconfined and confined aquifers is estimated to be 24 m. Modern faulting was attributed to ground-water withdrawal on the following bases (Holzer, 1980b). First, modern faulting postdated the beginning of water-level declines and associated subsidence. And second, precise monitoring of fault offset in 1977 and 1978 revealed that the fault moved only during periods of seasonal water-level decline in the summer; moreover, the magnitude of fault offset was greater during the year in which the seasonal low water level was lower. Holzer (1980b) concluded that the preexisting fault behaves as a partial ground-water barrier at least seasonally. On the basis of the measured compressibility of the aquifer system and the magnitude of the seasonal water-level difference, he interpreted that modern faulting probably is caused by differential compaction localized across the fault.

Other Areas in the Western United States

Earth fissures have been reported in more than 44 additional areas underlain by unconsolidated sediment in the western United States. In most of these areas, the fissures are evidently of natural origin because man has not altered the natural state of stress. Most of these fissures are associated with dry lakes or playas (Neal and others, 1968). Polygonal patterns are the typical mode of occurrence, although long straight or arcuate fissures, termed giant stripes (Neal and others, 1968), commonly are associated with the polygonal patterns. These polygonal-pattern-forming fissures are generally attributed to desiccation related to climatic changes. A few areas with only straight or arcuate fissures have also been described (for example, Robinson and Peterson, 1962, p. 7; Clark, 1972; U.S. Geological Survey, 1976, p. 27–31). Some of these are evidently of tectonic origin; for example, fissures in the Anza-Borrego Desert in southern California (Clark, 1972) were associated with a creeping, seismically active fault. The following paragraphs describe fissures that are suspected to be man-induced. These occurrences are discussed together, because either only isolated fissures have formed or documentation is meager.

Lofgren (1975) described a long straight earth fissure in the Raft River Valley, Idaho (Fig. 4, area 11), that formed on the boundary of a small man-induced subsidence bowl. The fissure was mapped in part on the basis of 1967 aerial photographs (B. E. Lofgren, 1976, oral commun.). Although the existence of the fissure is unquestioned, its mapped location is suspect because it coincides with a photolinearity representing an abandoned railroad grade. The fissure, which formed in the early 1960s, was attributed to ground-water withdrawal primarily on the basis of its similarity to man-induced fissures in other areas. The fissure formed in an area underlain by unconsolidated deposits, although basalt flows are interbedded in the aquifer system in the general area of the fissure. Water-level declines locally exceed 30 m; the maximum documented subsidence was 0.80 m from 1935 to 1974. From his inference that the fissure paralleled lines of equal water-level change, Lofgren (1975) hypothesized that the fissure was caused by horizontal seepage forces. The fissure, however, was perpendicular to lines of equal change in water level for the only published period of record, from 1952 to 1961 (Lofgren, 1975).

W. L. Burnham (1952, unpub. report) described a 600-m-long earth fissure, locally 5-m deep, that opened in the Yucaipa Valley, southern California, in January 1952 (Fig. 4, area 10). The fissure was approximately 5 km west of the town of Yucaipa. The Yucaipa Valley is a small structural trough partly filled with predominantly fine-grained unconsolidated alluvium. The valley has a long history of ground-water development; by 1952, ground-water levels in the unconsolidated alluvium had declined more than 35 m. Shallow materials exposed in the fissure walls consisted of clayey silt. Burnham speculated that the fissure was caused by long-term drying of the dewatered deposits during a period of subnormal rainfall. He performed no theoretical analysis but measured a large volume change in samples desiccated in the laboratory.

T. L. Holzer and M. M. Clark (1981, unpub. report for Soil Conservation Service) described an approximately 600-m-long, locally 2.3-m-deep, arcuate earth fissure in the Antelope Valley, southern California (Fig. 4, area 3); the fissure was 11 km east-northeast of the town of Lancaster. It was first noticed by area residents in 1978 after flooding of the fissure area; the fissure grew eastward after additional floods in 1980. The principal aquifer beneath the Antelope Valley is an unconsolidated alluvial unit that is heavily pumped. Beneath the fissure, the thickness of the unconsolidated deposits exceeds 350 m, and water levels have declined more than 75 m (Bloyd, 1967). A long history of land subsidence is documented by relevelings of a north-south-trending survey line that crosses Antelope Valley 9.6 km west of the fissure; more than 1.3 m of subsidence was measured along part of this line from 1955 to 1976 (R. O. Castle, 1981, written commun.). The fissure was attributed to water-level declines because of its similarity to other man-induced fissures and the absence of recent faulting or seismicity.

Fissures forming polygonal patterns have been attributed to ground-water withdrawal in three areas, principally on the basis of a temporal and areal correlation between the fissures and water-level declines. In two of these areas, in southern California, the fissures occur on playas—Rogers Lake playa (Fig. 4, area 7; Neal and others, 1968) and Lucerne Lake playa (Fig. 4, area 6; Fife, 1977); documentation is modest for both areas. Shifting of ponded surface water from the north to the south side of Lucerne Lake playa suggests possible subsidence (Fife, 1977). Cordova and Mower (1976) described fissures in polygonal patterns in the Milford area, southwestern Utah (Fig. 4, area 14). These fissures are in an area where near-surface materials consist of clayey silt and peaty materials. The fissures were attributed to horizontal contraction of the near-surface materials after they were dewatered by declining water levels.

An additional example of surface faulting possibly caused by ground-water withdrawal is modern dip-slip movement across an approximately 0.56-km-long segment of the Busch fault in the Santa Clara Valley north of Hollister, California (Fig. 4, area 4; Rogers, 1967). An earthquake on this fault in November 1974, however, indicates that the fault is seismically active, with left-lateral strike-slip movement beneath the aquifer system (Nason, 1976). The fault offsets unconsolidated sediment in which water-level declines exceed 30 m near the fault. Although releveling data are inadequate to describe the general configuration of subsidence within the valley area near the fault, at least 0.60 m of subsidence has been documented in the Hollister area (J. F. Poland, 1976, written commun.). The modern scarp, which was at least 0.17 m high in 1976, grew from 1960 to 1976 by aseismic creep at a rate of approximately 10 mm/yr (Holzer, 1977). Although the scarp occurs within 0.46 km of the Calaveras fault, a tectonically active strike-slip fault, Rogers (1967) proposed the modern offset on the Busch fault might be related to ground-water withdrawal because of the aseismic growth of the scarp and the absence of any strike-slip fault component. Fault creep measured from 1970 to 1975 by a tilt beam across the Busch fault (Nason and others, 1974; R. D. Nason, 1975, written commun.) was remarkably periodic; the average annual vertical displacement was 8.6 mm (Holzer, 1977, Fig. 3B). Negligible creep occurred until September or October of each year, and then most of the annual creep occurred during the next 3 months. The annual initiation of creep occurred at about the end of the pumping season, which preceded the beginning of seasonal rainfall, commonly in November.

MECHANISMS OF GROUND FAILURE

General Comments

The widespread association of ground failures with water-level declines and associated subsidence indicates that many, if not all, of the failures described here are man-induced. Such a conclusion has previously been drawn by many investigators of each of the affected areas. Mechanisms of formation, however, have been primarily speculative because of an absence of (1) information on local subsurface conditions and surface deformation and (2) rigorous theoretical investigations. Because earth fissures and surface faults generally have been investigated independently of each other, theories of ground failure also have focused narrowly on only one kind of failure. Although detailed investigations have not been conducted in each area affected by ground failure, only two mechanisms—localized differential compaction and horizontal contractions induced by capillary stresses in the zone above a declining water table—appear to have general significance; the latter mechanism is probably significant only in areas where earth fissures form polygonal patterns. Contributions to horizontal stresses and ground failure from regional differential compaction and horizontal seepage forces probably are minor in most areas.

Localized Differential Compaction

Recent investigations of subsurface conditions and surface

deformation in several areas near ground failures suggest that many such failures are related to localized differential compaction. The most comprehensive and detailed investigations have been in south-central Arizona, where gravity surveys are particularly well suited for determining subsurface conditions. These investigations have demonstrated that earth fissures are generally associated with zones where the aquifer system thins over ridges or steps on the underlying bedrock surface. Detailed studies reveal that fissures form above the loci of points of maximum convex-upward curvature on the bedrock surface. Jachens and Holzer (1979, 1982), on the basis of concepts developed by Lee and Shen (1969), have illustrated how differential compaction above such features causes surface stretching by bending of the overburden. According to their modeling results, horizontal strains are at maximum tension above the points of maximum convex-upward curvature on the bedrock surface. Points of maximum tension also coincide with the points of maximum convex-upward curvature in subsidence profiles (Lee and Shen, 1969). Although detailed subsurface studies have not been conducted in other areas, the localization of fissures near the points of maximum convex-upward curvature in subsidence profiles across fissures in south-central Arizona, Fremont Valley, California, and Las Vegas Valley, Nevada, suggests that localized differential compaction is the general failure mechanism (Holzer and Pampeyan, 1981). Horizontal extension, which approximately ranged from 100 to 700 microstrains per year, was measured at the points of maximum curvature.

Detailed investigations of subsurface conditions and surface deformation near the Pond-Poso Creek (California) and Picacho (Arizona) faults suggest that localized differential compaction can also cause modern surface faulting. Both of these modern scarps are associated with preexisting faults that are partial ground-water barriers. Man-induced seasonal water-level differences across the faults and inferred specific compaction of the sediment were sufficient to cause the observed scarp heights by localized differential compaction across the preexisting faults (Holzer, 1978b, 1980b). Reid (1973) and Kreitler (1977b) have proposed that such a mechanism may also apply to surface faulting in the Houston-Galveston, Texas, region, although the magnitude of water-level difference required to cause the offsets observed there does not appear to be compatible with available water-level data (see Gabrysch and Holzer, 1978; Kreitler, 1978). This result, however, does not preclude possible localized differential compaction. As was previously noted, faults in the Gulf Coast sometimes were active during deposition of the sediment that they offset. Therefore, the potential for differential thicknesses of compressible material exists across these faults. Though speculative, such lateral changes of thickness localized across preexisting faults might be sufficient to cause discrete differential compaction.

Capillary Stress

The formation of complex polygonal patterns by fissure systems in areas of water-level decline suggests that these fissures are large contraction cracks caused by a horizontally isotropic tensile stress field. By analogy to desiccation cracks, the probable source of such tension is a large, negative, capillary stress in the dewatered zone above a declining water table; such a mechanism was proposed by Neal and others (1968) to explain naturally occurring fissures that form giant polygons on playas. The feasibility of this mechanism rests on theoretical and experimental investigations of effective stresses in unsaturated soils. Skempton (1961), following Bishop (1959), proposed that the effective stress, $\bar{\sigma}$, for an unsaturated soil, in which pore-air pressure is in equilibrium with the atmosphere, may be expressed as:

$$\bar{\sigma} = \sigma - \chi u_w, \qquad (1)$$

where σ is the total stress, u_w is the pore-water pressure, and χ is an empirically determined parameter that depends on saturation. The parameter χ decreases from 1 at saturation to 0 as saturation decreases. According to equation 1, as pore-water pressures become increasingly negative above a declining water table, the effective stress increases. The ability of the soil to mobilize large effective stresses depend on χ. Increases in effective stress cause both horizontal and vertical contractions within the soil. Although Narasimhan (1979) demonstrated that volume changes in aquifer systems are possible by this mechanism, its operation under field conditions has not been rigorously tested. The occurrence of fissures in polygonal patterns in areas underlain by fine-grained playa and lacustrine deposits, however, suggests that effective stresses may be most efficiently mobilized in such deposits.

Regional Differential Compaction

Castle and Youd (1972) and Bouwer (1977) proposed that horizontal, centripetal, tensile strains around the margins of subsidence bowls related to ground-water withdrawal can cause faulting and fissuring. This mechanism was originally proposed to explain surface faulting associated with petroleum withdrawal (Yerkes and Castle, 1969). Although regional horizontal displacements in areas of ground-water withdrawal have not been systematically investigated, centripetal horizontal displacements have been measured in association with land subsidence over three oil fields (Grant, 1954; Whitten, 1961; Castle and Yerkes, 1976). These measurements demonstrate that bowl-like differential subsidence can induce radial tensile strains around the margin of a subsidence bowl. The positions of normal faults and tension cracks near the centers of subsidence bowls formed by ground-water withdrawal and the nonparallelism of failures with respect to regional-subsidence contours, however, suggest that this mechanism probably is not of major consequence in regions of ground-water withdrawal. The unimportance of this mechanism is particularly well illustrated for faults in south-central Arizona and Houston-Galveston (Figs. 10 and 13), which are not systematically located or oriented within the subsidence bowl, and for the earth fissures mapped in Figure 8. The diverse trends of intersecting fissures on the map (Fig. 8) preclude their formation by regionally controlled, unidirectional, horizontal displacements. In addition to these field observations, general theoretical considerations suggest that the broad regional extent of compacting aquifer systems relative to their thickness is not conducive to significant horizontal displacement by a regional compaction mechanism. Lee and Shen (1969) showed that horizontal displacements induced by differential compaction are proportional to the depth of the compacting zone and to the slope of the subsidence profile. Comparison of typical depths to the compacting zone and subsidence slopes in areas of ground-water withdrawal with corresponding parameters at those oil fields where horizontal displacements have been measured suggests that generally only minor horizontal displacements have been induced by regional subsidence caused by water-level declines.

Seepage Force

Another proposed mechanism of ground failure is horizontal strain caused by horizontal seepage forces (Lofgren, 1971); such forces are proportional to the horizontal component of the hydraulic gradient. According to B. E. Lofgren (1978, oral commun.), ground failures by this mechanism form concentric to the cone of depression and near its periphery. In addition, for those cones of depression that have radially expanded over time, the age of fissures should decrease radially outward. Although failures are more common in some regions around the periphery of the cone of depression, exceptions in these same areas commonly are noted. Perhaps even more significant are such areas as Las Vegas Valley and Houston-Galveston where failures bear no apparent systematic relation to the cone of depression. As noted by Domenico and others (1964), some of the earliest fissures in Las Vegas Valley formed in well fields, which, by this mechanism, should have been zones of compression.

Mechanics of Failure

The mechanics of subsidence-associated faulting has been examined in detail in only a few investigations. On the basis of geodetic data, Holzer and others (1979) inferred the history of deformation associated with the development of the Picacho fault. They concluded that scarp formation was preceded for approximately 13 years by differential subsidence across a 300-m-wide zone and that the scarp formed by the release of strain stored beneath the flexure (Fig. 36). This delay presumably was caused by a time lag in either (1) propagation of the subsurface rupture to the surface or (2) development of a discrete shear failure in the subsurface along the preexisting fault plane. The time lag between subsurface and surface deformation at the Picacho fault contrasts with the more rapid response of the Pond-Poso Creek fault (Holzer, 1980b). There, surface fault offset was seasonal and could be correlated with water-level fluctuation in the confined aquifer, the top of which is at a depth of approximately 76 m. The absence of observed water-level fluctuation in

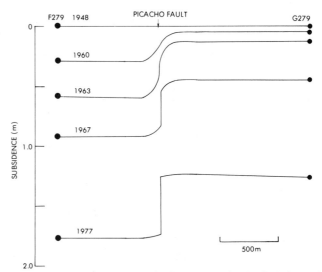

Figure 36. Idealized subsidence profiles across Picacho fault in south-central Arizona (Holzer and others, 1979).

the overlying water-table aquifer suggests that both subsurface failure below a depth of 76 m and its propagation to the surface was rapid.

As was previously noted, near-fault surface deformation above the footwall and hanging-wall blocks is commonly observed. The widths of these zones of deformation vary, but they commonly are more than several hundred meters wide. Presumably, the width of a zone of fault-associated deformation is proportional to the depth of fault rupture (Holzer and Thatcher, 1979).

Although the long-term growth of fault scarps by aseismic creep is documented by abundant field studies, the history of horizontal displacements across fissures after their initial appearance is poorly documented. Many fissures fill with sediment and appear after a time to become inactive (Holzer, 1977). Holzer and Pampeyan (1981), however, measured extension that persisted for 4 years, the duration of their investigation, across two fissures in south-central Arizona. They also measured ongoing localized differential compaction across one of the fissures. Jachens and Holzer (1982) inferred from fresh collapse along some fissures in their study area that extension had persisted for more than 24 years; they also inferred that the "active" fissures were located within zones of ongoing differential subsidence. Continuous monitoring for approximately 2 years of horizontal displacements across earth fissures in south-central Arizona by Boling and others (1980) indicated that movement was relatively smooth, with occasional sudden jumps; they also noted a tendency for some fissures to close after long dry periods.

PREDICTION OF GROUND FAILURE

The state of the art of ground failure prediction is based primarily on investigations of ground failures that have already occurred; successful predictions of ground failure have not yet been reported. The widespread association of ground failure with land subsidence caused by water-level declines in compressible sediment, however, suggests a basis for general areal predictions. The first step in evaluating the potential for ground failure should be an assessment of the potential for man-induced land subsidence (see D. C. Helm, this volume). Given such a potential, two approaches for predicting ground failure have been proposed: (1) determination of particular subsurface conditions conducive to failure and (2) monitoring of surface deformation for precursory signals. The first approach is applicable in areas not yet subject to water-level declines and associated subsidence, whereas the second approach requires that man-induced deformation be underway. In addition, if the mechanism of ground failure can be correctly anticipated, then it may be possible, given subsurface conditions and physical properties, to predict the magnitude of water-level decline required to cause failure.

Earth fissures caused by localized differential compaction can be anticipated by recognizing subsurface conditions conducive to such compaction. For example, in south-central Arizona, differential compaction commonly is related to variations in aquifer-system thickness caused by relief on the shallow buried bedrock surface at the base of the aquifer system. In that area, delineation of the configuration of the buried bedrock surface provides a useful predictive tool for the potential locations of most fissures. Other possible causes of localized differential compaction, both in Arizona and elsewhere, include preexisting faults that are partial ground-water barriers and local lateral changes in aquifer-system compressibility. Although examples of such preexisting fauls have been described, documentation of fissures associated with lateral changes in compressibility is modest (see Las Vegas Valley, Nevada subsection of this report). Detailed investigations of subsurface conditions beneath fissure systems suggest that the predicted locations of fissures can be quite specific. For example, Jachens and Holzer (1979, 1982) demonstrated that most of the fissures they studied in south-central Arizona directly overlie the loci of points of maximum convex-upward curvature on the underlying bedrock surface; their data permitted resolution of these loci to within a few dekameters.

Predictions of the time of occurrence of earth fissures or of the magnitude of water-level decline required to induce fissuring presumably requires specific data on the thickness and compressibility of subsurface materials and on the tensile strength of the surficial deposits. Few investigations have focused on this aspect of fissure prediction, and the feasibility of such predictions is still undemonstrated. Vertical compressibilities of aquifer systems in subsidence areas are commonly reported, although site-specific data probably would require detailed drilling and sampling. Most investigations of tensile strength have concentrated on compacted soils (for example, Leonards and Narain, 1963; Hasegawa and Ikeuti, 1966; Ajaz and Parry, 1975). In general, the drier the soil, the smaller the strain at which soils fail in tension. If this generalization applies to natural undisturbed soils, shallow-soil moisture may be an important aspect of fissure formation. Jachens and Holzer (1982), on the basis of modeling, estimated tensile strains at failure approximately ranged from 0.02% to 0.2% within a 10 km^2 area in south-central Arizona.

Earth fissures caused by tensions induced by capillary stress cannot be confidently predicted on the basis of current knowledge. As was previously suggested, some deposits, such as fine-grained ones, may be more susceptible to fissuring and, therefore, regional predictions may be feasible. Because the specific locations of these fissures do not appear to be controlled by special subsurface conditions, their locations probably are random. At best, then, the spacing of fissures may be predictable by a methodology similar to that proposed by Lachenbruch (1962) for thermal contraction cracks in permafrost.

The general association of surface faults with preexisting faults suggests that potential sites of surface faulting are predictable. The few detailed subsurface studies that have been reported, as well as the coincidence of surface scarps with preexisting scarps, suggest that modern offset is on preexisting fault planes. On this basis, site-specific predictions appear to be feasible. Recognition of which preexisting faults are susceptible to reactivation by ground-water withdrawal, however, requires, in part, knowledge of the mechanism of faulting. If this reactivation is by localized differential compaction, then, presumably, conditions conducive to localized water-level differences or to compressibility differences across the preexisting fault are required. Assessment of the magnitude of localized differential compaction may also be significant because failures presumably propagate upward from the aquifer system. The smaller the differential compaction, the less likely the subsurface failure will propagate to the surface.

An alternative approach for predicting ground failure in areas of ongoing water-level decline is to monitor surface deformation. As demonstrated by Holzer and Pampeyan (1981), straight to arcuate earth fissures occur near the points of maximum convex-upward curvature in subsidence profiles oriented perpendicular to the cracks. Thus, repeated leveling of closely spaced bench marks can be used to anticipate the locations of zones potentially susceptible to fissuring. In addition to vertical surveys, horizontal surveys can also be used to identify zones of extension. Horizontal surveys have the advantage, if the approximate strains at failure are known, of providing direct information on how close to failure the soil has been strained.

Monitoring of surface deformation may also be applicable to fault prediction. As argued by Holzer and others (1979), offset on the Picacho fault in south-central Arizona was preceded by differential subsidence over an approximately 300-m-wide zone. Although analysis of the deformation that is precursory to fault offset may be formidable, such analyses as that by Holzer and Thatcher (1979) may provide insight into the dynamics of the subsurface deformation. Monitoring across preexisting faults might at least permit exclusion of those faults that do not constitute a hazard by identifying those across which no differential subsidence is occurring. By contrast, differential subsidence across faults could provide a justification for further investigation (for example, Holzer, 1978a).

SUMMARY

Ground failure is associated with ground-water withdrawal and man-induced subsidence in at least 14 areas in the United States; the 2 largest affected areas are south-central Arizona and the Houston-Galveston, Texas, metropolitan region. Failures range from long tension cracks (earth fissures) to scarps (surface faults). Earth fissures, which commonly have lengths of several hundred meters, are visually the more spectacular because of their erosion by surface runoff into deep, wide, steep-walled gullies. These gullies, which commonly have widths of 1 to 2 m and depths of 2 to 3 m, are large enough to trap and injure livestock and other animals as well as to pose a potential hazard to humans. Horizontal separations across fissures, though commonly less than a few centimeters, are sufficient to damage rigid engineered structures. Economic losses from fissures have been relatively small because most fissures occur in agricultural areas and can usually be filled in and the land regraded. Scarps formed by surface faulting commonly are 1 km or more long; the longest scarp measured was 16.7 km. Vertical offsets of 0.5 m are common and constitute a significant hazard to engineered structures. Probably several million dollars of damage and losses of property value has been caused in the Houston-Galveston area by more than 86 active surface faults with an aggregate scarp length of more than 240 km.

Both types of failure occur by aseismic displacement, and examples of both types have remained active for decades after their initial appearance. The rate of growth of fault scarps by creep, which has been documented at many localities, ranges from 4 to 60 mm/yr. In areas where water levels fluctuate seasonally, comparable variations in fault movement have been reported. At several localities, faulting has ceased when water levels recovered significantly. In addition to the discrete failure, zones of surface deformation have been measured near both surface faults and earth fissures.

Localized differential compaction is suspected to be the principal mechanism of ground failure associated with water withdrawal, although this mechanism is speculative in the Houston-Galveston region. Such compaction causes straight to arcuate earth fissures by bending and stretching of the overburden above the differentially compacting zone. Variations in aquifer-system thickness, differential water-level declines across preexisting faults, and lateral variations in aquifer-system compressibility are potential causes of the differential compaction. Surface faults form when the differential compaction is discrete, as may occur across preexisting faults behaving as partial ground-water barriers. Fissures that form in polygonal patterns probably are caused by another mechanism—horizontal tension created by capillary water in the zone above declining water tables.

Ground failure can be predicted by two approaches: (1) recognition of appropriate subsurface conditions and (2) monitoring of surface deformation for precursory signals. Investigations of ground failures caused by localized differential compaction suggest that the locations of potential ground failures by this mechanism can be predicted by a detailed investigation of subsurface conditions; resolution of the conditions contributing to the differential compaction is required. Investigation of subsurface conditions adjacent to surface faults have indicated that in general modern offset is associated with preexisting faults. Because many of these surface faults coincide with preexisting fault scarps as well, modern movement probably is along preexisting fault planes. Ground failure by differential compaction in areas of ongoing water-level decline can also be predicted by resurveys of closely spaced bench marks. Earth fissures form at the points of maximum convex-upward curvature in subsidence profiles, where extension is greatest. Prediction of faulting on the basis of resurveys requires more sophisticated analysis.

ACKNOWLEDGMENTS

This report is the result of an investigation that was funded by the Reactor Hazards Program of the U.S. Geological Survey. I am grateful to the original program manager, Carl M. Wentworth, for bringing the problem to my attention and for his continued encouragement. Many professional colleagues have been generous with their time and their ideas and have contributed greatly to my understanding of this phenomenon; to them I am deeply grateful. They include: S. N. Davis, G. E., Figueroa, R. K. Gabrysch, D. C. Helm, R. C. Jachens, C. W. Kreitler, R. L. Laney, B. E. Lofgren, M. D. Mifflin, D. M. Morton, E. H. Pampeyan, R. O. Patt, J. F. Poland, R. H. Raymond, H. H. Schumann, E. R. Verbeek, C. C. Winikka, and R. F. Yerkes. Assistance in the field was provided, in part, by J. K. Boling, Jr., M. C. Carpenter, C. W. Roberts, and K. M. Williams. Reviews of the manuscript by J. F. Poland and R. F. Yerkes are gratefully appreciated. E. R. Verbeek reviewed the section on the Houston-Galveston, Texas, region.

REFERENCES CITED

Ajaz, A., and Parry, R.H.G., 1975, Stress-strain behavior of two compacted clays in tension and compression: Geotechnique, v. 25, no. 3, p. 495–512.

Anderson, S. B., 1978, Earth fissures in the Stewart area of the Willcox basin, Cochise County, Arizona [M.S. thesis]: Tucson, University of Arizona, 72 p.

Anderson, S. L., 1973, Investigation of Mesa earth-crack, Arizona, attributed to differential subsidence due to ground-water withdrawal [M.S. thesis]: Tempe, Arizona State University, 111 p.

Arizona Water Commission, 1975, Inventory of resources and uses: Arizona State Water Plan, Phase I, 224 p.

Bell, J. W., 1980, Results of leveling across fault scarps in Las Vegas Valley, Nevada, April, 1978–June, 1980: Nevada Bureau of Mines and Geology Open-File Report 80-7, 5 p.

—— 1981, Results of leveling across fault scarps in Las Vegas Valley, Nevada, April 1978–June 1981: Nevada Bureau of Mines and Geology Open-File Report 81-5, 7 p.

Bernard, H. A., Le Blanc, R. J., and Major, C. F., 1962, Recent and Pleistocene geology of southeast Texas, in Rainwater, E. H., and Zingula, R. P., eds., Geology of the Gulf Coast and central Texas: Houston Geological Society, p. 175–224.

Bishop, A. W., 1959, The principle of effective stress: Teknisk Ukeblad, v. 39, p. 859–863.

Bishop, A. W., and Blight, G. E., 1963, Some aspects of effective stress in saturated and partly saturated soil: Geotechnique, v. 13, no. 3, p. 177–197.

Bloyd, R. M., Jr., 1967, Water-resources of the Antelope Valley–East Kern Water Agency area, California: U.S. Geological Survey Open-File Report, Menlo Park, California, 73 p.

Boling, J. K., Carpenter, M. C., Johnson, N. M., and Davis, S. N., 1980, Measurement, prediction, and hazard evaluation of earth fissures and subsidence, south-central Arizona: National Technical Information Service Accession no. PB 80-183 908, 96 p.

Bouwer, H., 1977, Land subsidence and cracking due to ground-water depletion: Ground Water, v. 15, no. 5, p. 358–364.

Brown, S. G., and Schumann, H. H., 1969, Geohydrology and water utilization in the Willcox basin, Graham and Cochise Counties, Arizona: U.S. Geological Survey Water-Supply Paper 1859-F, p. F1–F32.

Burke, D. B., 1979, Log of a trench in the Garlock fault zone, Fremont Valley, California: U.S. Geological Survey Miscellaneous Field Studies Map MF-1028.

Castle, R. O., and Yerkes, R. F., 1976, Recent surface movements in the Baldwin Hills, Los Angeles County, California: U.S. Geological Survey Professional Paper 882, 125 p.

Castle, R. O., and Youd, T. L., 1972, Comment on 'the Houston fault problem': Association of Engineering Geologists Bulletin, v. 9, no. 1, p. 57–68.

Castle, R. O., Church, J. P., Yerkes, R. F., Manning, J. C., 1983 Historical surface deformation near Oildale, California: U.S. Geological Survey Professional Paper 1245, 42 p.

Clanton, U. S., and Amsbury, D. L., 1975, Active faults in southeastern Harris County, Texas: Environmental Geology, v. 1, no. 3, p. 149–154.

Clark, M. M., 1972, Collapse fissures of the Coyote Creek fault, in The Borrego Mountain earthquake of April 9, 1968: U.S. Geological Survey Professional Paper 787, p. 190–207.

—— 1973, Map showing recently active breaks along the Garlock and associated faults, California: U.S. Geological Survey Miscellaneous Geologic Investigations Map I-741, scale 1:24,000, 3 sheets.

Clark, M. M., Buchanan-Banks, J. M., and Holzer, T. L., 1978, Creep along parts of the Garlock fault: Possible relation to decline in ground-water levels [abs.]: Geological Society of America Abstracts with Programs, v. 10, no. 3, p. 100.

Cordova, R. M., and Mower, R. W., 1976, Fracturing and subsidence of the land surface caused by the withdrawal of ground water in the Milford area, Utah: U.S. Geological Survey Journal of Research, v. 4, no. 5, p. 505–510.

Davidson, E. S., 1973, Geohydrology and water resources of the Tucson basin, Arizona: U.S. Geological Survey Water-Supply Paper 1939-E, p. E1–E81.

Domenico, P. A., Stephenson, D. A., and Maxey, G. B., 1964, Ground water in Las Vegas Valley: University of Nevada, Desert Research Institute Technical Report 7, 53 p.

Domenico, P. A., Mifflin, M. D., and Mindling, A. D., 1966, Geologic controls on land subsidence in Las Vegas Valley, in Proceedings, Annual Engineering Geology and Soils Engineering Symposium, 4th, Moscow, Idaho, p. 113–121.

Eaton, G. P., 1972, Deformation of Quaternary deposits in two intermontane basins of southern Arizona, U.S.A.: International Geological Congress, 24th, Montreal, Proceedings, Section 3, Quebec, Harpell's Press, p. 607–616.

Eaton, G. P., Peterson, D. L., and Schumann, H. H., 1972, Geophysical, geohydrological, and geochemical reconnaissance of the Luke salt body, central Arizona: U.S. Geological Survey Professional Paper 753, 29 p.

Elsbury, B. R., Van Siclen, D. C., and Marshall, B. P., 1980, Engineering aspects of the Houston fault problem: Paper presented at ASCE, Texas and New Mexico Sections Fall Meeting, September 1980, El Paso, Texas (available from McClelland Engineers, Houston, Texas).

Feth, J. H., 1951, Structural reconnaissance of the Red Rock quadrangle, Arizona: U.S. Geological Survey Open-File Report, Tucson, Arizona, 32 p.

Fett, J. D., Hamilton, D. H., and Fleming, F. A., 1967, Continuing surface displacement along the Casa Loma and San Jacinto faults in San Jacinto Valley, Riverside County, California: Association of Engineering Geologists, Engineering Geology, v. 4, no. 1, p. 22–32.

Fife, D. L., 1977, Engineering geologic significance of giant desiccation polygons, Lucerne Valley Playa, San Bernardino County, California [abs.]: Geological Society of America Abstracts with Programs, v. 9, no. 4, p. 419.

Gabrysch, R. K., 1969, Land-surface subsidence in the Houston-Galveston region, Texas, in Proceedings, International Land Subsidence Symposium, 1st Tokyo, v. 1: International Association of Hydrological Sciences Publication 88, p. 43–54.

—— 1980, Approximate land-surface subsidence in the Houston-Galveston region, Texas, 1906–78, 1943–78, 1973–78: U.S. Geological Survey Open-File Report 80-338, scale 1:380, 160, 3 sheets.

Gabrysch, R. K., and Bonnet, C. W., 1975, Land-surface subsidence in the Houston-Galveston region, Texas: Texas Water Development Board Report 188, 19 p.

Gabrysch, R. K., and Holzer, T. L., 1978, Comment on 'Fault control of subsidence, Houston, Texas': Ground Water, v. 16, no. 1, p. 51–55.

Grant, U. S., 1954, Subsidence of the Wilmington oil field, California, in Jahns, R. H., ed., Geology of southern California: California Division of Mines Bulletin 170, Chapter 10, p. 19–24.

Gray, E. V., 1958, The geology, ground water, and surface subsidence of the Baytown–La Porte area, Harris County, Texas [M.S. thesis]: College Station, Texas A&M University, 66 p.

Guacci, G., 1979, The Pixley fissure, San Joaquin Valley, California, in Saxena, S. K., ed., Evaluation and prediction of subsidence: New York, American Society of Civil Engineers, p. 303–319.

Hackel, O., 1966, Summary of the geology of the Great Valley, in Bailey, E. H., ed., Geology of northern California: California Division of Mines and Geology Bulletin 190, p. 217–238.

Hardt, W. F., and Cattany, R. E., 1965, Description and analysis of the geohydrologic system in western Pinal County, Arizona: U.S. Geological Survey Open-File Report, Tucson, Arizona, 92 p.

Harrill, J. R., 1976, Pumping and ground-water storage depletion in Las Vegas Valley, Nevada, 1955–74: Nevada Department of Conservation and Water Resources, Nevada Water Resources Bulletin no. 44, 69 p.

Hasegawa, H., and Ikeuti, M., 1966, On the tensile strength test of disturbed soils, in Kravtchenko, J., and Sirieys, P. M., eds., Rheology and soil mechanics: Berlin, Spring-Verlag, p. 405–412.

Haynes, C. V., 1967, Quaternary geology of the Tule Springs area, Clark County, Nevada, in Wormington, H. M., and Ellis, D., eds., Pleistocene studies in southern Nevada: Nevada State Museum Anthropological Papers no. 13, p. 15–104.

Helm, D. C., 1978, Field verification of a one-dimensional mathematical model for transient compaction and expansion of a confined aquifer system, *in* Proceedings, Specialty Conference on Verification of Mathematical and Physical Models in Hydraulic Engineering, College Park, Maryland: New York, American Society of Civil Engineers, p. 189–196.

Holdahl, S. R., and Morrison, N. L., 1974, Regional investigations of vertical crustal movements in the U.S., using precise relevelings and mareograph data: Tectonophysics, v. 23, no. 4, p. 373–390.

Holzer, T. L., 1977, Ground failure in area of subsidence due to ground-water decline in the United States, *in* Proceedings, International Land Subsidence Symposium, 2nd, Anaheim, December 1976: International Association of Hydrological Sciences Publication 212, p. 423–433.

—— 1978a, Documentation of potential for surface faulting related to ground-water withdrawal in Las Vegas Valley, Nevada: U.S. Geological Survey Open-File Report 78-79, 21 p.

—— 1978b, Results and interpretation of exploratory drilling near the Picacho fault, south-central Arizona: U.S. Geological Survey Open-File Report 78-1016, 17 p.

—— 1979a, Elastic expansion of the lithosphere caused by ground-water depletion: Journal of Geophysical Research, v. 84, no. B9, p. 4689–4698.

—— 1979b, Leveling data—Eglington fault scarp, Las Vegas Valley, Nevada: U.S. Geological Survey Open-File Report 79-950, 7 p.

—— 1980a, Reconnaissance maps of earth fissures and land subsidence, Bowie and Willcox areas, Arizona: U.S. Geological Survey Miscellaneous Field Studies Map MF-1156, scale 1:24,000, 2 sheets.

—— 1980b, Faulting caused by ground-water level declines, San Joaquin Valley, California: Water Resources Research, v. 16, no. 6, p. 1065–1070.

—— 1981, Preconsolidation stress of aquifer systems in areas of induced land subsidence: Water Resources Research, v. 17, no. 3, p. 693–704.

Holzer, T. L., and Bluntzer, R. L., 1984, Land subsidence near oil and gas fields, Houston, Texas: Ground Water, v. 22, no. 4 (in press).

Holzer, T. L., and Davis, S. N., 1976, Earth fissures associated with watertable declines [abs.]: Geological Society of America Abstracts with Programs, v. 8, no. 6, p. 923–924.

Holzer, T. L., and Pampeyan, E. H., 1981, Earth fissures and localized differential subsidence: Water Resources Research, v. 17, no. 1, p. 223–227.

Holzer, T. L., and Thatcher, Wayne, 1979, Modeling deformation due to subsidence faulting, *in* Saxena, S. K., ed., Evaluation and prediction of subsidence: New York, American Society of Civil Engineers, p. 349–357.

Holzer, T. L., and Verbeek, E. R., 1980, Modern surface faulting in the Goose Creek oil field (Texas)—A reexamination [abs.]: Geological Society of America Abstracts with Programs, v. 12, no. 7, p. 449.

Holzer, T. L., Davis, S. N., and Lofgren, B. E., 1979, Faulting caused by ground-water extraction in south-central Arizona: Journal of Geophysical Research, v. 84, no. B2, p. 603–612.

Jachens, R. C., and Holzer, T. L., 1979, Geophysical investigations of ground failure related to ground-water withdrawal—Picacho basin, Arizona: Ground Water, v. 17, no. 6, p. 574–585.

—— 1982, Differential compaction mechanism for earth fissures near Casa Grande, Arizona: Geological Society of America Bulletin, v. 93, no. 10, p. 998–1012.

Jennings, M. D., 1977, Geophysical investigations near subsidence fissures in northern Pinal and southern Maricopa Counties, Arizona [M.S. thesis]: Tempe, Arizona State University, 95 p.

Johnson, N. M., 1980, The relation between ephemeral stream regime and earth fissuring in south-central Arizona [M.S. thesis]: Tucson, University of Arizona, 158 p.

Jorgensen, D. G., 1975, Analog-model studies of ground-water hydrology in the Houston District, Texas: Texas Water Development Board Report 190, 84 p.

Kinney, J. L., 1979, Laboratory procedures for determining the dispersibility of clayey soils: U.S. Bureau of Reclamation Report REC-ERC-79-10, 20 p.

Koehler, J. H., 1977, Ground water in the Koehn Lake Area, Kern County, California: U.S. Geological Survey Water-Resources Investigations 77-66, 24 p.

Konieczki, A. D., and English, C. S., 1979, Maps showing ground-water conditions in the lower Santa Cruz area, Pinal, Pima, and Maricopa Counties, Arizona—1977: U.S. Geological Survey Water-Resources Investigations Open-File Report 79-56, scale 1:125,000, 4 sheets.

Kovach, R. L., 1974, Source mechanisms for Wilmington oil field, California, subsidence earthquakes: Seismological Society of America Bulletin, v. 64, no. 3(I), p. 699–711.

Kreitler, C. W., 1977a, Faulting and land subsidence from ground-water and hydrocarbon production, Houston-Galveston, Texas, *in* Proceedings, International Land Subsidence Symposium, 2nd, Anaheim, December 1976: International Association of Hydrological Sciences Publication 121, p. 435–446.

—— 1977b, Fault control of subsidence, Houston, Texas: Ground Water, v. 15, no. 3, p. 203–214.

—— 1978, Reply *to* Comment *on* 'Fault control of subsidence, Houston, Texas': Ground Water, v. 16, no. 2, p. 126–128.

Lachenbruch, A. H., 1962, Mechanics of thermal contraction cracks and icewedge polygons in permafrost: Geological Society of America Special Paper 70, 65 p.

Laney, R. L., Raymond, R. H., and Winikka, C. C., 1978a, Maps showing waterlevel declines, land subsidence, and earth fissures in south-central Arizona: U.S. Geological Survey Water-Resources Investigations Report 78-83, scale 1:125,000, 2 sheets.

Laney, R. L., Ross, P. P., Littin, G. R., 1978b, Maps showing ground-water conditions in the eastern part of the Salt River Valley area, Maricopa and Pinal Counties, Arizona—1976: U.S. Geological Survey Water-Resources Investigations Report 78-61, scale 1:125,000, 2 sheets.

Lee, K. L., and Shen, C. K., 1969, Horizontal movements related to subsidence: American Society of Civil Engineers, Proceedings, Soil Mechanics and Foundations Division Journal, v. 94, no. SM6, p. 140–166.

Lee, W. T., 1905, Underground waters of Salt River Valley, Arizona: U.S. Geological Survey Water-Supply Paper 136, 196 p.

Lehner, P., 1969, Salt tectonics and Pleistocene stratigraphy on continental slope of Northern Gulf of Mexico: American Association of Petroleum Geologists Bulletin, v. 53, no. 12, p. 2431–2479.

Leonard, R. J., 1929, An earth fissure in southern Arizona: Journal of Geology, v. 37, no. 8, p. 765–774.

Leonards, G. A., and Narain, J., 1963, Flexibility of clay and cracking of earth dams: American Society of Civil Engineers, Proceedings, Soil Mechanics and Foundations Division Journal, v. 89, no. SM2, p. 47–98.

Lockwood, M. G., 1954, Ground subsides in Houston area: Civil Engineering, v. 24, no. 6, p. 48–50.

Lofgren, B. E., 1971, Significant role of seepage stresses in compressible aquifer systems [abs.]: EOS (American Geophysical Union Transactions), v. 52, no. 11, p. 832.

—— 1975, Land subsidence and tectonism, Raft River Valley, Idaho: U.S. Geological Survey Open-File Report 75-585, 21 p.

—— 1976, Land subsidence and aquifer-system compaction in the San Jacinto Valley, Riverside County, California—A progress report: U.S. Geological Survey Journal of Research, v. 4, no. 1, p. 9–18.

Lofgren, B. E., and Klausing, R. L., 1969, Land subsidence due to ground-water withdrawal, Tulare-Wasco area California: U.S. Geological Survey Professional Paper 437-B, p. B1–B103.

Los Angeles Department of Water and Power, 1974, San Joaquin nuclear project early site review report, Volume I: Los Angeles (City) Department of Water and Power.

Mabey, D. R., 1960, Gravity survey of the western Mojave Desert, California: U.S. Geological Survey Professional Paper 316-D, p. D51–D73.

Malmberg, G. T., 1964, Land subsidence in Las Vegas, Nevada: Nevada Department of Conservation and Natural Resources Ground-Water Resources Information Series Report 5, 10 p.

—— 1965, Available water supply of the Las Vegas ground-water basin, Nevada: U.S. Geological Survey Water-Supply Paper 1780, 116 p.

Maxey, G. B., and Jameson, C. H., 1948, Geology and water resources of Las

Vegas, Pahrump and Indian Spring Valleys, Clark and Nye Counties, Nevada: Nevada Water Resources Bulletin no. 5, 121 p.

Mindling, A. L., 1971, A summary of data relating to land subsidence in Las Vegas Valley: Reno, University of Nevada, Center for Water Resources Research, Desert Research Institute Report, 55 p.

Mindling, A. L., and John A. Blume and Associates, 1974, Effects of groundwater withdrawal on I-15 Freeway and vicinity in north Las Vegas, Nevada: Reno, University of Nevada, Center for Water Resources Research, Desert Research Institute, Project Report Series no. 33, 37 p.

Mitchell, J. K., 1976, Fundamentals of soil behavior: New York, John Wiley & Sons, 422 p.

Morton, D. M., 1977, Surface deformation in part of the San Jacinto Valley, southern California: U.S. Geological Survey Journal of Research, v. 5, no. 1, p. 117–124.

Murray, G. E., 1961, Geology of the Atlantic and Gulf coastal province of North America: New York, Harper, 692 p.

Narasimhan, T. N., 1979, The significance of the storage parameter in saturated-unsaturated ground-water flow: Water Resources Research, v. 15, no. 3, p. 569–576.

Nason, R. D., 1976, Active fault slippage at San Juan Bautista and Hollister: International Land Subsidence Symposium, 2nd, Guidebook, p. C1–C2.

Nason, R. D., Philippsborn, F. R., and Yamashita, P. A., 1974, Catalog of creepmeter measurements in central California from 1968 to 1972: U.S. Geological Survey Open-File Report 74-31, p. 246–250.

Neal, J. T., Langer, A. M., and Kerr, P. F., 1968, Giant desiccation polygons of Great Basin playas: Geological Society of America Bulletin, v. 79, no. 1, p. 69–90.

Page, R. W., 1973, Base of fresh ground water (approximately 3000 micromhos) in the San Joaquin Valley, California: U.S. Geological Survey Hydrologic Investigations Atlas HA-489, scale 1:500,000.

Pankratz, L. W., Ackermann, H. D., and Jachens, R. C., 1978a, Results and interpretation of geophysical studies near the Picacho fault, south-central Arizona: U.S. Geological Survey Open-File Report 78-1106, 17 p.

Pankratz, L. W., Hassemer, J. H., and Ackermann, H. D., 1978b, Geophysical studies relating to earth fissures in central Arizona [abs.]: Geophysics, v. 44, no. 3, p. 367.

Pashley, E. F., Jr., 1961, Subsidence cracks in alluvium near Casa Grande, Arizona: Arizona Geological Society Digest, v. 4, p. 95–101.

Patt, R. O., and Maxey, G. B., 1978, Mapping of earth fissures in Las Vegas Valley, Nevada: Reno, University of Nevada, Water Resources Center Desert Research Institute Publication 41051, 19 p.

Peirce, H. W., 1975, Rumbles and rattles: Arizona Bureau of Mines Field Notes, v. 5, no. 2, p. 5, 8.

—— 1976, Tectonic significance of basin and range thick evaporite deposits: Arizona Geological Society Digest, v. 10, p. 325–339.

Peterson, D. E., 1962, Earth fissuring in the Picacho area Pinal County, Arizona [M.S. thesis]: Tucson, University of Arizona, 35 p.

—— 1964, Earth fissuring in Picacho area, Pinal County, Arizona: U.S. Geological Survey Open-File Report, Tucson, Arizona, 52 p.

Péwé, T. L., 1978, Terraces of the lower Salt River Valley in relation to the Late Cenozoic history of the Phoenix Basin, Arizona, in Burt, D. M., and Péwé, T. L., eds., Guidebook to the geology of central Arizona: Arizona Bureau of Geology and Mineral Technology Special Paper 2, p. 1–14.

Poland, J. F., Lofgren, B. E., Ireland, R. L., and Pugh, R. G., 1975, Land subsidence in the San Joaquin Valley, California, as of 1972; U.S. Geological Survey Professional Paper 437-H, p. H1–H78.

Pratt, W. E., and Johnson, D. W., 1926, Local subsidence of the Goose Creek oil field (Texas): Journal of Geology, v. 34, no. 7, p. 557–590.

Proctor, C. V., Jr., 1973, Late Pleistocene deposition, upper Texas coastal plain [abs.]: Geological Society of America Abstracts with Programs, v. 5, no. 3, p. 276.

Proctor, R. J., 1962, Geologic features of a section across the Casa Loma fault, exposed in an aqueduct trench near San Jacinto, California: Geological Society of America Bulletin, v. 73, no. 10, p. 1293–1296.

—— 1974, New localities for fault creep in southern California—Raymond and Casa Loma faults [abs.]: Geological Society of America Abstracts with Programs, v. 6, no. 3, p. 238.

Raymond, R. H., Winikka, C. C., and Laney, R. L., 1978, Earth fissures and land subsidence, eastern Maricopa and northern Pinal Counties, Arizona, in Burt, D. M., and Péwé, T. L., eds., Guidebook to the geology of central Arizona: Arizona Bureau of Geology and Mineral Technology Special Paper 2, p. 107–113.

Reid, W. M., 1973, Active faults in Houston, Texas [Ph.D. thesis]: Austin, University of Texas, 122 p.

Riley, F. S., 1969, Analysis of borehole extensometer data from central California, in Proceedings, International Land Subsidence Symposium, 1st, Tokyo, v. 2: International Association of Hydrological Sciences Publication 89, p. 423–431.

Robinson, G. M., and Peterson, D. E., 1962, Notes on earth fissures in southern Arizona: U.S. Geological Survey Circular 466, 7 p.

Rogers, T. H., 1967, Active extensional faulting north of Hollister near the Calaveras fault zone: Seismological Society of America Bulletin, v. 57, no. 4, p. 813–816.

Ross, P. P., 1978, Maps showing ground-water conditions in the western part of the Salt River Valley area, Maricopa County, Arizona—1977: U.S. Geological Survey Water-Resources Investigations Report 78-40, scale 1:125,000, 2 sheets.

Schumann, H. H., 1974, Land subsidence and earth fissures in alluvial deposits in the Phoenix area, Arizona: U.S. Geological Survey Miscellaneous Investigations Map I-845-H, scale 1:250,000, 2 sheets.

Schumann, H. H., and Poland, J. F., 1969, Land subsidence, earth fissures, and ground-water withdrawal in south-central Arizona, U.S.A., in Proceedings, International Land Subsidence Symposium, 1st, Tokyo, v. 1: International Association of Hydrological Sciences Publication 88, p. 295–302.

Scott, J. D., and Moore, R. T., 1976, The Palo Verde nuclear power station: Arizona Bureau of Mines Field Notes, v. 6, no. 3-4, p. 1–7.

Sharp, R. V., 1967, San Jacinto fault zone in the Peninsular ranges of southern California: Geological Society of America Bulletin, v. 78, no. 6, p. 705–730.

Skempton, A. W., 1961, Effective stress in soils, concrete, and rocks, in Proceedings, Conference on Pore Pressure in Soils: London, Butterworths, p. 4–16.

Smith, G.E.P., 1938, The physiography of Arizona valleys and the occurrence of ground water: Tucson, University of Arizona, Agricultural Experiment Station Technical Bulletin 77, 91 p.

—— 1940, The ground-water supply of the Eloy district in Pinal County, Arizona: Tucson, University of Arizona, Agricultural Experiment Station Technical Bulletin 87, 42 p.

Stulik, R. S., and Twenter, F. R., 1964, Geology and ground water of the Luke area, Maricopa County, Arizona: U.S. Geological Survey Water-Supply Paper 1779-P, p. P1–P30.

Sturgal, J. R., and Irwin, T. D., 1971, Earthquake history of Arizona and New Mexico, 1850–1966: Arizona Geological Society Digest, v. 9, p. 1–39.

Taylor, T. V., 1907, Underground waters of coastal plain of Texas: U.S. Geological Survey Water-Supply Paper 190, 73 p.

Turcan, A. N., Jr., Wesselman, J. B., Kilburn, C., 1966, Interstate correlation of aquifers, southwestern Louisiana and southeastern Texas, in Geological Survey Research 1966: U.S. Geological Survey Professional Paper 550-D, p. D231–D236.

U.S. Geological Survey, 1976, Field trip to Nevada test site: U.S. Geological Survey Open-File Report 76-313, 64 p.

Van Siclen, D. C., 1966, Active faulting in the Houston area, in Comprehensive study of Houston municipal water system for the City of Houston, Phase I, Basic Studies: Consulting report by Turner, Collie, and Braden to the City of Houston, p. 42–48.

—— 1967, The Houston fault problem: American Institute of Professional Geologists, Texas Section, Annual Meeting, 3rd, Dallas, Proceedings, p. 9–31.

Verbeek, E. R., and Clanton, U.S., 1978, Map showing faults in the southeastern Houston metropolitan area, Texas: U.S. Geological Survey Open-File Report 78-797, scale 1:24,000.

—— 1981, Historically active faults in the Houston metropolitan area, Texas, *in* Etter, E. M., ed., Houston area environmental geology—surface faulting, ground subsidence, hazard liability: Houston Geological society Special Publication, p. 28-68.

Verbeek, E. R., Ratzlaff, K. W., and Clanton, U.S., 1979, Faults in parts of north-central and western Houston metropolitan area, Texas: U.S. Geological Survey Miscellaneous Field Studies Map MF-1136, scale 1:24,000.

Weaver, P., and Sheets, M. M., 1962, Active faults, subsidence, and foundation problems in the Houston, Texas, area: Geological Society of America, 74th Annual Meeting, Guidebook, p. 254-265.

White, N. D., 1963, Analysis and evaluation of available hydrologic data for San Simon basin, Cochise and Graham Counties, Arizona: U.S. Geological Survey Water-Supply Paper, 1619-DD, p. DD1-DD33.

Whitten, C. A., 1961, Measurements of small movements in the earth's crust: Academiae Scientiarum Fennicae Annales, Series A, III, Geologica-Geographica, v. 61, p. 315-320.

Winikka, C. C., and Wold, P. D., 1977, Land subsidence in central Arizona, *in* Proceedings, International Land Subsidence Symposium, 2nd, Anaheim, December 1976: International Association of Hydrological Sciences Publication 121, p. 95-103.

Yerkes, R. F., and Castle, R. O., 1969, Surface deformation associated with oil and gas field operations in the United States, *in* Proceedings, International Land Subsidence Symposium, 1st, Tokyo, v. 1: International Association of Hydrological Sciences Publication 88, p. 55-66.

—— 1976, Seismicity and faulting attributable to fluid extraction: Engineering Geology, v. 10, no. 2-4, p. 151-166.

MANUSCRIPT ACCEPTED BY THE SOCIETY APRIL 18, 1984

Printed in U.S.A.

Organic soil subsidence

John C. Stephens, Consulting Geohydrologist
(Formerly Agricultural Administrator, USDA, ARS)
1111 N.E. 2nd Street
Ft. Lauderdale, Florida 33301

Leon H. Allen, Jr., Soil Scientist
Soil and Water Unit, USDA. ARS
Gainesville, Florida 32611

Ellen Chen, Research Associate
Institute of Food and Agricultural Sciences
University of Florida
Gainesville, Florida 32611

ABSTRACT

Organic soil subsidence occurs mainly with drainage and development of peat for agriculture. Subsidence occurs either from densification (loss of buoyancy, shrinkage, and compaction) or from actual loss of mass (biological oxidation, burning, hydrolysis and leaching, erosion, and mining). Densification usually occurs soon after drainage is established. Slow, continuous loss of mass is due mainly to biological oxidation. Erosion is minor except in specific sites. Mining losses vary greatly and depend upon direct removal of the materials.

Subsidence rates are determined mainly by type of peat, depth to water table, and temperature. Subsidence losses have been carefully measured in several locations (e.g., the Florida Everglades), and predictions of future subsidence developed in 1950 have proved reliable.

Peat drainage and subsidence have several consequences: loss of plant rooting depth where the substrate is unfavorable (stony, acidic, saline), increased pumping for drainage, instability of roads and other structures, increase in nutrient outflows, colder surface temperature during winter nights, and increase of CO_2 flux to the global atmosphere

The water table for organic soils should be held as high as crop and field conditions allow to reduce subsidence.

Computer models offer methods for refining oxidation rate processes and prediction of subsidence losses where adequate calibration data are available. Remote sensing offers a method of assessing organic soil area and drainage changes. These new technologies should improve our assessment, and guide our management, of organic soil resources.

INTRODUCTION

We present the nature and extent of organic soils, the history, observed rates of sinkage, and causes of organic soil subsidence, the present state of the art in predicting such soil losses, known methods of control, and, finally, a look into the future of organic soil subsidence studies.

Nature and Classification of Organic Soils

Peat and muck soils have unique biological, physical, and chemical properties. Some of the common drainage and agricultural management practices used on mineral soils are not suitable for use on peat soils. Biologically, the organic soils support hosts of micro-organisms that are largely responsible for the formation and alteration of the peat. In an anoxic environment the action of anaerobic micro-organisms break down the parent plant structure and creates peat which accumulates because biomass production is greater than biomass decomposition. Aerobic organisms are largely responsible for decomposition of organic soils under drained conditions.

Peat and muck soils have a dark color, low bulk density, and an absorbent sponge-like texture. They have low albedo for solar radiation and high emissivity for thermal radiation. Compared to mineral soils, they have a high heat capacity when wet, and low heat conductivity. They burn at relatively low temperatures, have a high cation exchange per unit volume, and a high buffer capacity that strongly resists changes in acid reactions. Peat soils tend to be acid but range from pH 3.5 to 8.

Organic soil deposits are classified, under the FAO soil classification system, as Histosol or Gleysols. Histosols must contain at least 12 to 18 percent organic carbon by weight and exceed 30 to 40 cm in depth. Gleysols are mineral soils with gleyed horizons that may be moderately high in organic carbon but contain less than the minimum required for Histosols.

This paper deals only with the Histosols, which are commonly known as peat and muck. Peat and muck develop from the vegetation of bogs, marshes, and swamp forests, or from sphagnum and other mosses, by the action of micro-organisms, under waterlogged conditions. Several taxonomies have been developed and updated in different parts of the world, based on either soil profile properties or botanical composition. One classification defined peat as a soil that has less than 50 percent of mineral matter on a dry weight basis; muck as a soil having 50 to 80 percent of mineral matter; and mineral soils as those having less than 20 percent organic matter. Another usage separates peat and muck based on state of decomposition without regard to organic matter content. Peat soils are usually only partially decomposed and retain a fibrous or granular nature. The mucks are thoroughly decomposed and are finely textured, uniform, amorphous, and black. The principal types of peat, based on botanical origin, are sedimentary, fibrous, woody, and moss. The first three are formed in basins or on poorly drained land. They are known as the low-moor peats. The sphagnum moss types may develop on higher land and are commonly called high-moor peats.

Sedimentary peats are derived primarily from submerged succulent open water plants such a naiads and pond weeds that contain relatively small proportions of cellulose or hemi-cellulose materials. These may be mixed with fecal material from aquatic animals (about one-third on average), with algae and dead micro-organisms, with fallen pollen and leaves from the higher plants together with water or wind-borne mineral sediments. Sedimentary peats are colloidal in particle size, i.e., the clay of organic soils. Drained sedimentary peats seldom yield first-class agricultural soils. Fibrous peats are composed of the remnants of sedges, reeds, and related marsh plants that grow in shallow water of marshes. The fibrous peats contain a reasonably high proportion of cellulose. The water-holding capacity of the fibrous peats is high, and the water transmissibility is satisfactory for drainage and water management for agriculture. They are less acid than the moss peats and have a more desirable texture for tillage than the sedimentary peats. They contain from 2 to 3.5 percent nitrogen. Most of the drained fibrous peats yield good agricultural soils. Woody peats are formed from the climax vegetation in swamp deposits. They develop from the residue of trees and shrubs that occupy the forest floor of the swamp. Woody peat has a lower water-holding capacity than fibrous peat but when drained has a loose granular and blocky structure through which water moves readily. This soil has excellent tilth and usually is rated intermediate between fibrous and sedimentary peat as an agricultural soil.

Moss peats are formed in northerly latitudes where a cool moist climate favors their development. Extensive areas occur in northern Canada and Europe. This peat is formed principally from sphagnum mosses and associated vegetation, which depend upon rainfall, dews, and fogs for moisture and nutrients. The spagnum peats can hold up to 15 or 16 times their weight of water and absorb moisture to the extent that they actually lift the water table above the surrounding country. They build up by layers and sometimes reach a thickness of 30 ft. Generally, they present a dome-shape surface profile. They are extremely acid, pH 3.5 to 4, and are not readily decomposed. Sphagnum peat is a poor medium for the growth of micro-organisms causing decomposition, and the nitrogen content is low. Even when drained, it is a poor soil for the growth of higher plants unless the low pH is raised by the heavy use of lime, and a complete fertilizer is added, which includes the micro-nutrient elements. Sphagnum deposits in Europe and Canada may be used for forest production and as pastures. Some is used in Russia and northern Europe for fuel. Other uses are for the improvement of other soils, such as soil mixes for horticultural use, for packing plants and flowers, as an absorbent litter for stables and poultry houses, and for related uses where its high water-holding capacity is of value. Dawson (1956), Farnham and Finney (1965), Aandahl and others (1974), McCollum and others (1976), and Lucas (1982) review organic soils in further detail.

Extent of Organic Soils

Histosols cover an estimated 21.0 million hectares in the United States (Farnham in Armentano, 1980a; and Lucas, 1982). This includes about 10.9 million hectares in Alaska. Land areas of Histosols are considerably larger than estimates of peat (Moore and Bellamy, 1974). Histosols in the contiguous United States are located mostly in the cool, temperate, humid regions of Maine, Massachusetts, New York, and New Jersey; and in the Great Lakes states of Michigan, Wisconsin, Minnesota, Illinois, and Indiana. The northern glaciated land areas have about 70 percent of the total peat and muck land in the contiguous United States. Deposits also exist in the warm, temperate, humid region within the southeastern coastal plain swamps of Virginia, North Carolina, and Georgia, and along the coastal marsh tidelands of Louisiana and Texas. Other deposits in the Pacific coastal area include

TABLE 1. FOUR ESTIMATES OF THE WORLD'S PEAT LAND RESOURCES (IN 10^6 HA.)

Country	Davis and Lucas (1959)	Moore and Bellamy (1974)[1]	Farnham[2]	Lucas (1982)
U.S.S.R.	70.8	71.5	150.0	150.0
Canada	112.5	129.5	112.0	112.0
United States	56.7[3]	7.5	21.0	21.0
Finland	8.1	10.0	9.7	10.0
Sweden	6.1	1.5	7.8	7.0
E./W. Germany	3.0	1.6	5.2	1.6
Great Britain/Ireland	3.2	1.8	5.2	2.6
Poland	2.4	1.5	3.4	1.5
Norway	3.0	3.0	3.0	3.0
Indonesia/Malaysia	1.4	0.74	2.4	18.9[5]
All Others	1.0[6]	1.9[6]	2.2[6]	15.0
Total	268.2	230.5	321.9	342.6

[1] Exploitable peat reserves mainly.
[2] Based on a presentation by R.S. Farnham of the University of Minnesota in Armentano (1980a). See also Heikurainen (1964).
[3] Includes 44.6 x 10^6 ha in Alaska.
[4] Indonesia only.
[5] Includes Sumatra, Kalimatan, Sarawak, Brunei, Malaya, and Papua.
[6] Incomplete global estimates.

the California Delta peats formed where the climate is hot and dry and the marsh and valley deposits in Washington and Oregon. Florida, including the Everglades and related localities, contains about 12 percent of the deposits in the contiguous United States. The Everglades, containing more than 810,000 hectares, is the largest parcel of peat and muck soils in the contiguous United States. The bulk of peat and muck lands, however, are in small scattered pockets of 1 hectare to several hundred hectares. The deposits are valuable for crop production, forestry, natural water treatment, water reservoirs and wildlife refuges, and as a source of organic materials.

Worldwide estimates of the extent of Histosols vary widely. Four estimates of peatland resources are given in Table 1. These estimates are likely incomplete for the less accessible parts of the world. Major uncertainties concern the extent of Histosols in southeast Asia, Canada, Europe, and the U.S.S.R. Farnham, in Armentano (1980a, 1980b), believes that the estimate given by Moore and Bellamy (1974) in Table 1 should be about 40 percent larger, based on larger estimates by Heikurainen (1964) and other sources. Lucas (1982) compiled global estimates of Histosols that are about 49 percent greater than the estimates of Moore and Bellamy (1974).

HISTORY AND RECORDED RATES OF SINKAGE

Netherlands

The oldest records of subsidence appear to be those of the old polders in western Netherlands as related by Schothorst (1977). The low-moor peat soils in these old polders were reclaimed in a period between the 9th and 13th centuries. Initially, the elevation of these peat soils was reportedly equal to or somewhat above sea level. At low tides, excess water could then be discharged by means of sluices into the sea or into rivers. This system of gravitational discharge was possible until the beginning of the 16th century when the surface had subsided to such an extent that excess water had to be discharged artificially by means of windmills. Although throughout the centuries only shallow seasonal drainage was applied, the soil surface nevertheless subsided to about 1 to 2 m below sea level over a period of 8 to 10 centuries. After steam pumping stations began controlling water levels all year long, about 1870, the process of subsidence was accelerated. During the period 1877 to 1965, the polder water level was lowered by 0.5 m. During the same period, the surface subsided an equal amount. The rate of subsidence was 6 mm per year in spite of shallow drainage, which was only 0.1 to 0.2 m below the surface.

Specific records kept of the polder Zegvelderbroek, situated west of Utrecht, show that before drainage in winter was made possible by steam pumps, the total subsidence in the past had amounted to at least 1.5 m—assuming that the initial elevation was equal to or somewhat above sea level. According to historical sources, the polder was reclaimed around 1,000 A.D. Thus, subsidence during the subsequent 9 centuries amounted to approximately 1.7 mm per year. After water control in winter was introduced, however, the subsidence accelerated from 1.7 to 6 mm per year. More detailed data obtained from an experimental field laid out in 1952 for water control in the subject polder showed that a 0.4 m drawdown in the ditch water level over a period of 20 years resulted in a total surface subsidence of 23 cm. In the first 2 years, the subsidence proceeded very rapidly constituting 44 percent of the total for the 20-year period. Subsequently the subsidence rate decreased to a constant 7 mm per year, which is approximately equal to the subsidence rate of the entire polder.

England

The history of drainage, which began in 1652, for the low-moor peats of the English Fens, is largely a story of troubles caused by lowering of the land surface. The story of the Fenlands has been marked by alternate cycles of improved drainage with increased subsidence and consequent higher water tables. First, there was gravity drainage. This was followed by pumpage using windmills, then steam engines, and finally diesel engines. All of these methods greatly increased the rate of water removal but failed to provide a satisfactory solution to the problem of drainage. Frustratingly, the more the water table was lowered by more effective drainage, the more rapidly the peat surface continued to sink; thus, the achievements of one generation became the problems of the next (Darby, 1956). Measurements at the renowned Holme Post, which was solidly imbedded into the underlying clay at Holme Fen about 1848, provides one of the oldest authenticated records of subsidence in existence. This Post shows the lowering of the ground level around the post of 3.14 m in the 84-year period up to 1932 (Fowler, 1933). Thompson (1957) reports that the declining surface elevations continued long after the Fens had become well drained. This subsidence persisted to the date of his last examination in 1957 at a rate variously estimated from a fraction of cm to more than 2.5 cm per year. These historical observations and records of subsidence at Holme Fen are being continued (Hutchinson, 1980). Before drainage began in the 17th century, it is estimated that the peat lands stood 1.5 to 2.1 m above the silt areas. Now the peat is about 2.05 m lower than these areas. Thompson concluded that the continuance of peat subsidence will in time result in the complete elimination of all peat deposits in the region. During a visit to the Fens in 1969, the senior author observed that the peat which had been under continuous drainage had subsided to the point where special plows were needed to incorporate the underlying material in order to provide sufficient depth for cultivation of sugar beets.

Russia and Norway

Skoropanov (1961, 1962) gives an account of peat subsidence in Belorussia at the Minsk Bog Experiment Station. These soils were developed as sedge, sedge cane, and wood cane peat and are classed as low-moor peats with a pH of 5.6. Minsk, U.S.S.R., is near Latitude 54° N. Frost-free days for peat bog soils at Minsk average 123 days. In 1961 thickness of the Minsk peat deposit did not exceed 1.0 m, whereas it was 2.0 m at the beginning of the cultivation in 1914. Research under Belorussian conditions showed that oxidation accounted for 14 percent of the decrement on two plots which had been cultivated for 46 years. Skoropanov contends that oxidation should not be considered as destruction of soil but instead represents the process of soil formation. On the other hand, according to Ivitskii (1962), deep depression of the water table intensifies decomposition of the organic matter by mineralization and that although this process is an integral aspect of soil fertility excessive drainage is undesirable. He recommends that the drainage norm for intertilled crops and grains below bog soils be at least 60 cm but not over 175 cm for the pre-sowing period. During the growing season, he recommends an optimum norm of 120 cm to 130 cm with maximum depth not exceeding 220 cm to 260 cm.

In Norway the principal problem entailed in peatland conservation has been soil destruction brought about by the irrational digging of peat for fuel (Løddesøl, 1947). Subsidence also occurs, however, under agricultural drainage in Norway. Løddesøl, former director of the Norwegian Bog Association, Oslo, Norway, reported that a loss of 1.5 m in 65 years occurred for cultivated bog soils at Stend Agricultural School at Hordaland County. From these and other observations on 33 parishes, the average loss of about 20 mm per year was shown to occur (Løddesøl, 1949).

U.S.A.

In the organic soils of the Sacramento-San Joaquin Delta of California, the average rate of subsidence has been approximately 8 cm a year with no indication that this rate is decreasing (Weir,

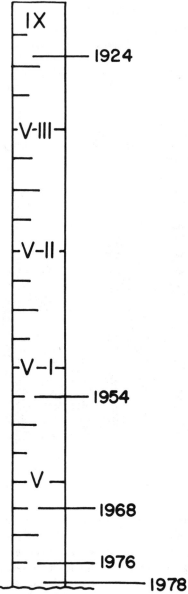

Figure 1. This concrete monument sketch shows a subsidence loss of 1.8 m, occurring between 1924 and 1978 at the Everglades Experiment Station, Belle Glade, Florida. (After Shih and others, Belle Glade, AREC Research rept. EV-1979-1, Mimeo.)

1950). These soils are tule reed peats which have subsided from 1.8 m to 2.4 m with drainage between 1922 and 1950. At the latter date, most of the area was between 3.0 and 3.6 m below sea level and protected by dikes. In Michigan, losses in surface elevation on organic soils as great as 21.5 cm over a 5-year period have been measured, with an average 5-year loss of 9 cm from 13 sites (Davis and Engberg, 1955; Davis and Lucas, 1959). Near Hennepin County, Minn., the total settling of a newly drained peat bog varied from 30 to 60 cm over a 6-year period, with the variation being approximately proportional to water-table depths (Roe, 1936).

In the Everglades of Florida (Stephens and Speir, 1969;

Figure 2. The subsidence of soil around this house built on piling about 1924 at the Everglades Experiment Station is shown in this recent photograph.

Stephens, 1969), the arable organic soils have an average subsidence rate, after initial settlement, of 2.54 cm per year. This rate is continuing (Shih and others, 1979a). Concrete monuments with their tops originally flush with the ground surface and set directly on the solid underlying rock some 57 years ago now protrude approximately 1.8 m above ground and furnish visual evidence of soil loss. Figure 1 shows losses that have occurred between 1924 and 1978. At Belle Glade and other Everglades towns in Florida, where homes are built on pilings, a new doorstep to the entrance may be needed about every 10 years as the ground subsides (Figure 2). In Indiana and adjacent states, experiments at Purdue University have shown that organic soils subside at about half the rate of those for Florida (Jongedyk and others, 1953).

Global Summary

Worldwide, subsidence rates have been found to vary from less than 1 to more than 8 cm per year. Table 2 shows data from a survey by L. H. Allen, Jr., University of Florida, and J. M. Duxbury, Cornell University. It was compiled for a workshop report on the role of organic soils in the world carbon cycle, (Armentano, 1980a), relating measured rates of subsidence of organic soils for specific sites in different areas.

CAUSES OF ORGANIC SOIL SUBSIDENCE

Disregarding the mining of organic soil for use as a fuel or for agricultural and industrial purposes, organic soil subsidence

TABLE 2. MEASURED RATES OF SUBSIDENCE OF ORGANIC SOILS FOR SPECIFIC SITES IN DIFFERENT AREAS[1]

Location of Site	Annual Subsidence Rate (cm yr^{-1})	Cumulative Subsidence (cm)	Time Period (yr)	Average Depth to Water Table (cm)
California Delta (2 sites)	2.5 - 8.2	152 - 244	26	
Louisiana (estimated)	1.0 - 5.0			
Michigan	1.2 - 2.5	7.6 - 15	5	
New York	2.5	150	60	90
Indiana	1.2 - 2.5	7.6 - 15	6	
Florida Everglades (2 sites	2.7 2.7 - 4.2	147 19 - 29	54 7	90 60
Netherlands (2 sites)	0.7 1.0 - 1.7	70 6 - 10	100 6	10 - 20 50
Ireland	1.8			90
Norway	2.5	152	65	
England	0.5 - 5.0	325 (by 1932) 348 (by 1951)	84 103	
Israel	10			
U.S.S.R. (Minsk bog)	2.1	100	47	

[1]Data from survey by L.H. Allen and J.M. Duxbury. Prepared for Armentano (1980a, 1980b).

has occurred mainly because of drainage and the development of peat deposits for growing food and fiber crops. Thus, we will focus on subsidence processes resulting from agricultural drainage. Under natural conditions of peat formation, water acts as the preservation agent for peat by excluding oxygen.

Causality Classes

Under drainage, six causes of organic soil subsidence are: 1) shrinkage due to desiccation; 2) consolidation, or loss by the bouyant force of groundwater, or loading, or both; 3) compaction, normally with tillage; 4) wind and water erosion; 5) burning; and 6) biochemical oxidation. The first three causes, drying, consolidation, and compaction, increase soil density only and do not cause a loss of soil mass. Such densification is largely a nonrecurring phenomenon. Wind and water erosion or burning may occur occasionally in certain bog deposits and cause significant soil loss. However, these types of losses can usually be controlled. Biochemical oxidation, on the other hand, is a long-term process and continues as long as temperature, pH, and aeration are conducive to microbial oxidation of the organic matter. Thus, the major causes of subsidence can be divided into two main categories: a) physical, which increases soil density and reduces volume only; and b) biochemical, which causes a loss in soil substance and can eventually lead to the loss of the bog deposit (Stephens and Stewart, 1976).

Laboratory Studies

The biochemical aspect of peat oxidation was demonstrated by laboratory studies at Rutgers University as early as 1930 (Waksman and Stevens, 1930; Waksman and Purvis, 1932). Peat types, such as low-moor, high-moor, forest, and sedimentary differ in their chemical composition, in the nature of the microbial flora inhabiting them, and in the rate of attack by micro-organisms. Thus, peats vary considerably in the rate of decomposition depending on different environmental conditions and management. Waksman found for samples of Florida low-moor peat (those with pH from neutral to basic) that about 15 percent of the dry total weight was decomposed at 28° C in 18 months, most of which could be accounted for as CO_2 gas evolved. The optimum moisture content for decomposition was 50 to 80 percent of the total moist peat. Above and below this moisture range, the rate rapidly diminished. Wet and dry cycles greatly stimulated peat decomposition as compared with constant moisture. Broadbent (1960) also observed this effect with California organic soils. In related studies, bacteria were found to be most numerous in drained low-moor peats and less numerous in extremely acid high-moor peats. However, when the acid peats were limed, manured, and put under cultivation, the microbial population increased to about that of the low-moor peats under similar drainage; decomposition rates also increased to that of the low-moor peats. Cold climate retards the activity of micro-organisms. Waksman found that organisms causing decomposition were perceptibly active only when soil temperatures remained above 5° C. Jenny (1930) noted that soil microbial activity generally doubled for each 10-degree increase in temperature. Thus, soil temperature is a factor in determining subsidence rates.

Laboratory studies of the role of micro-organisms in the subsiding of Histosols were made at the University of Florida Agricultural Research and Education Center at Belle Glade by Tate (1976, 1979a, 1979b, 1980a, 1980b), Terry (1980), Terry and Tate (1980), and Duxbury and Tate (1981). Tate traced the microbial reaction sequence involved in peat soil subsidence and in the production of nitrates. He points out that the process usually produces more nitrates than growing plants can use and that the excess is probably volatilized by denitrification, or is removed in drainage water. He suggests that raising the water table will decrease the rate of soil subsidence; hence, decrease the amount of inorganic nitrogen formed, as well as inhibit nitrification. Accordingly, preservation of the organic soils should also help in preserving the quality of surrounding lakes and streams.

Field Studies

Investigations of organic soil subsidence in the Netherlands (Schothorst, 1977), show that compression losses can be reversible or irreversible. In the Dutch polders, a high rate of surface subsidence in summer is largely caused by elastic compression of the peat below groundwater level due to high evaporation rates, which reduces the buoyancy effect. In the winter as the water table rises and the buoyancy effect is recovered, the soil subsidence also recovers almost 100 percent. Thus, in temperate climates where there is a marked difference in water-table levels between winter and summer, it is important that measurements be made at the same season each year to obtain true soil losses. Seasonable subsidence measurements for a 6-year study in three Netherlands polders, with ditch water-level depths that average approximately 65 cm, showed an average annual sinkage rate of 0.53 cm. Schothorst computed the components of loss to be as follows: compression 28 percent (subject to elastic rebound and recovery); irreversible shrinkage 20 percent; and oxidation 52 percent. He found that deeper drainage increased subsidence rates and that higher irreversible shrinkage and oxidation goes with higher levels of organic matter. In the future he expects compression and irreversible shrinkage to gradually decrease, but oxidation to continue at a more or less constant rate until a new, lowered ditch water-level will be necessary.

Typical subsidence rates and patterns for drained organic soils in sub-tropical climates are illustrated by the subsidence line studies in the Florida Everglades south of Lake Okeechobee. The first profile elevations along subsidence lines were surveyed in 1913. In the early 1930's lines were added at the Everglades Experiment Station where land-use and treatment could be controlled (Clayton, 1936, 1938; Allison, 1956; and Neller, 1944). Altogether, 15 lines were established. About 11 are still being surveyed (Shih and others, 1979a). Those abandoned have been negated by construction or have subsided until the organic cover has disappeared and the underlying mineral soil exposed. The original subsidence lines were usually 1 mile long. Elevations were established on permanent bench marks set in bedrock near the point of origin. All elevations were converted to mean-sea-level data and the elevation loss determined by resurveying the lines about every 5 years. Ground elevations were measured every 25 ft along the line, and then averaged to obtain mean elevation of the soil.

Plotted surface elevations vs time show very fast sinkage occurred following initial drainage because of shrinkage and loss of the buoyant force of water. Then, the first tillage caused additional subsidence by compaction. After 5 years of cultivation, the density of the top 45 cm increased to about double that of the brown fibrous peat underneath. The top layer was changed into a black, mucky, amorphous mass with a marked decrease in hydraulic conductivity. As drainage continued, the subsidence rate leveled off to a more or less steady rate, dependent on water-table depth. After the initial rapid shrinkage, due to desiccation or tillage or both, the slow, steady subsidence was largely due to oxidation, which in turn was related to the amount of oxygen in the drained zone.

A graph of the characteristic sequence of observed subsidence in the organic soils is shown for three profile lines (A, B, C, Figure 3). Line A, normal to the North New River Canal just below the old South Bay Lock near the 1-mile post is a peaty muck soil that had gravity drainage before pumps were installed in 1927. Sinkage was rapid during the first 5 years, beginning in 1914. As the ground level fell, gravity drainage was less effective and the subsidence rate decreased. In 1927 the subsidence rate again increased because of pumped drainage and initial tillage. Since 1926 the land generally has been planted to sugar cane and truck crops, and since 1938 the subsidence rate has rather consistently held at 1.60 cm per year. Line B, located 85 m north of the Bolles Canal, a major drain at Okeelanta, is also on peaty muck soil; it had gravity drainage until 1942 when pumps were installed. Historically, this line has had better gravity drainage than Line A. Truck crops were planted sporadically prior to 1953 when the entire area was put into pasture and then converted to sugar cane in 1965. Since 1953 the subsidence rate has been 1.75 cm per year, which is not significantly different from that of Line A for the same period.

Both Lines A and B show typical subsidence patterns: high initial subsidence, a decreasing rate as the ground sank and drainage was impaired, and the consequent increase in the subsidence rates after pump drainage. Line C, located on Everglades peat at the Everglades Experiment Station near Belle Glade, Florida, represents land that has already had its initial rapid subsidence after drainage. It has been continously planted to truck crops since 1934 and probably represents the average subsidence rate (2.44 cm per year) for peat soils planted to truck crops. The subsidence rate of Lines A and B is about two-thirds of that for C. This difference is due to the higher mineral content of the peaty muck soils, or to better pump drainage at the Everglades Station, or both.

It has been observed that grazed sod fields sink slower than adjacent tilled fields under the same drainage. At the Everglades Experiment Station, the annual rate of sinkage in 1950 for the prior three decades was 2.13 cm per year for St. Augustine grass pasture, and 2.67 cm per year for land in truck crops. However, when the entire soil profiles were examined for bulk density and when the higher pastureland elevation was corrected for lower soil density, the difference of loss in soil material between pasture and crop land was not significantly different (Stephens, Craig, and Allison, 1952). A recent analysis by Shih and others (1979b) shows that these trends still hold.

Wind and water erosion have been implicated in causing significant organic soil losses in unprotected fields in Quebec, Parent and others (1982).

PREDICTING SUBSIDENCE RATES

A reliable estimate of the resultant subsidence rate is usually the most important need in determining the economic feasibility of reclaiming organic soils. As previously stated, densification of these soils (caused by drying, consolidation, and compaction) is largely a nonrecurring event and does not cause loss of soil mass. It occurs soon after draining or loading and may be estimated by conventional soil physical formulas, compaction tests, or even short-term observations. Oxidation, on the other hand, is a long-term process that continues as long as temperature, pH, and

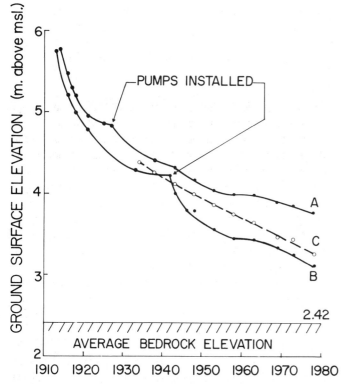

Figure 3. Sequence of observed subsidence of organic soils in the Florida Everglades after initial drainage, circa 1912.

aeration are conducive to biological oxidation of the soil matter, which can eventually cause the loss of the bog deposit.

Acid peats with a low pH are neither conducive to the growth of micro-organisms that cause a breakdown in oxidation of organic soils nor to the growth of higher plant life common to most agricultural production. A few specialized crops such as cranberries and blueberries do well on acid peats, but these are relatively unimportant in the total agricultural economy. As previously cited, Waksman observed that when acid peats were treated with lime, manured, and put under cultivation the micropopulation increased to about the same as that for low-moor peats under similar drainage. Thus, the two principal variables with which we are concerned in determining the rate of subsidence for agricultural soils are a) depth of drainage, and b) soil temperature.

Effect of Depth of Drainage

The relationship between depth of drainage and rate of subsidence has been well established at specific sites where the soil properties are known. The use of controlled water-table plots, where results are available for a period of years sufficient to provide statistically significant data, has proven to give the best guide for computing subsidence rates at representative locations. In all cases, it has been found that the lower the water table the greater the subsidence rate.

At the University of Florida, Everglades Agricultural Experiment Station in 1934, a field of Everglades peat was divided into eight blocks, each 30.5 × 73.4 m (100 by 240 ft) and provided with a system of ditches, underdrains, and check dams so that the water table could be held at any depth desired (Allison and Clayton, 1934; Stephens and Johnson, 1951). Water levels ranging from 30.5 to 91.4 cm (12 to 36 in) between blocks were established in 1936 and held at the same depths until 1943. Surface levels were surveyed and soil samples were taken annually from each block to determine elevation losses and any compaction changes. The rate of subsidence was found to be dependent upon the depth to the water table—the higher the water table the lower the soil loss. Density and mineral content determination showed that all soil losses took place above the water table. By collecting the soil gasses, Neller (1944) found that the production of carbon dioxide (CO_2), end-product of oxidation, was directly related to the amount of soil loss.

At the Purdue University Muck Crop Experiment Station near Walkerton, Indiana, on controlled plots similar to those at the Everglades Experiment Station, the rates of subsidence were found to be about half those for Florida (Jongedyk, 1950, 1953). This difference can be explained when water-table treatments are considered. Prescribed water tables were held year round in the Florida experiments, while in the Indiana experiments the lowered water tables were held only during the crop year from May to September. Furthermore, during the winter season, the Indiana plots were exposed to freezing temperatures. The significant re-

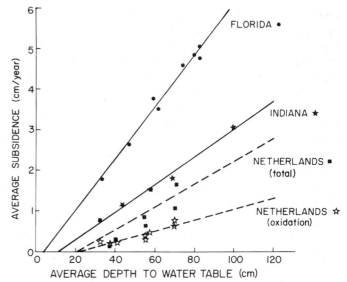

Figure 4. Comparative subsidence rates of organic soils in Indiana and the Florida Everglades versus water-table depth. Organic soil data for western Netherlands superimposed from Schothorst, 1977. The two lines shown for the Netherlands show total subsidence and subsidence attributable to biological oxidation. The linear regression equations are (a) Florida: $Y = 0.0643X - 0.259$; (b) Indiana: $Y = 0.0344X - 0.429$; (c) Netherlands (total subsidence): $Y = 0.0281X - 0.581$; (d) Netherlands (oxidative subsidence): $Y = 0.0134X - 0.291$; where Y is the predicted subsidence in cm per year, and X is the average depth to water table in cm.

sults from the Florida and Indiana water-table studies on subsidence rates are summarized in Figure 4.

The annual subsidence rate of low-moor peat soils for the Zegvelderbroek water-table experimental plot in the western Netherlands, (Schothorst, 1977), is also shown on Figure 4. This soil is similar to the Florida and Indiana peats in organic matter content. The Netherlands rates differ from those in Florida and Indiana; however, the rates are not strictly comparable since the Florida and Indiana losses were primarily from oxidation, whereas the Netherlands rates included initially higher sinkage rates for the first three years of drainage in addition to oxidation losses. Schothorst noted that subsidence occurred somewhat more rapidly during the initial years of the six-year study.

Coefficients for linear regression of annual subsidence rates vs depth to water table are given in the caption of Figure 4.

Effect of Soil Temperature

Although known for years that subsidence of organic soils was faster in warmer regions than in cooler climates, as witnessed by the Florida vs Indiana water-table studies, it has been only in the past decade that much progress has been made in relating soil temperatures to subsidence.

Laboratory studies, (Knipling and others, 1970; Volk, 1973), determined the rate of CO_2 evolved from peats as a function of soil temperature. Both studies showed that CO_2 evolution increased with water-table depth and with temperature. Knipling found that from 10° to 20°C the CO_2 rate of change was smaller than that from 20° to 60° C, but that generally CO_2 evolution rates doubled for each subsequent 10° increase. Volk found that CO_2 losses were directly proportional to water-table depth and to temperature, and were also a function of organic content and bulk density. He also found that each 10° C increase about doubled the CO_2 evolution, within normal soil temperature ranges. Further, for a given water-table depth and temperature, he found that the amount of CO_2 involved increased as the percentage of organic matter increased. Volk ascribed the reduction in biochemical activity for the higher density mucks, as contrasted to peats, to the higher mineral content of the soil fraction that formed a clay-organic complex less accessible to the micropopulation.

In studying the effects of climate on soil subsidence, Stephens and Stewart (1976) compared the rates of CO_2 released, as determined by the laboratory studies of Knipling and Volk, to the measured rates of soil subsidence at the Everglades Experiment Station water-table plots. By converting subsidence rates for peat soil solely due to CO_2 evolution, they found that for these soils (bulk density 0.18 g/cm^3) a CO_2 flux density of about 1.0×10^{-8} g/cm^2/sec would be required to produce an elevation loss of 1 cm annually, assuming that the subsidence rates were due solely to oxidation of carbon to CO_2. The measured annual subsidence rate for the 60 cm depth water-table plot, where the annual soil temperature was 25° C, was 3.76 cm per year (Figure 4). The laboratory CO_2 evolved at the 60 cm depth at 25° C was 1.9×10^{-8} g/cm^2/sec, which computes to a subsidence loss of 2.00 cm per year for the lab studies, or which accounts for approximately 53 percent of the measured loss in elevation for the water-table plots. However, the CO_2 evolution accounts only for the loss due to microbial respiration. It does not account for the loss of material hydrolized by microbial action, which is removed by leaching under field conditions, or for compression or compaction. For the water-table plots, Neller (1944) judged that only a small part of the subsidence could be attributed to compaction since he found no increase in bulk density over a period of six years. However, his density studies were too few to be conclusive. We estimate that the increase in density during the years for which Neller reported (1935–41) did not exceed 10 to 15 percent. Thus, the loss of hydrolized material by microbial action, which is removed by leaching, could be as much as 25 to 30 percent according to the experimental data cited. Whatever the exact contribution of the leachates from hydrolized organic material to biochemical subsidence, it is the same group of organisms that evolves CO_2 gas, and thus may react in the same manner to a change in soil temperature.

We recently attempted to estimate carbon losses from the Everglades Agricultural Area in the drainage water pumped into the conservation area based on data of Lutz (1977) and/or in the drainage water pumped into Lake Okeechobee based on data of Dickson and others (1978). Davis (1981) data showed that the C/N ratio of drainage water was about 7. A typical areal export rate of N was 36 kg/ha per year (Dickson and others, 1978). Multiplied by 7, this would yield an areal export rate of C of about 250 kg/yr. Assuming a 2.5 cm per year subsidence rate, and a C density of 0.1 g/cm^3, we would expect a loss of about 25,000 kg C per hectare. Therefore, the water transported C leaving the Everglades Agricultural Area appears to be only of the order of 1 percent of the total amount of C lost as a result of subsidence. However, it is possible that losses of both gaseous C and N could occur between the point where leachates leave the fields and the point where they are pumped out of the Everglades Agricultural Area. Nevertheless, these data suggest that losses of hydrolized materials are small compared to decomposition to CO_2 and other gases.

Linking Drainage and Soil Temperature

In developing a mathematical model linking drainage depth and soil temperature that could be used to estimate subsidence of low-moor peats in different climates, Stephens and Stewart (1976) applied the Arrhenius law to both water-table and laboratory findings. (Arrhenius showed that the logarithm of the velocity coefficient, k, of a chemical reaction is linearly related to the reciprocal of the absolute temperature, T.)

With the assistance of Victor Chew, biometrician, USDA, ARS, Gainesville, Florida, they developed the "Stephens-Stewart-Chew" basic subsidence equation, which follows:

$$S_T = (a + bD) e^{k(T-T_o)}, \qquad (1)$$

where

S_T = biochemical subsidence rate at temperature, T.
D = depth of watertable
e = base of the natural logarithm
k = reaction rate constant
T_o = threshold soil temperature where biochemical action becomes perceptible
a and b are constants

Using the empiricism that S_T multiplies to $Q_{10} \cdot S_{(T+10)}$ where Q_{10} is the change in reaction rate for each 10° C rise in temperature, then

$$S_{(T+10)} = Q_{10} \cdot S_T = (a+bD) e^{k \cdot [(T+10)-T_o]} \qquad (2)$$

Dividing eq. (2) by eq. (1):

$$\frac{S_{(T+10)}}{S_T} = Q_{10} = e^{10k} \qquad (3)$$

By rewriting (2) to express k in terms of Q_{10},

$$k = 1/10 \ln Q_{10}, \qquad (4)$$

and substituting (4) into (1),

$$S_T = (a + bD) \; Q_{10}^{(T-T_o)/10} \quad (5)$$

From laboratory studies, the value of Q_{10} was assumed to be 2.0 and the value of T_o to be 5° C. The relation found between drainage depth, soil temperature, and annual subsidence rates at the Everglades Experiment Station water-table plots was chosen to find the value of constants a and b where measured annual soil temperature was 25° C. Thus, from Figure 4, when D = 40 cm, S = 2.29 cm/yr, and D = 80 cm, S = 5.00 cm/yr. Substituting these values into (5) and solving the simultaneous equation to evaluate a and b, we find that a = −0.1035 and b = 0.0169. Then from (5)

$$S_x = (-0.1035 + 0.0169 \, D) \, (2)^{(T_x - 5)/10} \quad (6)$$

Stephens and Stewart recommended that equation (6) be used to estimate the biochemical subsidence rate for low-moor organic soils at locations where the annual average soil temperature at the 10-cm depth is T_x.

For example, equation (6) was used as follows to estimate the annual subsidence rate—exclusive of compaction or other bulk density change—at the Lullymore Experimental Station in the Irish Republic for the arable low-moor soils where the average annual soil temperature is 8.5° and the water-table depth is held at 90 cm:

$$\begin{aligned} S_L &= (-0.1035 + 0.0169 \times 90) \, (2)^{(8.5 - 5.0)/10} \\ S_L &= 1.4175 \times 2^{0.35} \\ S_L &= 1.4175 \times 1.27 = 1.80 \text{ cm/yr} \end{aligned} \quad (7)$$

This indicates that if the Everglades Histosols were transposed to Lullymore, the rate of decomposition would be only 32 percent of the rate in south Florida where T = 25°.

If the Everglades soils were in a more tropical climate where the average soil temperature was 30° C, for example, again from (6)

$$\begin{aligned} S_x &= (-0.1035 + 0.0169 \times 90) \, (2)^{(30-5)/10} \\ S_x &= 8.02 \text{ cm/yr} \end{aligned} \quad (8)$$

which is a rate 43 percent greater than the decomposition rate for south Florida.

For convenience in estimating S_x for any selected temperature, T_x, the results from equation (6) were plotted at water-table depths, D, of 30, 60, 90, and 120 cm, which has been reproduced as Figure 5.

Stephens and Stewart cited limitations of the mathematical model due to limited field and lab data, but also pointed out that if future studies indicated the values of Q_{10} and T_o should be different from those assumed, the mathematical procedure for computations would still be valid. Only the constants for a and b would change. For instance, where $Q_{10} = 1.5$ and $T_o = 0°$ C, then a = −0.1523, and b = 0.0246.

When using equation (6) or the graph, Figure 5, remember

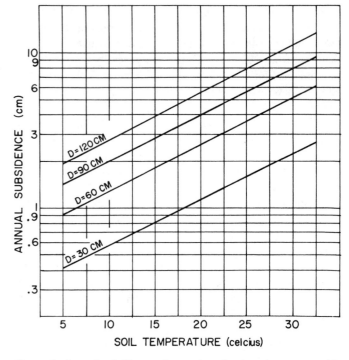

Figure 5. Annual subsidence of organic soils at various water-table depths and soil temperatures.

these values were developed from organic soils with a mineral content of less than 15 percent and a bulk density of approximately 0.22 g/cm^3. With muck soils of increased mineral content, and higher bulk density, the expected subsidence rates would be from one-half to three-fourths of those shown by the subject graph, or equation, depending on the increase in mineral content.

CONTROLLING SUBSIDENCE

Presently, organic soils cannot be used for crops that require drainage without paying the cost of subsidence. The rate can be slowed by proper water control and good land management. Crop yield studies indicate a water table of 30 to 60 cm for sod, and one of 60 to 90 cm is desirable for most truck and field crops on peat (Snyder and others, 1978). Thus, even with optimum water-table control for good production, subsidence will continue at an undesirable rate. On seasonally cropped lands, flooding during the idle season could lengthen the life of organic soils. Growing water-tolerant crops such as rice or kenaf could add many years to the life of peatlands. However, several agronomic, economic, and institutional problems have to be solved before many of the water-tolerant crops can be grown commercially.

The value of plowing under litter or cover crops to prolong the life of organic soils is debatable. A study in 1968 was made using radiocarbon age determinations to ascertain the rate of peat formation in the upper Everglades (McDowell and others, 1969). These age determinations indicate that peat formation began during late Hypsithermal time, about 4,400 years ago. About 500 to

1,000 years were required to build up an 8 cm layer of basal, mucky peat composed of a mixture of marl and organic matter. Then about 3,500 to 4,000 years ago, coincident with a rise in sea level or the climatic optimum, or both, plant growths were preserved as fibrous peats with little mineral admixture, and peat soils developed relatively rapidly. By 1914, peat had developed to a depth of about 3.7 m at the sample sites, which represents an average peat development of about 9 cm per century. Age determinations of moss and forest peat in the Netherlands polders show an accumulation rate of 10 cm per century (Van der Molen, 1981). In Scotland, radiocarbon dating and pollen graphs show that basin peats have accumulated at rates of from 8.0 to 9.6 cm per century during the last 2,500 years since the beginning of the wetter Atlantic Climatic Period, while the moss hill peats were accumulating at a much slower rate of 4.8 cm per century (Durno, 1961). Initially, it was concluded that in Florida turning under cover crops would be trivial, since the 9 cm of peat, which required 100 years to develop, would be lost by oxidation in about two years with a drainage depth of 70 cm. On the other hand, some scientists believe that organic soil deposits might recover faster than radioactive dating of natural strata would indicate. They suggested that managed flooding might restore deposits faster, because of no misfortunate losses due to desiccation during natural drought cycles that could lead to intermittent biologic oxidation, or to fire, or other catastrophic losses.

In any event, barring a fortunate and now unforeseen breakthrough in the science of organic soil conservation, subsidence will continue on drained, cropped Histosols. Meanwhile, several steps may be taken to obtain the maximum agricultural use of these soils: 1) provide adequate water control facilities for keeping water tables as high as crop and field requirements will permit; 2) make productive use of drained lands as soon as feasible; and 3) intensify research studies to develop practices to prolong the life of the soils.

A LOOK AT THE FUTURE

New Mathematical, Biochemical, and Agronomic Investigations

Agricultural scientists have started an intensive team effort, including field, laboratory, and mathematical approaches, aimed at better defining the specific causes of biochemical subsidence and gathering better insights into the interrelationship of the causes. In connection with these studies, Browder and Volk (1978) developed a computer systems model that simulates the interaction of climatic and biologic factors on the biochemical oxidation of Histosols following drainage. This model conserves the mass balances of carbon and nitrogen as they are transformed in a series of complex rate processes. Predictions, based in part on estimates of rate process data, gave results reasonably consistent with historical records. Simulations were designed to show the effects of temperature, depth to water table, and organic composition on CO_2 evolution. Simulation results showed that temperature and water-table depth have the greatest effect, which is similar to earlier experimental conclusions. It is hoped that such studies will help researchers develop a better understanding of how the organic soil system works and identify those factors that have the greatest impact on subsidence, so that additional research studies can be concentrated in those areas.

Studies have been initiated to try to determine the biological and enzymatic factors involved in biochemical oxidation of Histosols (Duxbury and Tate, 1981; Tate, 1976, 1979a, 1979b, 1980a, 1980b; Terry, 1980; and Terry and Tate, 1980). So far, these studies have not given successful clues for reducing subsidence of drained organic soils.

Alternative crops that can be grown under flooded or waterlogged conditions offer one possibility of reducing or eliminating subsidence. Morton and Snyder (1976) reviewed the potential of numerous aquatic crops such as watercress, water spinach, swamp fern, vegetable fern (pako), water dropwort or water celery, ottelia, asiatic pickerel weed, bengok, jungle rice, rice, wild rice, Chinese water chestnuts, swamp potato, lotus root, taro, and dasheen. Snyder and others (1977) reviewed the economic potential of incorporating rice into Everglades vegetable production systems. Crops that can be grown in flooded or waterlogged conditions offer the best opportunity of preserving and utilizing the organic soil resource.

If subsidence continues, many of the developed organic soils will have to be returned to a continuously saturated state. Deep organic soils such as in the Sacramento-San Joaquin Delta of California are already below sea level, and the peat will not support unlimited levee loads. Shallow organic soil such as in the Everglades Agricultural Area of southern Florida will eventually leave rock exposed. In either case, either a wetlands agriculture will have to adapt or the areas designated for other nonagricultural use.

Subsidence and the Global Carbon Cycle

One question that has been raised recently is the role of organic soils in the global carbon cycle. Based on measurements of subsidence rates (~2.54 cm/year), knowledge of soil properties (bulk density $\simeq 0.23$ g/cm^3; % organic matter $\simeq$ 88%, % C of organic matter $\simeq$ 56%), and knowledge of the area drained (~3.1 $\times 10^5$ ha), we computed the carbon released to the atmosphere annually from the Everglades Agricultural Area of Florida to be about 9×10^{12} g carbon. Since the recent rate of release of carbon from fossil fuels is about 5×10^{15} g, oxidation subsidence in this area is equivalent to about 0.18 percent of that released by fossil fuels. In order to meet the carbon-equivalent of the global annual fossil fuel usage, a 14 m depth equivalent of the Everglades Agricultural Area organic soil would be required. For the Sacramento-San Joaquin Delta, assuming the same soil properties, a subsidence rate of ~7.6 cm per year, and a drained area of ~1 $\times 10^5$ ha, a carbon release of about 7.6×10^{12} g annually was

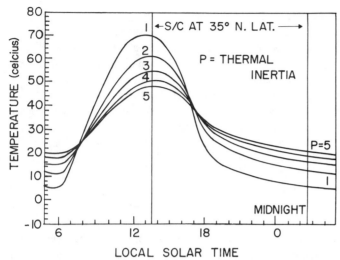

Figure 6. Illustration of diurnal surface temperature variation as a function of thermal inertia (from Heat Capacity Mapping Mission (HCMM) Data Users Handbook for Applications Explorer Mission-A (AEM), Goddard Space Flight Center, NASA, Greenbelt, MD, Dec. 1978, Sec. Rev. Oct. 1980). The unit value of P is $41.86\ kJ \cdot m^{-2} \cdot s^{-1/2} \cdot {}^\circ K^{-1}$.

computed, which is equivalent to about 0.15 percent of the carbon released annually from fossil fuels.

A workshop called "The role of organic soils in the world carbon cycle" estimated that the current annual release of carbon from organic soils falls within the range of 0.03 to 0.37×10^{15} g, which is equivalent to 1.3 to 16 percent of the annual increase of carbon in the atmosphere (Armentano, 1980a). If half of the released carbon remains in the atmsophere, then organic soils were estimated to contribute 0.6 to 8.0 percent of the annual rise in CO_2. Uncertainties in the data indicate that the actual release could lie outside this range. Evaluation of organic soil areas, drainage status, and biological oxidation rates are needed on a global scale to reduce the uncertainties.

Remote Sensing Potentials

Remote sensing offers much promise for assessing both the area and water status of organic soils now and changes that may occur in the future. Aerial photographic interpretation has been used for several years to assess muskeg areas in Canada (Korpijaakko and Radforth, 1968). Scientists at the Department of Peat and Forest Soils, Macauley Institute for Soil Research, Aberdeen, Scotland, have made significant advances in peat surveys by application of remote sensing techniques using Landsat satellite imagery and airborne sensor devices. Information can now be obtained on land use, peat and forest resources, crop conditions, water deficit or excess, and several other surface features (Stove and Hulme, 1980). Chen and others (1979, 1982) showed that drained organic soils could be clearly distinguished from undrained organic soils in Florida based on nocturnal surface temperatures from Geostationary Operational Environment Satellite (GOES) data. A combination of global soils map, visual

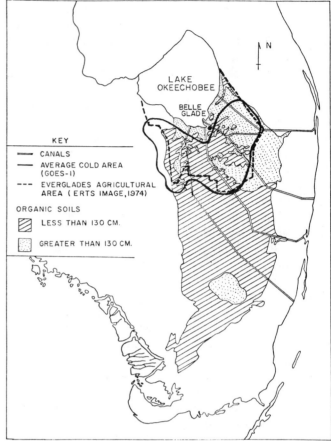

Figure 7. Comparison of Everglades Agricultural Area and Satellite Sensed Cold Area. (Adapted from Chen and others, 1979.)

sensors (reflected solar radiation), and thermal sensors (emitted radiation) could be calibrated to delineate more clearly the areas of organic soil on a global basis. The satellite systems are now available to do the job. The data analysis packages are available to discriminate various surface conditions. It only remains for someone, or some group, to apply the technology: first in an experimental phase, to determine how accurately organic soil surfaces can be distinguished from other surfaces, and then in an operational phase, to determine the areas and drainage status of the soils. Already, on a local scale, some remote sensing technology is being developed using reflected solar radiation and microwave energy. In 1980, a unit called Remote Sensing was formed within the Macauley Institute for Soil Research in Scotland to utilize information that can be obtained from satellite imagery and enhanced by radar coverage from the newer U.S. satellites and the European space programs. Newly acquired electronic stereoplotters, digitizers, and computer hardware are expected to greatly upgrade the efficiency of present Macauley Institute equipment for remote sensing research (West and others, 1981).

Not only areas, but also the drainage condition of organic soils could be estimated from remote sensing based on thermal inertia (sometimes called conductive capacity) of the surface. Thermal inertia = $(\lambda C)^{1/2}$, where λ = thermal conductivity and C

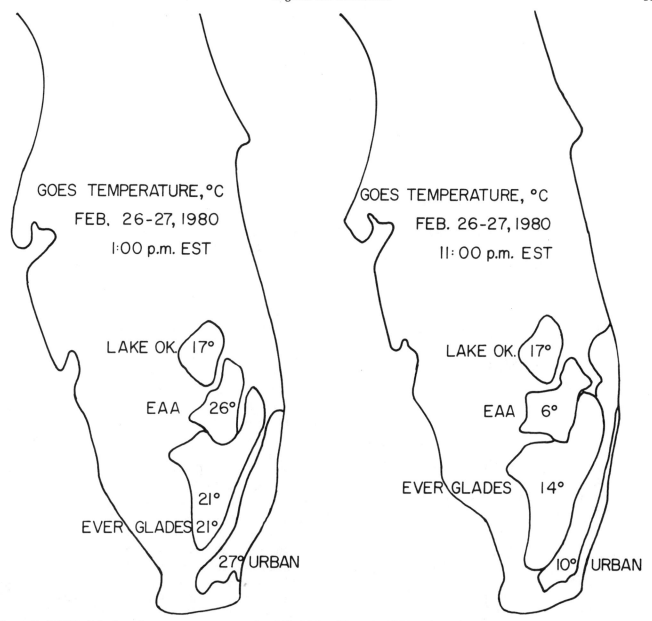

Figure 8. GOES derived surface temperature maps for 1:00 P.M. EST (early afternoon) on February 26, 1980.

Figure 9. GOES derived surface temperature maps for 11:00 p.m. EST (late evening) on February 26, 1980.

= heat capacity. Thermal sensors can be used in twice-daily polar orbiting overpasses, or from hourly data from the GOES, to detect differences in surface temperature from typical early afternoon maxima to predawn minima. Drained organic soils have a low thermal inertia because of (a) low heat capacity due to low water content, and (b) low thermal conductivity of the remaining organic matter. On the other hand, undrained or flooded organic soils have a different (greater) thermal inertia.

A case in point will be discussed. Figure 6 shows a family of lines of thermal inertia. This figure illustrates that surfaces with low thermal inertia would be expected to have higher early afternoon temperatures and lower predawn temperatures than surfaces with high thermal inertia.

Figure 7 shows the southern end of Florida, with four identifiable areas:
1. Lake Okeechobee
2. Drained Everglades Agricultural Area (EAA)
3. Undrained Everglades with shallow organic soils
4. Southeastern Urbanized or Developed area.

A fifth area of mineral soils in Hendry County, just west of the EAA, was not marked on Figure 7.

The next series of three figures, (Figures 8, 9, and 10) show Geostationary Operational Environmental Satellite derived surface temperature maps for three times on February 26-27, 1980, during clear sky conditions. (The maps are distorted slightly by elongation because it was dawn from a computer printer symbol-

TABLE 3. GOES SURFACE TEMPERATURES IN DEGREES CELCIUS IN SOUTH FLORIDA
AT THREE TIMES DURING A DIURNAL CYCLE ON FEBRUARY 26-27, 1980

Time (EST)	Hendry Co.	Lake Okeechobee	EAA	Everglades	Urban/developed
1:00 p.m.	23	17	26	21	27
11:00 p.m.	9	17	6	14	10
5:00 a.m.	8	16	5	13	9
Range[1]	15	1	21	8	18

[1]Difference in surface temperature between 1:00 p.m. EST February 26 and 5:00 a.m. EST February 27, 1980.

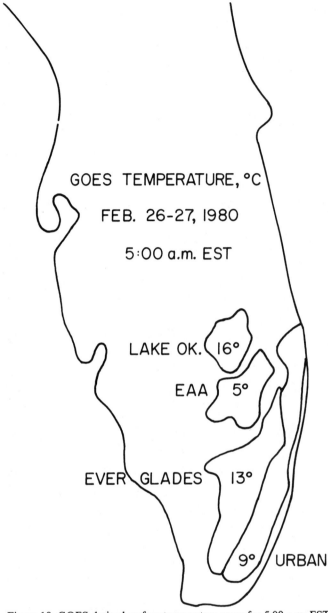

Figure 10. GOES derived surface temperature maps for 5:00 a.m. EST (predawn) on February 27, 1980.

coded map.) The three time periods shown are 1:00 p.m. (early afternoon), 11:00 p.m., and 5:00 a.m. (predawn). The surface temperature information at each time in each of the above identified areas is summarized in Table 3.

During early afternoon on February 26 (1:00 p.m.), the highest temperatures were in the Urban/Developed strip along the Southeast Coast (27° C). The area would be characterized by lower evapotranspiration, but with considerable thermal inertia. The next highest mid-day temperatures were in the Everglades Agricultural Area (26° C). This area is drained and had low thermal inertia in the surface layers. Over the whole area, evapotranspiration can be moderately high because only a part of the sugarcane is harvested by February 26. The next highest temperatures (for comparison here, but not illustrated) were in Hendry County just west of the EAA (23° C). This is a rural area of mineral soil with higher thermal inertia than the EAA. The Everglades surface temperature was lower at 21° C, and Lake Okeechobee was lowest at 17° C. The EAA has higher thermal inertia because of water at or near the surface. Lake Okeechobee has the highest thermal inertia of all, and, of course, has the ability to mix the waters.

By 11:00 p.m., the surface temperature patterns have changed completely. Lake Okeechobee had the highest temperatures (17° C), whereas the EAA had the lowest (6° C). Of the land surfaces, the Everglades (undrained) had changed the least, dropping from 21° C to 14° C, which indicates that it had a much higher thermal inertia than the drained EAA. The mineral soil surface of Hendry County appeared to have a lower thermal inertia than did the undrained Everglades, but greater than the EAA, actually about midway between. The Urban/Developed area had a large temperature drop that was partly due to high temperatures of nonevaporating surfaces during the day.

There may have been a slight latitudinal gradient of temperatures, but the maps show clearly that there was a wide difference in surface thermal responses of the undrained and the drained organic soils.

Also, this example shows how in the future we may be able to detect and quantify areas of drained and undrained organic soils on other parts of the earth. The technology and capability is just around the corner. Predicting rates of subsidence and contribution to the global carbon cycle will be more difficult, but we anticipate that much better estimates can be obtained in the future.

ACKNOWLEDGMENTS

The GOES information was interpreted from thermal data maps obtained through the National Earth Satellite Service (NESS) under the auspices of J. D. Martsolf, Fruit Crops Department, University of Florida.

REFERENCES CITED

Aandahl, A. R., and others, eds., 1974, Histosols, their characteristics, classification, and use: Soil Science Society of American Special Publication Number 6, Madison, Wisconsin, 136 p.

Allison, R. V., 1956, The influence of drainage and cultivation on subsidence of organic soils under conditions of Everglades reclamation: Soil and Crop Science Society of Florida Proceedings, v. 16, p. 21–31.

Allison, R. V., and Clayton, B. S., 1934, Water control investigations: 1934 Annual Report of the Everglades Station, Florida Agricultural Experiment Station, Gainesville, Florida, p. 94–95.

Armentano, R. V., ed., 1980a, Carbon Dioxide Effects Research and Assessment Program-004, The role of organic soils in the world carbon cycle: Report of a Workshop, Indianapolis, Indiana, May 7-8, 1979, U.S. Department of Energy, CONF-7905/35, UC-11, February 1980, 35 p.

—— 1980b, Drainage of organic soils as a factor in the world carbon cycle: Bioscience, v. 30, p. 825–830.

Broadbent, F. E., 1960, Factors influencing the decomposition of organic soils of the California delta: Hilgardia, v. 29, p. 587–612.

Browder, J. A., and Volk, B. G., 1978, Systems model of carbon transformations in soil subsidence: Ecological Modelling, v. 5, p. 269–292.

Chen, E., and others, 1979, Satellite-sensed winter nocturnal temperature patterns of the Everglades agricultural area: Journal of Applied Meteorology, v. 18, p. 992–1002.

—— 1982, Delineation of cold-prone areas using SMS/GOES thermal data: Effects of soil and water: Journal of Applied Meteorology, v. 21 (in press).

Clayton, B. S., 1936, Subsidence of peat soils in Florida: U.S. Department of Agriculture Bureau of Agricultural Engineering, Report No. 1070, (mimeograph).

—— 1938, Subsidence of Florida peat soil: Transactions of the International Society of Soil Science Commun., Zurich, ser. B, v. 6, p. 840–843.

Darby, H. C., 1956, The draining of the fens (second edition): Oxford, Cambridge University Press, 314 p.

Davis, F. E., 1981, Personal communication of South Florida Water Management District data.

Davis, J. F., and Engberg, C. A., 1955, A report on investigation of subsidence of organic soils: Michigan Agricultural Experiment Station, Quarterly Bulletin, v. 37, p. 498–505.

Davis, J. F., and Lucas, R. E., 1959, Organic soils, their formation, distribution, utilization, and management: Special Bulletin 425, Michigan Agricultural Experiment Station, 156 p.

Dawson, J. E., 1956, Organic soils: Advances in Agronomy, v. 8, p. 377–401.

Dickson, K. G., Federico, A. C., and Lutz, J. R., 1978, Water Quality in the Everglades Agricultural Area and its impact on Lake Okeechobee: Technical Publication number 78-3, South Florida Water Management District, West Palm Beach, Florida, 132 p.

Durno, S. E., 1961, Evidence regarding the rate of peat growth: Journal of Ecology, v. 49, p. 347–351.

Duxbury, J. M., and Tate, R. L., III, 1981, The effect of soil depth and crop cover on enzymatic activities in Pahokee muck: Soil Science Society of America Journal, v. 45, p. 322–328.

Farnham, R. S., and Finney, H. R., 1965, Classification of organic soils: Advances in Agronomy, v. 17, p. 115–162.

Fowler, G., 1933, Shrinkage of the peat covered fenlands: Geographical Journal, London, v. 81, p. 495–505.

Given, P. H., and Dickinson, C. H., 1975, Biochemistry and microbiology of peats, in Paul, E. A., and McLaren, A. D., eds., Soil Biochemistry: New York, Marcel Dekker, v. 3, p. 123–212.

Heikurainen, L., 1964, Improvement of forest growth on poorly drained peat soils, in Romberger, J. A., and Mikola, P., eds., International review of forestry research, vol. 1: New York, Academic Press, p. 39–113.

Hutchinson, J. N., 1980, The record of peat wastage in the East Anglican Fens at Holme Post, 1848-1978 A.D.: Journal of Ecology, v. 68, p. 229–249.

Ivitskii, A. I., 1964, Maximum and optimum norms of bog drainage: Gidrotehnika i Melioratsiya, No. 12, 33–42. (Translated from Russian for USDA and NSF in Jerusalem, Israel, 1968. Available from U.S. Department of Commerce Clearing House for Federal Scientific and Technical Information, Springfield, Virginia, 22151.)

Jenny, H., 1930, A study on the influence of climate on the nitrogen and organic matter content of the soil: Research Bulletin 152, Missouri Agriculture Experiment Station.

Jongedyk, H. A., and others, 1950, Subsidence of muck soils in northern Indiana: Station Circular no. 366, Purdue University Agricultural Experiment Station, in cooperation with U.S. Department of Agriculture-Soil Conservation Service, 11 p.

—— 1953, Methods and effects of maintaining different water tables in muck soils: Final report 1941-50, Muck Drainage Studies, Soil Conservation Service-U.S. Department of Agriculture, and Purdue University Agricultural Experiment Station (Mimeograph).

Knipling, E. B., Schroeder, V. N., and Duncan, W. G., 1970, CO_2 evolution from Florida organic soils: Soil and Crop Science Society of Florida Proceedings, v. 30, p. 320–326.

Korpijaakko, E. O., and Radforth, N. W., 1968, Development of certain patterned ground in Muskeg as interpreted from aerial photographs: Proceedings of the 3rd International Peat Congress, Quebec, p. 69–73.

Løddesøl, Aaslv, 1947, Soil Destruction in Norway, address delivered at the Annual Meeting, Norwegian Peat Society, March 10, 1947; English translation by Paul Johnson, Safetykk Av Norsk Geografisk Teddsskrift, bind xl, h5-6, A.W. Brøggers Boktrykker A/S-Oslo.

—— 1949, Soil conversation problems in Norway: Experience paper prepared for Section Meetings, Land Resources 1(b) on Methods of Soil Conservation for the United Nations' Economic and Social Council, E 1 Consf. 2/Sec/W.87, 23 March, 1949.

Lucas, R. E., 1982, Organic Soils (Histosols) Formation, Distribution, Physical and Chemical Properties, and Management for Crop Production: Research Report 435 from the Michigan State University Agricultural Experiment Station and Cooperative Extension Service, East Lansing, in cooperation with the Institute of Food and Agricultural Sciences, Agricultural Experiment Stations, University of Florida, Gainesville, 77 p.

Lutz, J. R., 1977, Water quality and nutrient loadings of the major inflows from the Everglades Agricultural Area to the conservation areas, Southeast Florida: Technical Publication 77-6, South Florida Water Management District, West Palm Beach, Florida, 40 p.

McCollum, S. H., Carlisle, V. W., and Volk, B. G., 1976, Historical and current classification of organic soils in the Florida Everglades: Soil and Crop Science Society of Florida Proceedings, v. 35, p. 173–177.

McDowell, L. L., Stephens, J. C., and Stewart, E. H., 1969, Radiocarbon chronology of the Florida Everglades peat: Soil Science Society of America Proceedings, v. 33, p. 743–745.

Moore, P. D., and Bellamy, D. J., 1974, Peatlands: London, Elek Science, 221 p.

Morton, J. F., and Snyder, G. H., 1976, Aquatic crops vs organic soils subsidence: Proceedings of the Florida State Horticultural Society, v. 89, p. 125–129.

Neller, J. R., 1944, Oxidation loss of lowmoor peats in the field with different water tables: Soil Science, v. 58, p. 195–204.

Parent, L. E., Millette, J. A., and Mehuys, G. R., 1982, Subsidence and erosion of a Histosol: Soil Science Society of America Journal, v. 46, p. 404–408.

Roe, H. B., 1936, A study of depth of ground-water level on yields of crops grown on peat lands: Bulletin 330, University of Minnesota Agricultural Experiment Station, 32 p.

Schothorst, C. J., 1977, Subsidence of low moor peat soil in the western Netherlands: Institute for Land and Water Management Research, Technical Bulletin 102, Wageningen, The Netherlands, p. 265–291.

Shih, S. F., and others, 1979a, Variability of depth to bedrock in Everglades organic soils: Soil and Crop Science Society of Florida Proceedings, v. 38, p. 66–71.

—— 1979b, Subsidence related to land use in the Everglades Agricultural Area:

Transactions of the American Society of Agricultural Engineering, v. 22, p. 561–568.

Skoropanov, S. G., 1961, State of cultivation and fertility of peat bog deposits: Chapter 7 in Reclamation and Cultivation of Peat Bog Soils, Minsk, p. 142–154. (Translated from Russian for USDA and NSF in Jerusalem, Israel, 1968. Available from U.S. Department of Commerce Clearing House for Federal Scientific and Technical Information; Springfield, VA 22151.)

—— 1962, Drainage norms for peat bog soils: Gidrotehnika i Melioratsiya, No. 1, 3440. (Translated from Russian for USDA and NSF in Jerusalem, Israel, 1968. Available from U.S. Department of Commerce Clearing House for Federal Scientific and Technical Information; Springfield, VA 22151.)

Snyder, G. H., and others, 1977, The economic potential for incorporating rice in Everglades vegetable production systems: Proceedings of the Florida State Horticultural Society, v. 90, p. 380–382.

—— 1978, Water table management for organic soil conservation and crop production in the Florida Everglades: Florida Agricultural Experiment Station Bulletin 801, 22 p.

Stephens, J. C., 1969, Peat and muck drainage problems: Journal of the Irrigation and Drainage Division, American Society of Civil Engineering, v. 95, p. 285–305.

Stephens, J. C., Craig, A. L., and Allison, R. V., 1952, Water controlled investigations: Annual Report of the Everglades Station, University of Florida Agricultural Experiment Station, p. 205.

Stephens, J. C., and Johnson, L., 1951, Subsidence of organic soils in the upper Everglades region of Florida: Soil Science Society of Florida Proceedings, v. 9, p. 191–237.

Stephens, J. C., and Speir, W. H., 1969, Subsidence of organic soils in the U.S. of America: International Association of Hydrological Sciences, Colloque De Tokyo, Publication 89, p. 523–534.

Stephens, J. C., and Stewart, E. H., 1976, Effect of climate on organic soil subsidence: In Land Subsidence Symposium, Proceedings of 2nd International Symposium on Land Subsidence, Anaheim, California, December 1976, International Association of Hydrological Sciences, Publication 121, p. 649–655.

Stove, G. C., and Hulme, P. D., 1980, Peat resource mapping in Lewis using remote sensing techniques and automated cartography: International Journal of Remote Sensing, v. 1, p. 319–344.

Tate, R. L., III, 1976, Nitrification in Everglades Histols: a potential role in soil subsidence: In Land Subsidence Symposium, Proceedings of 2nd International Symposium on Land Subsidence, Anaheim, California, December 1976, International Association of Hydrological Sciences, Publication 121, p. 657–663.

—— III, 1979a, Microbial activity in organic soils as affected by soil depth and crop: Applied and Environmental Microbiology, v. 37, p. 1085–1090.

—— 1979b, Effect of flooding on microbial activities in organic soil: carbon metabolism: Soil Science, v. 128, p. 267–273.

—— 1980a, Effect of several environmental parameters on carbon metabolism in Histosols: Microbiological Ecology, v. 5, p. 329–336.

—— 1980b, Microbial oxidation of organic matter of Histosols: Advances in Microbial Ecology, v. 4, p. 169–201.

Terry, R. E., 1980, Nitrogen mineralization in Florida Histosols: Soil Science Society of America Journal, v. 44, p. 747–750.

Terry, R. E., and Tate, R. L., III, 1980, Denitrification as a pathway for nitrate removal from organic soils: Soil Science, v. 129, p. 162–166.

Thompson, K., 1957, Origin and use of the English Peat Fens: Scientific Monthly, v. 85, p. 68–75.

Van der Molen, W. H., 1981, Personal Communication: Agricultural University, Wageningen, The Netherlands.

Volk, B. G., 1973, Everglades Histosol Subsidence: 1. CO_2 evolution as affected by soil type, temperature, and moisture: Soil and Crop Science Society of Florida Proceedings, v. 32, p. 132–135.

Waksman, S. A., and Purvis, E. R., 1932, The influence of moisture upon the rapidity of decomposition of lowmoor peat: Soil Science, v. 34, p. 323–336.

Waksman, S. A., and Stevens, K. R., 1930, Contribution to the chemical composition of peat: V. The role of microorganisms in peat formation and decomposition: Soil Science, v. 28, p. 315–340.

Weir, W. W., 1950, Subsidence of peat lands of the Sacramento-San Joaquin Delta, California: Hilgardia, v. 20, p. 37–56.

West, T. S., and others, 1981, 1979–1980 Annual Report, number 50: The Macauley Institute for Soil Research, Aberdeen, Scotland, AB92QS.

Manuscript Accepted by the Society April 18, 1984

Coal mine subsidence—eastern United States

Richard E. Gray
Robert W. Bruhn
GAI Consultants, Inc.
Monroeville, Pennsylvania 15146

ABSTRACT

Underground coal mining has occurred beneath eight million acres of land in the United States, two million acres of which have been affected by subsidence. Most of this mining has taken place in the eastern half of the United States (east of the 100th meridian) where thousands of acres in urban areas are threatened by subsidence.

Early mining was not as efficient as today. Unrecovered coal pillars, often of variable size and spacing, remain to support the overlying strata for an indefinite period of time after mining has ceased. Roof collapse, crushing of pillars, or punching of pillars into the floor is now resulting in sinkhole or trough subsidence tens or even hundreds of years after mining.

In areas of active mining, where nominal total extraction is practiced, subsidence is essentially contemporaneous with mining. Limited observational data on ground movements over total extraction mines—room and pillar and longwall—suggest subsidence over deep longwall mines in Europe is similar in general respects, but different in detail, to subsidence in the United States.

Ground deformations resulting from subsidence have often been assumed to cause damage to structures in terms of simple tension and compression transferred by friction and adhesion to the undersides of foundations. Differential settlement, intensified pressure on subgrade walls, and other modes of soil-structure interaction are of equal significance in the eastern United States.

INTRODUCTION

Underground mining of coal can result in subsidence of the ground surface and cause:

1. Damage to structures;
2. Modification to surface drainage;
3. Modification to the groundwater regimen.

Coal is found in 37 states and is mined underground in 22 states (HRB Singer, Inc., 1980). Underground mining produced 40 percent of the 830,000,000 tons of U.S. coal mined in 1980 (Keystone, 1981). Underground coal mining in the United States is estimated to eventually cover 40 million acres with eight million already undermined (HRB Singer, Inc., 1977). Subsidence due to mining has affected more than two million acres in 30 states, 1.9 million acres from bituminous coal mining and the remainder primarily from anthracite mining (Johnson and Miller, 1979). Subsidence from other metal and nonmetal underground mines has affected only 17,000 acres (Johnson and Miller, 1979).

Most underground coal mining has occurred in the eastern United States, where the U.S. Bureau of Mines estimates that 398,600 acres in urban areas in 18 states are threatened by subsidence (Table 1) (Johnson and Miller, 1979). Pennsylvania and West Virginia contain 60 percent of the total threatened areas. The U.S. Bureau of Mines estimates that the total cost to stabilize the estimated 398,600 urban acres located above abandoned mines potentially subject to subsidence is almost $12 billion (1978 costs). Areas undermined by efficient, total extraction methods offer little potential for future subsidence.

This paper reviews the extent of underground coal mining in the eastern United States,[1] methods of mining as they relate to subsidence, the mechanisms of subsidence over active and aban-

---

[1]The eastern United States for the purposes of this paper is defined as that area located east of the meridian passing through central Kansas. Included in this region are the Eastern, Interior, and Gulf Coal Provinces.

TABLE 1. POTENTIAL SUBSIDENCE IN EASTERN URBAN AREAS[1]

State	Acres Threatened[2]	Estimated Cost to Prevent Subsidence, Million Dollars[3]
Alabama	11,700	351
Arkansas	1,300	39
Illinois	41,800	1,254
Indiana	10,900	327
Iowa	3,800	114
Kansas	2,900	87
Kentucky	37,200	1,116
Maryland	2,900	87
Michigan	400	12
Missouri	2,500	75
Ohio	21,800	654
Oklahoma	1,700	51
Pennsylvania, bituminous	97,000	2,910
Pennsylvania, anthracite	54,400	1,632
Tennessee	5,000	150
Texas	800	24
Virginia	13,400	402
West Virginia	89,100	2,673
Totals	398,600	11,958

[1]Estimated undermined acres in urban areas that have not been affected by subsidence.
[2]Acres threatened for individual states are estimates based on the amount of underground production from each state through 1975.
[3]Based on an average of $30,000 per acre.
Ref: Modified from Johnson and Miller (1979).

doned mines, partial mining methods to prevent subsidence, and similarities and differences between subsidence in the eastern United States and Europe.

DEVELOPMENT OF EASTERN COAL FIELDS

Although explorers discovered coal in the Appalachian and Illinois regions at least as early as the 1670's, the early American economy was based primarily upon agriculture, and the utilization of coal was minimal. Ironworks and forges operating during that period were fueled with charcoal.

As America became more industrialized following the Revolution, anthracite discovered in eastern Pennsylvania in 1791 became important in the smelting of iron ore (Gregory, 1980).

After the Civil War (1865), coke produced from bituminous coal became increasingly important in iron processing, causing coal production to grow at an accelerated rate, particularly in western Pennsylvania (Gregory, 1980). In the early 18th century, most coal was imported from England and Nova Scotia. However, the total quantity used yearly prior to 1776 did not exceed 9,000 tons. In 1800, only two cities, Pittsburgh, Pennsylvania, and Richmond, Virginia, were using coal to any extent for domestic purposes. The amount of coal used in Philadelphia, New York, and Boston did not exceed the amount of wood until many years later (Eavenson, 1942). Although coal mining started about 1700 in Virginia (first mined commercially near Richmond in 1745) and mining of the famous Pittsburgh seam started in 1759, few reports on coal production were made prior to 1870 and accurate records were not kept until after 1885 (Eavenson, 1942).

Figure 1 shows the location of coal fields in the eastern United States. These fields are grouped into the Eastern Coal Province, the Interior Coal Province, and the Gulf Coal Province. Table 2 summarizes information on mining in each coal province. Detailed descriptions of seams by state are contained in the Keystone Coal Industry Manual (1981).

The bulk of underground mining has occurred in Pennsylvania, West Virginia, Kentucky, Ohio, and Illinois (Table 2). In 1917, underground mining accounted for 99 percent of total U.S. coal production. By 1969, underground mining produced about 66 percent of U.S. coal and accounted for 89 percent of total cumulative U.S. production (Averitt, 1970). In 1981, underground mining accounted for only 39 percent of U.S. coal production (Department of Energy, 1982). Subsidence is a potential problem in all areas of underground mining and, as indicated by Table 1, potentially most severe in the five states (listed above) having the most underground mining.

METHODS OF UNDERGROUND MINING

Methods of extraction vary depending upon subsurface conditions and sensitivity of the surface to mine-related disturbances. The coal mining industry employs two basic methods of extraction: longwall mining (popular in European coalfields) and room and pillar mining (used extensively in the United States) (Gray and others, 1974). Either system can be, and usually is, extensively modified for local mining conditions.

Room and Pillar Mining

Room and pillar mining originated as a method of extracting as much coal as possible while still providing roof control by means of coal pillars. During the 18th and 19th centuries, mines were small hand-operations under shallow cover. Coal was cheap and the spacing, size, and regularity of pillars were somewhat arbitrary. Coal pillars were left in place as a matter of convenience and safety to the miners. Increased production by the mid- to late-19th century brought mechanization and ventilation requirements to mines that necessitated a systematic arrangement of pillars, but still resulted in considerable coal being left underground. Mining often extended to where the overburden was only 25 feet thick. Early extraction ratios, the proportion of coal removed, averaged 30 to 40 percent. Since coal deposits were widespread and accessible, little effort was made to improve extraction ratios.

In the latter part of the 19th century, total extraction mining was initiated to achieve greater production of the coal that was becoming increasingly valued for its coking properties by the steel industry and was first implemented in partial extraction mines of the day. The distinction from partial extraction mines was that the long, narrow pillars left between rooms during first mining were now extracted in a second stage of mining. Subsidence of the ground surface in a properly executed operation took place contemporaneously with pillar extraction.

For a long while, wide rooms and narrow pillars (10 to 15 feet wide) such as those shown in Figure 2, were employed in

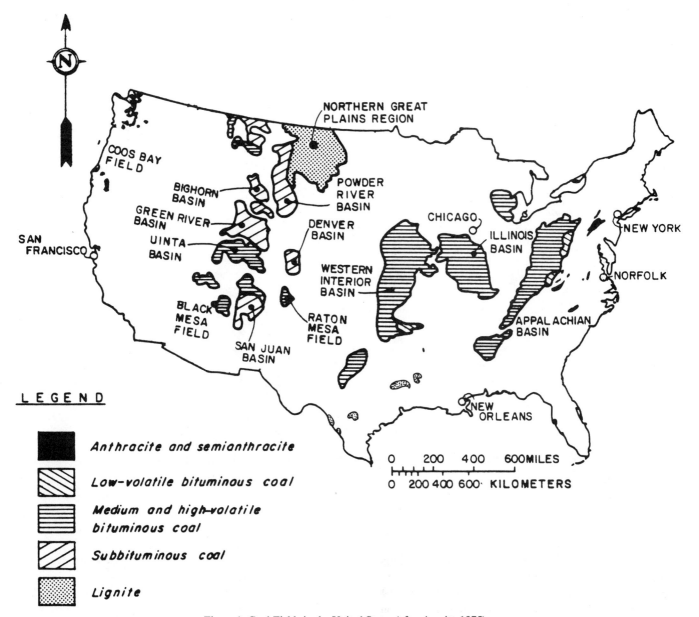

Figure 1. Coal Fields in the United States (after Averitt, 1975).

total extraction mines because it was believed that more lump coal could be produced by room mining than by extracting the pillars left between rooms. However, by the 1920's, block systems of mining came into favor, wherein square or rectangular pillows 50 to 100 feet on a side (Figure 3) were separated by narrow rooms and entries, reducing roof deterioration, roof falls, and support problems during pillar extraction (Paul and Plein, 1935). From 1948 to 1952, most remaining mines of the old pattern were converted to the block system as continuous mining machines were introduced on a large scale. Subsequently, as break lines controlling failure of the mine roof parallel to the pillar faces replaced angled break lines, the transition to the relatively efficient pillar extraction methods of today was essentially complete.

Most modern room and pillar mining has been modified from the original form and is classified as either panel mining or partial extraction mining. The configuration of main entries depends upon the depth and orientation of the coal seam and may consist of drifts, slopes, or vertical shafts. Commonly, in bituminous coal fields, main entries consisting of four to nine parallel headings protected by chain pillars are advanced to the far boundary of the property. Three to five parallel headings also protected by chain pillars are then driven at right angles from the main entries at intervals of a few thousand feet to provide further access to the mine property. Additional series of parallel entries on this orthogonal pattern further subdivide the property into panels, from which coal is extracted by driving rooms separated by pillars into the coal followed by removal of the pillars themselves. The size and design of rooms and pillars depend upon

TABLE 2. EASTERN COAL PROVINCE MINING SUMMARY

Coal Province	Coal Region and Fields	Coal Rank	Coal Age	Location	Approx. Total Production Tons	Comments
Eastern	Rhode Island	Meta-Anthracite	Pennsylvanian	Rhode Island	-	Several seams have been locally mined - Rocks are structurally complex and affected by igneous and metamorphic activity.
	Pennsylvania Anthracite Region 4 Fields- Northern Eastern Middle Western Middle Southern	Anthracite	Pennsylvanian	Pennsylvania	4.6 Billion	Extensive mining with sites having as many as eleven mined seams. Dips greater than 60° are common around the rims of the four synclinal fields. Current production largely from surface operations.
	Atlantic Coast Region 3 Fields - Richmond Farmville Deep River		Triassic	Virginia Virginia N. Carolina	8 Million	Mining was initiated in the early 1700s in the Richmond Field. Production has been mostly for local use.
	Appalachian Region	Bituminous	Mississippian to Permian	Alabama Maryland West Va. Ohio Kentucky (East) Virginia Tennessee Georgia Pennsylvania	8.7 Billion 3 Billion 2.8 Billion 1.4 Billion 9.3 Billion	Many seams which generally dip less than 2°. Mining was initiated in the mid-1700s. More excellent coking coal than anywhere in the world.
Interior	Michigan Basin	Bituminous	Pennsylvanian	Michigan		Local mining near Saginaw Bay.
	Eastern Region	Bituminous	Pennsylvanian	Illinois Indiana (S.W.) Kentucky (West)	4.5 Billion 1.5 Billion	Coals are of lower rank and quality than in the Appalachian Region - seams are flat-lying. Extensive underground mining occurred in Illinois from the 1870s until the 1930s. Strip mining then became significant and now accounts for nearly half of the annual 60 million ton coal production. Almost all of Indiana's current production is by strip mining.
	Western Region	Bituminous	Pensylvanian and Permian	Iowa Missouri Kansas	330 Million	Coals are not as high quality as in the Eastern Region. Seams are relatively flat. All current production is from surface mines.
		Bituminous to Anthracite	Pennsylvanian	Oklahoma Arkansas	105 Million	Folding has increased rank of some coal to anthracite. Most production is bituminous coal. Beds dipping up to 65° have been mined but most production has been from coals dipping less than 5°.
Gulf		Lignite	Tertiary	Texas Arkansas Louisiana Mississippi Alabama		Flat-lying beds. Production is almost all by surface mining.

Ref: Keystone Coal Industry Manual, 1981
U.S. Bureau of Mines Mineral Yearbook, 1976
Ashley, 1928

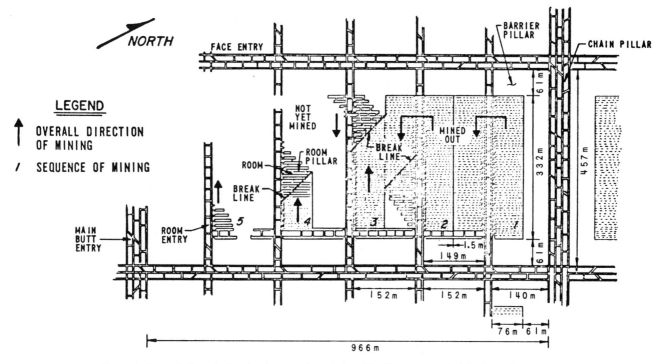

Figure 2. General plan of mine development by narrow pillar line common in Allegheny County, Pa.; Half advance-Half retreat method of pillar extraction with short break line (after Paul and Plein, 1935).

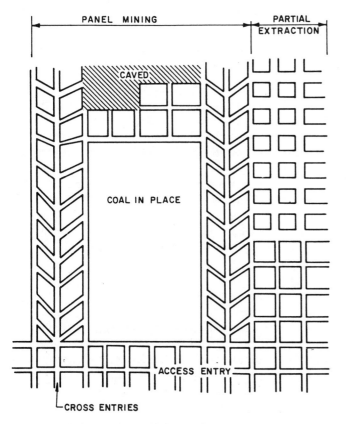

Figure 3. Example of room and pillar mining (Gray and others, 1974).

local conditions and the working design of the mine (Kauffman and others, 1981).

Partial mining is utilized where subsidence prevention is required or where subsurface conditions are such that extraction of pillars would result in roof support problems at mine level. Where partial mining is practiced, a pattern of coal pillars is left in place to provide overburden support either in selected portions of the mine or over the entire mine property. In the bituminous coal fields of western Pennsylvania, selective protection is conventionally provided for gas and oil wells and mine shafts and rather commonly for surface developments. Partial mining with 55 to 65 percent extraction is commonly practiced in eastern Ohio and West Virginia's northern panhandle to avoid "squeezes" at mine level brought about by strong limestone roof strata that are not easily broken and consequently present difficulties in roof support if total extraction were to be attempted. In the flat prairie lands of the midwest where subsidence can result in ponding of water on valuable farmland, state legislation or restricting legal agreements in certain areas dictate the use of partial mining. In all cases, pillars are designed on the basis of geologic conditions, mining depth, and local factors. Average extraction ratios in the United States range from 40 to 60 percent for this system (Cassidy, 1973).

Total extraction mining results in nearly complete removal of coal in a panel. Total extraction is accomplished by systematically removing the coal pillars from one end of the panel to the other, at the same time allowing the mine roof to collapse to the floor. The sequence and rate of pillar extraction are critical in order to maintain a safe distance between the area of collapsing overburden and the development of new entries being mined. Extraction reaches 80 to 90 percent of the seam (Cassidy, 1973).

Longwall Mining

Longwall mining techniques are designed for a one-stage operation in which complete extraction is achieved concurrent to closure of the mined area. This technique is widely used outside of the United States because the limited distribution and accessibility of coal reserves has resulted in mining at considerable depth where room and pillar techniques experience significant roof control problems. Longwall mining was tried in the United States prior to 1900, but the labor costs involved with manually moving the supports led to its abandonment. Following the development of self-advancing props, longwall mining was reintroduced to the United States in 1960 (Poad, 1977). Longwall or modified longwall mining is gaining popularity in this country but accounts for less than 15 percent of the present total underground coal production (Conroy, 1980).

The development of main entries for access and ventilation is very similar to the development phase in room and pillar mining. If retreat mining is to be used, groups of two or three parallel entries are driven perpendicular to the main entry, on either side of the proposed extraction area. These entries are separated by roof support pillars and serve as access, haulage, and ventilation entries (Figure 4). The width of the extraction area varies from 600 to 800 feet, while the length of extraction, controlled by the logistics of the mining operations or the limits of the mining area, may be on the order of 3,000 feet (Anonymous, 1965).

When these secondary entries have been advanced to the desired lengths, they are joined by an entry driven across the extraction area to form an extraction face. This face is then equipped with a row of hydraulic roof supports, a coal conveyor, and a machine to break the coal from the face. The system is designed to support only the coal face and allow caving behind the supports in the mined-out area, with the support system and conveyor automatically advanced as mining proceeds. Mining progresses towards the main entry and achieves a high percentage extraction. Pillars supporting the entries are generally not recoverable with longwall mining. In Europe, single entries are used with longwall mining. In the United States, multiple entries are required by law and extraction percentages are comparable with those obtained by room and pillar panel mining. The use of single entry systems, which can recover 90 to 95 percent of the seam, have been investigated by the Bureau of Mines (Poad, 1977).

Advancing longwall mining utilizes the same basic design, but with the mining proceeding away from the main entry. Access and haulage entries must be developed concurrently with the advance of the extraction face. Advancing methods have the advantage of immediate resource recovery to offset development costs, but tend to experience difficulty in maintenance of haulageways affected by subsiding overburden. Retreat mining

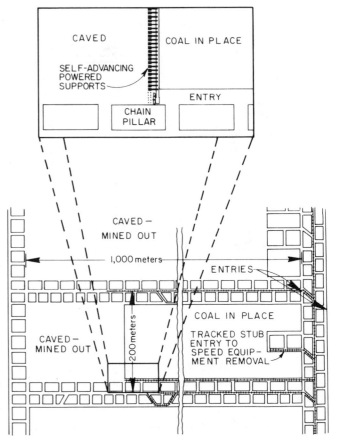

Figure 4. Example of longwall mining (Coal Age, Feb. 1965).

requires more extensive initial development, but results in no subsidence between the face and the main entry, and thus all access entries are constantly surrounded by unmined, stable coal.

Although longwall mining achieves high extraction rates within the extraction area, it has several disadvantages that have delayed its adoption over present room and pillar techniques. The equipment used on the longwall face is not adaptable for driving entries or for selective mining on a small scale. Thus, two equipment systems are required. For the same reason, selective ground surface support with this system involves either the modification of the mine design or the use of other equipment in those areas requiring support (Hunter, 1972).

"Stope" Mining in Steeply Dipping Seams

A modified room and pillar type of mining has been used in mining steeply dipping coal seams in the anthracite region of Pennsylvania (Jones and Hunt, 1954). A haulageway entry (gangway) and a ventilation entry are driven horizontally along the strike of the dipping coal seam. From the haulageway, inclined shafts or chutes are driven upwards to working rooms (breasts or chambers), as illustrated in Figure 5. Cross drifts connect the ventilation entries with the rooms. Coal pillars are left for support between the mine rooms along the strike of the coal bed.

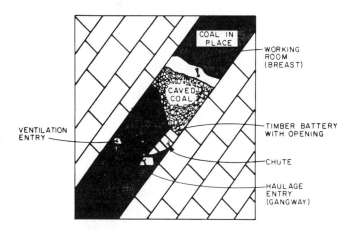

Figure 5. Example cross-section of "stope" mining in steeply dipping seams (Gray and others, 1974).

The allowable dimensions of the rooms and pillars depend upon the condition of the coal, roof rock and floor rock, and the thickness of overburden.

A timber battery controls the flow of coal into the chute for loading. The rooms are kept nearly full of loose mined and caved coal to provide a place for the miners to stand while working at the face. Several modifications of this method have been used for mining thin seams or for multiple seams. As mining progresses, pillars are mined or "robbed" in one or more operations. The amount of extraction varies depending upon conditions, but in some cases reaches nearly complete extraction when the rooms are backfilled, allowing mining of the pillars.

SUBSIDENCE ASSOCIATED WITH ACTIVE MINING

Subsidence is a time-dependent deformation of the ground surface resulting from readjustment of the overburden above a mine. Although the vertical components of movement are usually largest, horizontal movements and the resulting strains and displacements are often most significant in causing surface damage (Bräuner, 1973). Some movements take place during mining and some after, depending on the type and extent of mining, the thickness and character of the overburden and the mine floor, and other details of the site. Areally, the movements cover from a few square feet to many acres and vertically, from a few inches to several feet.

With active mines, the amount and areal extent of subsidence can be reasonably estimated and surface developments designed to minimize the impact. Also, partial mining can be utilized so that coal pillars left in place will support the ground surface; or construction of surface structures can be delayed until subsidence is complete.

Total Extraction

The following discussion deals with essentially flat-lying seams and moderate surface topography, with mining deep enough—approximately 100 feet—that sinkholes do not form. Mining a small area underground initially results in small deformations of the mine roof and imperceptible movements at the ground surface. As the area of extraction increases, measurable movements occur at the ground surface. Although this subsidence involves primarily downward vertical movements, upward vertical and horizontal movements may also occur. The surface expression of these movements is a large, shallow depression in the shape of a trough or basin. In most cases, the surface area affected by subsidence exceeds the area of the seam extracted. The following parameters relate geometry of the mined area to the geometry of the surface area affected by subsidence. These parameters, which have been extensively studied in Europe, constitute a set of subsidence principles that are basic to the understanding of subsidence movements.

Critical Width. Of fundamental importance to an understanding of the basic mechanics of subsidence is the concept of critical width. Given a large enough extraction area, there exists a critical width of extraction at which subsidence at the center of the trough will reach a maximum possible value. Extraction of an area of less than this critical width, subcritical extraction, produces a subsidence trough with vertical movements less than the maximum possible value. Alternately, extraction of the seam over an area of width larger than the critical width results in an essentially flat-bottomed subsidence trough with the flat portion of the area subsiding a vertical distance equal to the maximum possible value. These conditions are illustrated in Figure 6.

In addition to being dependent upon a critical width of

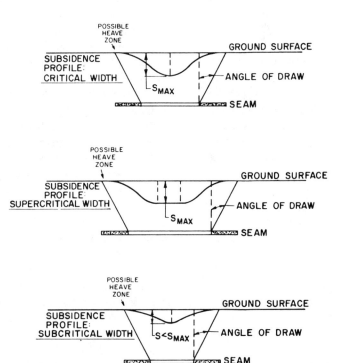

Figure 6. Critical/supercritical/subcritical widths (Gray and others, 1974).

extraction, the maximum possible vertical movements are also a function of the mined thickness and the method of support remaining within the extraction area. This relationship is simply expressed as the maximum vertical movement being equal to the seam thickness multiplied by a coefficient depending upon the type of support remaining and local geologic conditions.

Width/Depth Ratio. The width of the extraction area divided by the depth of the seam beneath the surface is defined as the width/depth ratio. Width is used to define a horizontal dimension of the extraction area. A rectangular extraction area will have a width/depth ratio corresponding to each of its two plan dimensions.

Studies have shown that for maximum subsidence to occur at a particular point on the ground surface, the width/depth ratio of the extraction must exceed a certain minimum or critical value in both plan dimensions. It follows that as the depth of the seam being mined increases, the size of the critical extraction area increases. For specific local conditions, the critical width/depth ratio is generally considered a constant. In Europe and Great Britain, its value has been empirically determined to lie between 1.0 to 1.4 (Wardell, 1969). Values near 1.0 are associated with thick sandstone while values of 1.4 are more common where the overburden is predominantly interbedded shales and siltstones without thick sandstones. Karmis and others (1981) report a value of 1.2 while Peng and Cheng (1981) suggest a value of 1.1 in the Northern Appalachian Coal Field where thick sandstones are common. Critical dimensions have not been reported for the Interior Coal Province.

Angle of Draw. The angle of inclination of a line connecting the edge of the extraction area with the limit of the surface area affected by subsidence is called the angle of draw, or limit angle. British and United States' convention is to measure the angle of draw from a vertical reference line while European measurements are from the horizontal.

Since the surface area affected by subsidence is generally larger than the area of the seam extracted, angles of draw are usually positive. Observed values in Europe and Britain, measured from the vertical, vary between 19 and 45 degrees (Abel and Lee, 1980). United States' data are limited. Angles of draw on the order of 10 to 25 degrees have been reported in Southwestern Pennsylvania (Newhall and Plein, 1936; Maize and Greenwald, 1939; Abel and Lee, 1980). Reported values from Illinois range from 12 to 26 degrees (Bauer and Hunt, 1981).

The angle of draw, in the United States, appears generally to range from 20 to 35 degrees, and most commonly from 20 to 28 degrees. The angle of draw for pillar retreat panels may be a few degrees higher than for longwall panels, but the data are not definitive. Kohli and others (1980) report that angles of draw in excess of 28 degrees for longwalls in the Northern Appalachian Coal Field were generally measured adjacent to gob areas mined by the room and pillar method. Adamek and Jeran (1981) suspect that locally the angle of draw in the Northern Appalachian Coal Field may be as small as 12 to 17 degrees, owing to the bridging of competent strata in the upper overburden interval.

This is based on a comparison of measured and predicted ground profiles using Bal's theory.

Observed differences in the angle of draw for similar mining conditions have generally been explained by differences in subsurface conditions. Large angles of draw have been observed in areas containing thick deposits of unconsolidated materials, while smaller angles of draw are reported in areas where many thick, competent rock strata are present above the mine (Zwartendyk, 1971).

In the United States, overburden composition by rock type generally has not been found to bear a strong relationship to angle of draw, although Abel and Lee (1980) claim a relationship exists between angle of draw and percent sandstone and limestone. The general absence of a trend between angle of draw and rock type may be due to within-mine lithologic departures from boring log data, the failure to account for the position of the strata in the rock column, as well as poor field data.

Little specific information concerning soil cover or near surface rock strata is available for most reported subsidence cases. Because soil is commonly less than 20 feet thick in the Northern Appalachian Coal Field, its presence is not expected to have been a major influence overall on angles of draw that have been reported.

Surface Movements. The following parameters describe distortions of the ground surface associated with subsidence. These distortions are the effects of particular importance from the standpoint of damage to structures.

Vertical Displacement. Vertical displacements occur in all portions of a subsidence trough but achieve maximum values only above the central portion of the extraction area. Damage resulting from vertical movements is primarily a function of differential movements rather than the magnitude of displacement. Relative vertical movements are illustrated by the subsidence profile (Figure 6).

Tilting. Differential vertical movements within the subsidence trough result in a tilting of the ground surface. If this tilt is uniform, it produces little surface damage except for the resulting nonlevel surface. Differential tilt, however, results in a curved surface with a much higher potential for surface damage. These vertical movements are illustrated by the curved and sloping portions of the subsidence profile (Figure 6).

Horizontal Displacement and Strain. Associated with subsidence trough development are horizontal displacements of the ground surface. Differential horizontal displacement rather than uniform horizontal displacement produces damage to surface structures. The relative magnitude of displacement is proportional to the maximum vertical displacement.

Horizontal displacements during subsidence result in a shortening or lengthening of the ground surface. These displacements are conventionally expressed in terms of strain or change in length per unit length. Idealized distributions of horizontal displacements and strains are illustrated for critical, subcritical, and supercritical extraction ratios in Figure 7, 8, and 9, respectively. Tensile and compressive strains correspond to concave down-

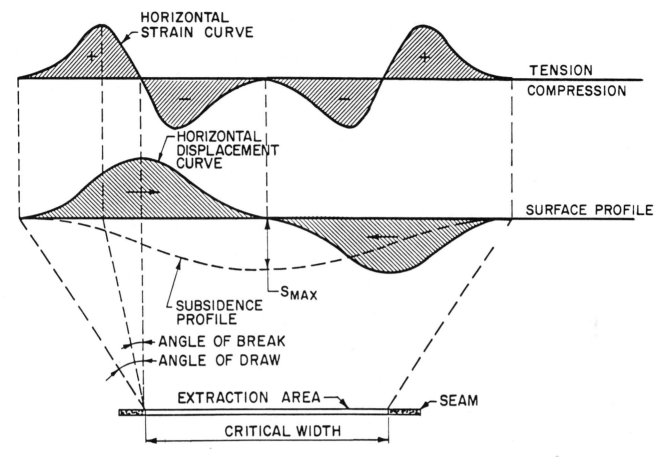

Figure 7. Strain and horizontal displacement-Critical width (Gray and others, 1974).

ward and concave upward portions of the subsidence profile, respectively. It has been shown that these strains are proportional to the curvature of the profile, with the maximum strain occurring where curvature is maximum (Zwartendyk, 1971). The transition point of the trough profile is the point of maximum horizontal displacement which undergoes no strain, since no profile curvature occurs at this point. A point of maximum extension exists within the area of tensile strain, indicating maximum differential horizontal displacement, and is often associated with extensive surface damage.

Time Effects. Almost no movement is associated with the advance phase of mining in a retreat pillar operation. The previously discussed concept of the subsidence trough is an idealized representation of the final state of the ground surface after mining-induced movements have ceased. The trough does not form during one period of vertical displacement, but evolves from the subcritical stage through the critical state to the supercritical as the mined-out area is progressively enlarged. If mining activity ceases before this evolution is complete, the shape of the trough will be determined by the parameters prevailing at the cessation of mining. This evolution is manifested as the migration of one end of the trough, in the direction of mining, accompanied by the tensile and compressive strains described in Figures 7, 8, and 9. Where nominal total extraction is practiced, subsidence is essentially contemporaneous with mining. Wardell (1953) reported that on the average, 95 percent of the total subsidence occurs while the mine face is within the critical area. Subsidence development curves for four pillar retreat mines in southwestern Pennsylvania (Figure 10) indicate:

1. Subsidence is commonly 90 percent complete and progressing at a much diminished rate when the face has advanced 0.75h to 1.0h beyond the monitoring point.
2. Subsidence is virtually complete when the face has advanced 1.25h to 1.75h beyond the monitoring point (Bruhn, and others, 1982).

Very similar characteristics have been reported for longwall mines in the Northern Appalachian Coal Field (Cheng and Peng, 1981; Kohli and others, 1980). Seam depth and rate of face advance control the time-dependent parameters (Cheng and Peng, 1981). Residual movements are generally complete within two years after mining.

Subsidence Experience—Total Extraction

Subsidence was not considered a significant problem in the early years of United States coal mining since coal was mined in largely agricultural areas (Gray and Meyers, 1970). The earliest surface subsidence investigations date to the early part of this

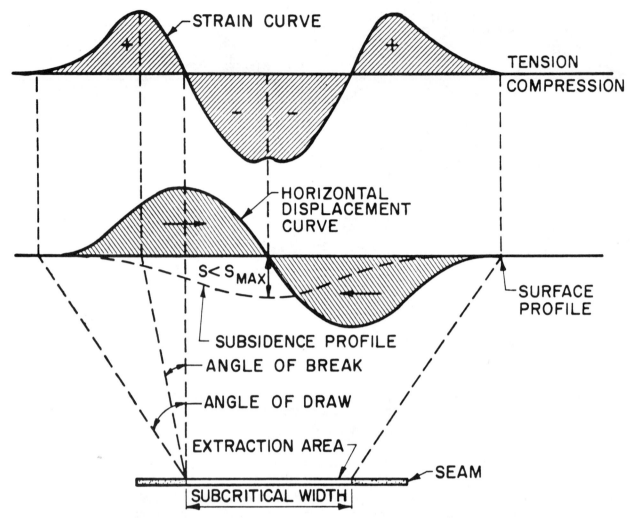

Figure 8. Strain and horizontal displacement-Subcritical width (Gray and others, 1974).

century when Young and Stoek (1916) described subsidence-related problems in several coal fields. In 1927, Herbert and Rutledge published accounts of mine subsidence at several locations in Illinois. From 1931 to 1934, subsidence monitoring in western Pennsylvania included observations on damage to structures (Newhall and Plein, 1936). In 1935, Greenwald and others began a series of systematic measurements of subsidence above mines in the Pittsburgh Coal (Greenwald and others, 1937; Maize and Greenwald, 1939; Maize and others, 1941).

A number of researchers have made notable strides in recent years towards improving the state of the art in subsidence engineering. In connection with the testing of a three-dimensional, elastic-plastic finite element analysis, Dahl and Choi (1975) monitored surface subsidence above a room and pillar mine in southwestern Pennsylvania and above a longwall operation in northern West Virginia. Perhaps the most interesting conclusions drawn by these researchers are that surface topography plays a minor role in affecting subsidence and that subsidence is essentially time independent for longwall mining. Kohli and others (1980) working in the same area agree topography is not a significant factor. Conroy and Gyarmaty (1982) report that in Ohio horizontal movement is influenced by topographic effects. The general applicability of these conclusions to other geologic and mining conditions in the United States has not yet been firmly established.

In a cooperative agreement between Old Ben Coal Company and the U.S. Bureau of Mines, subsidence monitoring was undertaken at the first modern successful longwall demonstration operation in Illinois, at the Old Ben Mine No. 24 near Benton in Franklin County. Conroy (1978 and 1980) and Wade and Conroy (1980) report the findings from mining of the first of two panels.

Investigations of differential strata movements and water table fluctuations during longwall operations at the Bethlehem Steel Corporation's Somerset Mine No. 60 in southwestern Pennsylvania, conducted by Columbia University under a contract with the Bureau of Mines, are of interest primarily for the findings on the influence of mining on groundwater levels. In this study, the fracture of impermeable shale layers, caused by passing of the longwall face, was related to sudden drops in water level in a

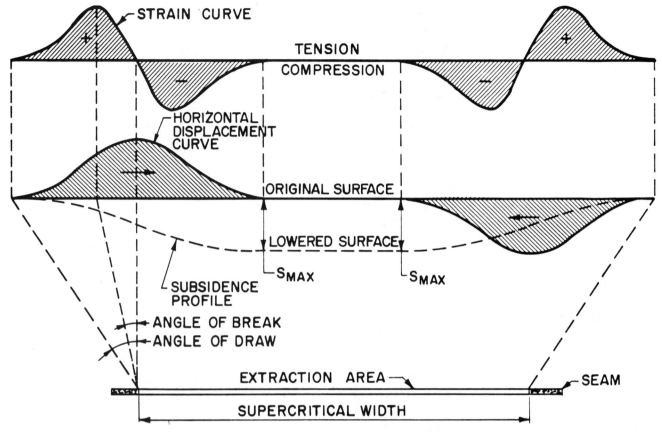

Figure 9. Strain and horizontal displacement-Supercritical width (Gray and others, 1974).

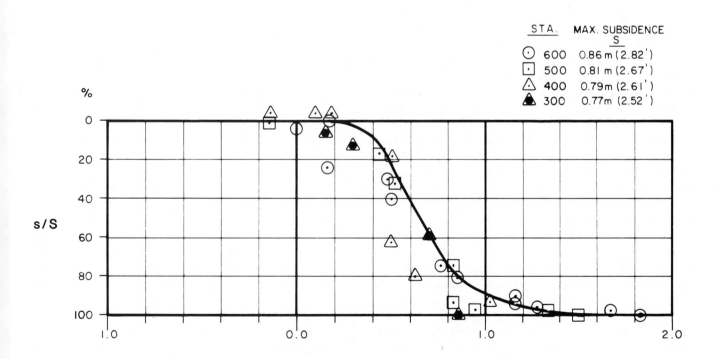

Figure 10. Typical subsidence development curve, Southwestern Pennsylvania.

sandstone aquifer above the impermeable shale (Barla and Boshkov, 1978). A similar phenomenon might occur in the coal measures of other mining districts, even though the stratigraphy might be different.

A comprehensive research effort in the area of subsidence above longwall and room and pillar mining in the Northern Appalachian Coal Region is that executed under contract between the Department of Mining Engineering, West Virginia University, and the U.S. Bureau of Mines. The work involves a compilation and analysis of subsidence data collected by individual coal companies during the past 20 years (presumably from mines in the Pittsburgh Coal seam) (Kohli and others, 1980). Preliminary analyses of the longwall data indicate the following:

1. The angle of draw ranges from 21 to 30 degrees with a few exceptions.
2. The subsidence factor ranges from 0.22 to 0.7 and increases with seam depth.
3. The subsidence profiles can be approximated and predicted by a simple profile function in most cases.
4. The time dependent or residual subsidence is less than 13 percent of the total subsidence.
5. Surface topography or surface slopes do not seem to have any effect on surface subsidence.
6. In multipanel mining, interpanel effect will add from 15 to 33 percent to the maximum possible subsidence obtained in single panel mining.

Although the ground surface survey data on which the above conclusions are based are not as complete as would be desired, the regional nature of the study and the collective, observational approach to the problem are clearly the most direct routes to usable, practical solutions to improved technology in subsidence engineering. Cheng and Peng (1981) analyzed the data from Kohli and others (1980) plus some new data using the approach of the eastern European countries; i.e., identification of mathematical functions for subsidence profile, slope curvature, horizontal displacement and strain, the relationship among functions and time effects.

A summary of known subsidence surveys conducted above active mines in the Eastern and Interior Coal Provinces is shown in Table 3. In a few of these cases, underground rock mechanics studies, involving in-mine instrumentation, have also been conducted (Federal No. 2 Mine, Old Ben Mine No. 24, Somerset No. 60 Mine). Mine level convergence measurements have been made in a few of the older room and pillar mines (Crucible Mine, Montour No. 10 Mine).

Although observational data on ground movements over total extraction mines—room and pillar and longwall—in the United States is limited, they suggest that subsidence over deep longwall mines in Europe is similar in general respects, but different in detail to subsidence in the eastern United States (Munson and Eichfeld, 1980). Available data on subsidence parameters for mines in the Northern Appalachian Coal Field and Illinois Basin are summarized in Table 4 (Bruhn and others, 1982).

Munson and Eichfeld (1980) compared subsidence measurements in the Illinois Basin with European prediction methods. The National Coal Board methods (National Coal Board, 1975) correctly predicted the total amount of subsidence for the subcritical trough investigated, but predicted a much smaller maximum strain even though the geology of the Illinois Basin is probably the most similar of U.S. coal mining areas to that of the United Kingdom (O'Rourke and Turner, 1979). The significant difference in strain is attributed to prior multiple seam mining in the United Kingdom, which has "softened" the ground compared with the single seam U.S. mine in virgin ground (Munson and Eichfeld, 1980).

The Eastern Coal Provinces contain directional, residual horizontal stresses that may exceed the vertical stress by factors of

TABLE 3. GROUND SURVEYS CONDUCTED ABOVE ACTIVE MINES IN THE EASTERN AND INTERIOR COAL PROVINCES

Room and Pillar Mining
1. Bethlehem Mine No. 60 (Bethlehem Mines Corporation), Washington County, Pennsylvania. Date of Survey: 1967-69. Source: Pennsylvania DER-MSR.
2. Clyde Mine (Republic Steel Corporation), Washington Township, Greene County, Pennsylvania. Date of Survey: 1968-69. Source: Pennsylvania DER-MSR.
3. Crucible Mine (Crucible Fuel Company), Cumberland Township, Greene County, Pennsylvania. Date of Survey: 1937-38. Source: U.S. Bureau of Mines RI 3452, 1939.
4. Humphrey No. 7 Mine (Consolidation Coal), near Mt. Morris, Greene County, Pennsylvania. Date of Survey: 1970-71. Source: Proc. 15th Symposium on Rock Mechanics, Sept. 1973, ASCE, 1975 (Dahl and Choi, 1973).
5. Maple Creek Mine (U.S. Steel Corporation), Somerset Township, Washington County, Pennsylvania. Date of Survey: 1970-74. Source: Pennsylvania DER-MSR.
6. Mathies Mine (Mathies Coal Company), North Strabane Township, Washington, Washington County, Pennsylvania. Date of Survey: 1969-72. Source: Pennsylvania DER-MSR.
7. Montour No. 10 Mine (Pittsburgh Coal Company), South Park Township, Washington County, Pennsylvania. Date of Survey: 1936. Source: Bureau of Mines RI 3355, 1937.
8. Nemacolin Mine (Buckeye Coal Company), Cumberland Township, Greene County, Pennsylvania. Date of Survey: 1938-40. Source: U.S. Bureau of Mines RI 3562, 1941.
9. Republic Mine (Republic Steel Corporation), Redstone Township, Fayette County, Pennsylvania. Date of Survey: 1931-32. Source: Transactions AIME, Volume 110, pp. 58-94, 1936.
10. Mine not identified (Bethlehem Mines Corporation), Cambria County, Pennsylvania. Date of Survey: Unknown. Source: Pennsylvania DER-MSR (Ebensburg Reservoir).
11. Two mine operations (Mining Companies Unknown), Northern Appalachian Coal Field. Dates of Surveys: Unknown. Source: Kohli, Peng, and Thill "Surface Subsidence Due to Underground Longwall Mining in the Northern Appalachian Coalfield." Preprint AIME Annual Meeting, Las Vegas, Nevada, 1980.

Longwall Mining
1. Federal No. 2 Mine (Eastern Associates Coal Company), Monongalia County, West Virginia, Date of Survey: 1973-76 (approx.). Source: USBM Contract H0230012 Comprehensive Ground Control Study of a Mechanized Longwall Operation.
2. Gateway Mine (Jones and Laughlin Steel Corporation), Morgan Township, Greene County, Pennsylvania. Date of Survey: 1972-75. Pennsylvania DER-MSR.
3. Old Ben Mine No. 24 (Old Ben Coal Company), Benton, Franklin County, Illinois. Date of Survey: 1976-77. Source: Coal Mine Subsidence, Session No. 71, ASCE National Convention, Pittsburgh, Pennsylvania, April 24-28, 1978 (Conroy, 1978; Dames and Moore, 1975).
4. Somerset No. 60 Mine (Bethlehem Mines Corporation), Somerset Township, Washington County, Pennsylvania. Date of Survey: 1976-78. Source: Department of Energy Contract No. ET-76-C-01-9041 (Barla and Boshkov, 1978).
5. 17 Mine Operations (various coal companies), Northern Appalachian Coal Field. Dates of Surveys: Unknown. Source: Kohli, Peng, and Thill, 1980.

Shortwall Mining
1. Valley Camp No. 3 Mine (Valley Camp Coal Company), West Finley Township, Washington County, Pennsylvania. Date of Survey: 1973-75. Source: Pennsylvania DER-MSR.

TABLE 4. PRINCIPAL SUBSIDENCE PARAMETERS
NORTHERN APPALACHIAN COAL FIELD

Subsidence Factor[1],	$S/m = 0.72$	Panel of critical or supercritical dimensions
	$S/m = 0.72 \sqrt{n_1 n_2}$	Panel of subcritical dimensions
Panel Dimensions[2]	$w/h = 1.1$; $L/h = 1.1$	Critical
	$n_1 = 0.9^{w/h}$; $n_2 = 0.9^{L/h}$	Subcritical
Maximum Slope of Ground Surface	$i_o = 1.85 m/h$	
Maximum Curvature	$k_o = 15.9 m/h^2$	
Maximum Horizontal Strain	$\varepsilon_o = 0.71 m/h$	
Angle of Draw[3]	$\alpha = 22°$	
Angle of Break[3,4]	$\beta = -3°$	
Angle of Complete Mining[3]	$\gamma = 25°$	

[1]Maximum subsidence, S, per unit of mined height, m, for critical or supercritical panel dimensions.
[2]w is panel width, L is perpendicular distance from subsidence profile to nearest panel edge; h is overburden thickness.
[3]Measured to the vertical.
[4]Negative angle denotes inclination over mined-out panel.
Based on Cheng and Peng, 1981.

1.5 to 4 (Ganow, 1975; Haimson and Fairhurst, 1969). This horizontal stress field undoubtedly accounts for some of the observed differences in subsidence (Jeremic, 1981). European profile functions provided a satisfactory comparison with the Illinois data. O'Rourke and Turner (1979) report that subsidence (vertical movement) in the Appalachian Basin and western United States mines is significantly less than predicted by the National Coal Board experience. They attribute the difference to variations in lithology. Tandan and Powell, 1982, show that the ratio of maximum subsidence to the extraction thickness can be expressed in terms of the width-to-depth ratio by a simple exponential equation containing a coefficient that varies with lithology. Adamek and Jeran (1981) also relate differences in subsidence profiles to variations in stratigraphy. Abel and Lee (1980) review the impact of geologic variations on subsidence. It is interesting to note that in 1898, Sopworth (1898) indicated a relationship between lithology and angle of draw.

Figure 11 presents maximum subsidences that have been reported over longwall and pillar retreat panels in the United States. As per National Coal Board convention, the maximum settlement has been normalized relative to the thickness of the coal seam. Data have been compiled by O'Rourke and Turner (1979), Kohli and others (1980), Karmis and others (1980), Abel and Lee (1980), and von Schonfeldt and others (1979).

The amount of subsidence becomes greater as the width of the panel increases relative to the mine depth, and the limiting subsidence observed for all types of mining in the United States is slightly in excess of 0.7 compared with 0.9 in the United Kingdom.

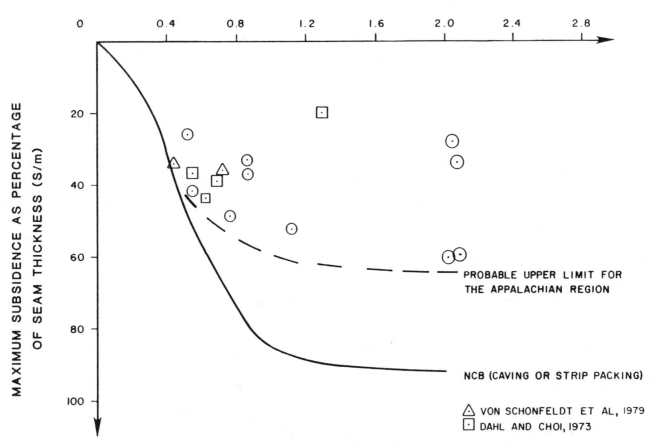

Figure 11. Influence of panel width to depth ratio on the maximum subsidence factor (Karmis and others, 1981).

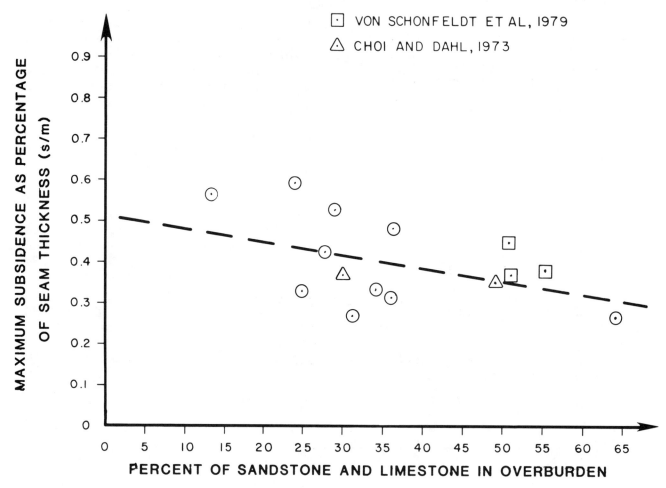

Figure 12. Influence of sandstone and limestone in overburden on the maximum subsidence factor (Karmis and others, 1981).

Subsidence from both longwall and pillar retreat mining in the United States tends to be substantially less than in the United Kingdom, even when the movements have been adjusted to virgin conditions. The lesser subsidence in the United States is thought due to the greater proportion of sandstone and limestone in the overburden.

Abel and Lee (1980), Peng and Cheng (1981), and Karmis and others (1980) have all suggested the amount of subsidence decreases with increase in the proportion of strong rock units in the overburden. Figure 12 shows the effect of percent sandstone and limestone on the subsidence factor in the Northern Appalachian Coal Region.

The data are much too variable to establish any more than broad bounds for expected subsidence on the basis of lithology.

Multiple Seam Mining

U.S. subsidence experience is largely with single seam mining except for the Pennsylvania Anthracite Region. However, multiple seam mining is becoming somewhat more common. Stemple (1956) and Britton (1980) discuss problems encountered in multiple seam mining. Subsidence experience with total extraction mining of multiple seams should be reasonably predictable as we obtain data on regional subsidence parameters. Following subsidence related to mining of the first seam, future subsidence movements may more closely compare with observations from the United Kingdom. Multiple seam mining beneath a seam which has been only partially mined by room and pillar methods could produce unusual subsidence patterns. Research on this situation is required.

Total Surface Support and Partial Extraction

Subsidence can be prevented by not mining the coal beneath surface areas requiring protection. Because of the angle of draw, the plan area of this coal support increases with depth of the coal seam below ground surface and may be a significant amount of coal even at shallow depths. The area to be left unmined will exceed the area to be protected in all cases since the surface area undergoing subsidence will extend over the unmined area as illustrated in Figure 13.

While the structure or area to be protected undergoes no

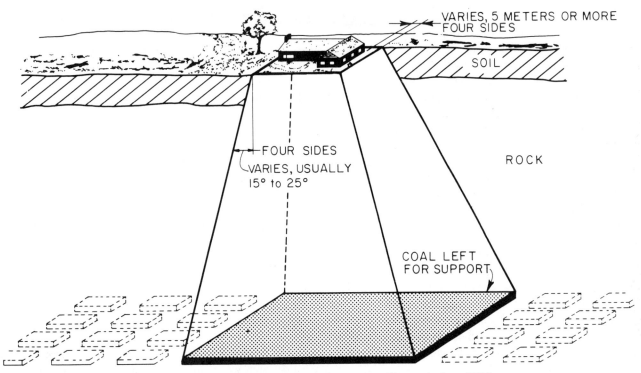

Figure 13. Mineral abandonment for surface support (Gray and others, 1974).

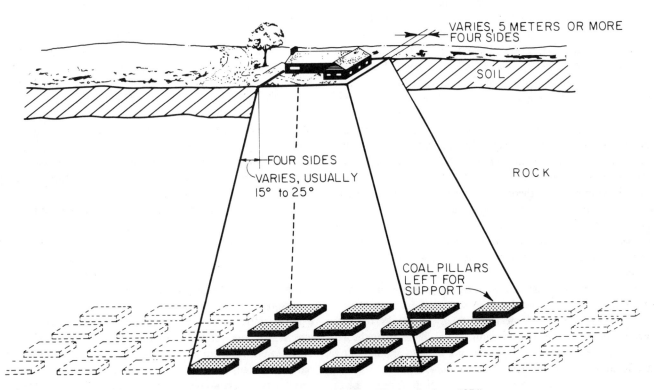

Figure 14. Partial extraction for surface support (Gray and others, 1974).

movements, the flanks of the protected area are subjected to tension and compression commensurate with their position at the margin of the subsidence trough. Partial extraction systems designed for subsidence prevention allow the removal of a portion of the coal without surface subsidence (Figure 14).

Pennsylvania's Bituminous Mine Subsidence and Land Conservation Act of 1966 provides protection for public buildings, dwellings used for human habitation, and cemeteries above active mines. Using the partial support method indicated in Figure 14, unmined pillars of coal equivalent in area to 50 percent of the support area are left in place to prevent subsidence (Cortis, 1969; Bise, 1980). Alexander (1979) reported that of the 17,000 structures which had been protected up to 1979, only 1.6 percent incurred some damage which was repaired at a cost of $1.4 million.

Large coal pillars to be left for surface protection must be accounted for in mine design. Areas left unmined in a longwall operation may require the redesign of panel layout and the use of room and pillar equipment around protective pillars. Permanent pillars left in mines utilizing room and pillar methods generally disrupt mining to a lesser extent but may require a modified layout of access and haulage entries to avoid conflicts with the support area.

The average axial stress borne by the coal pillars is considered to be the overburden weight multiplied by the ratio of the mined area within the support area to the plan area of the coal pillars. The load on an individual pillar is considered to be the average stress for the mined area multiplied by the individual pillar area (Coates, 1965-66; Holland, 1962). Hustrulid (1976) and Bieniawski (1981) present excellent reviews of coal pillar strength. In comparing available pillar strength to applied loads on an individual pillar, a factor of safety of 2.0 is usually used for design of permanent pillar support (Holland, 1962). In cases where particularly sensitive structures exist at the surface, or where the mine is to be subsequently flooded (saturation reduces the strength of most sedimentary rocks), higher factors of safety are sometimes used.

Experiments and theories to determine the strength of the mine roof have generally been formulated from the mining viewpoint of attempting to prevent localized roof failure rather than considering the problem of roof strength being compatible with overburden load applied to individual pillars (Adler and Sun, 1968).

The strength of the mine floor can be estimated using bearing capacity theories developed in soil mechanics for shallow foundations or from rock mechanics theory (Vesic, 1976; Speck, 1979; Adler and Sun, 1968; Ganow, 1975). Floor strengths may vary considerably from one portion of a mine to another due to variations in geology or flooding of the mine.

The size of the area in which a significant portion of the seam must be left unmined increases as the depth of the seam beneath the surface increases. Since the stress on individual pillars increases, the percentage of the coal seam which must be left unmined also increases.

SUBSIDENCE OVER ABANDONED MINES

Although there are similarities between subsidence over abandoned and active mines, there are some important differences. For example, with abandoned mines often no maps or other information on subsurface conditions or ground movements are available. As a result, even with an extensive subsurface investigation, prediction of the amount of future subsidence and its areal extent are often largely an educated guess based upon experience.

Two types of subsidence features are recognized above abandoned mines—sinkholes and troughs (Figure 15). In Illinois, these features may be called pits and sags, respectively.

A sinkhole is a depression in the ground surface that occurs from collapse of the overburden into a mine opening (a room or an entry). Boundaries between ground surface and walls of the sinkhole are often abrupt; and because diameter generally increases with depth, the sinkhole in profile may resemble a bottle with the cap removed. Erosion of soil at the sinkhole's periphery may increase the diameter near ground surface to create an hourglass profile.

A trough is a shallow, often broad, dish-shaped depression that develops when the overburden sags downward into a mine opening in response to the crushing of mine pillars or the punching of pillars into the mine floor. Figures 16 and 17 show a structure damaged by trough subsidence 110 feet above an abandoned mine in the Pittsburgh Coal.

Sinkholes generally develop where the cover above a mine is less than 100 feet. This appears to be largely due to: (1) the shape of the cavity developed by caving of the roof above the mine opening, and (2) the width of the rooms common in coal mines. Competent strata above the coal will limit sinkhole development (Figure 15). In areas of thick soil, such as Illinois, sinkholes have developed 164 feet above a mine (Bauer and Hunt, 1981).

Troughs develop where a pillar or pillars fail by crushing or punching into the mine roof or floor (Figure 15). Pillar strength may be affected by groundwater fluctuations or weathering. Stress concentrations may cause spalling of pillars; and the resulting smaller pillars, if unable to carry the loads, will fail. Failure of one pillar results in a redistribution of stresses by arching of the mine roof, which may cause adjacent pillars to fail by crushing or punching. Punching occurs where a soft seam of rock, generally claystone, is located immediately above or below the coal.

Subsidence troughs associated with abandoned mines typically resemble those above active mines. They are, however, almost always elliptical in plan rather than rectangular, since they often do not conform to panel boundaries, and usually measure only a few tens of feet to a few hundreds of feet in diameter. Trough diameters above abandoned mines in the Northern Appalachian Coal Field commonly measure 1.5 to 2.5 times the overburden thickness, reflecting the limit to which the overburden can bridge over local crushed pillars or roof failures before sagging into the distressed area (Gray and others, 1977).

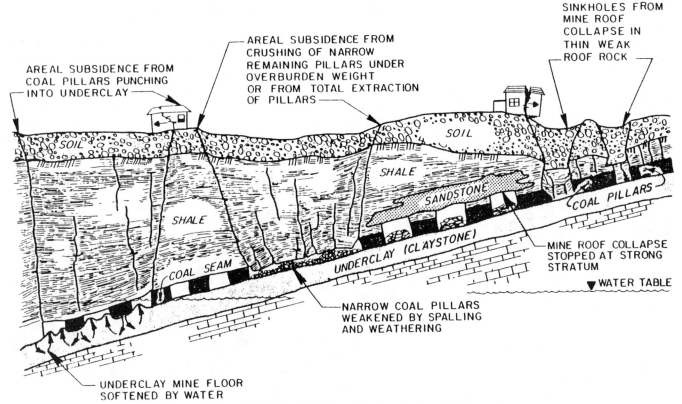

Figure 15. Modes of subsidence (Bruhn and others, 1978).

Similar trough diameter/mine depth ratios have been found for several subsidence cases reported in Illinois. This is interesting in that soil composed 40 to 50 percent of the overburden in Illinois, but only 10 to 20 percent in the Northern Appalachian Coal Field.

Subsidence over abandoned mines may occur many years after mining and may go unnoticed unless it is in an urban area. The most extensive study of subsidence experience over abandoned mines has been conducted for the Pittsburgh Coal seam (Gray and others, 1977). Hunt (1978) has summarized experiences over room and pillar coal mines in Illinois. Of the 354 subsidence incidents identified in the Pittsburgh Coal study, 71 percent were in highly urbanized Allegheny County, Pennsylvania, having an area equal to only 6 percent of the Pittsburgh Coal. The distribution of subsidence sites by overburden thickness is shown in Figure 18.

Nearly 81 percent of the subsidence incidents took place where the overburden (rock and soil) is less than 100 feet thick, and nearly 59 percent occurred where the overburden is less than 50 feet thick (Bruhn and others, 1978). Most sinkholes developed where the cover is less than 50 feet thick, and most troughs developed where the cover above the abandoned mine is more than 50 feet thick.

Subsidence has occurred as early as a decade after mining and as late as a century (Figure 19). More than half the subsidence incidents took place 50 or more years after mining (Figure 20). The time of occurrence of subsidence is undoubtedly governed by the rate of deterioration of the rock strata and coal pillars, and by other factors, which sometimes include robbing of pillars by small operators years after initial mining. This represents a complex interaction of phenomena that prohibits convenient prediction of the time of subsidence. The frequency of reports of sinkhole development appears to reflect the quantity of precipitation experienced in the preceding three to eight months (Figure 21) (Bruhn and others, 1978). Note, for example, that Point F' on the curve representing cumulative subsidence incidents follows eight months after the corresponding point (F) on the cumulative precipitation curve. Wildanger and others (1980) report that sinkholes in St. David, Illinois, have tended to appear after rainstorms.

Reported incidents of subsidence-damaged structures tend to be fewer as overburden thickness increases (Figure 18 and 22), supporting the contention that the likelihood of subsidence tends to reduce with increased interval above mine level. Sites located 60 or more feet above mine level avoid the majority of sinkholes, which constitute about 95 percent of all reported subsidence incidents and are responsible for more than half the reported incidents of damage to structures (Bruhn and others, 1981).

A subsidence incident documented at a site where cover ranged from 230 to 450 feet is noteworthy in that it covered 40 acres and damaged more than 40 structures. It represents the single most extensive subsidence incident recorded above an

Figure 16. Subsidence damage to office building above abandoned mine in the Pittsburgh Coal, Allegheny County, Pa.

abandoned mine in the Pittsburgh Coal region and supports the view that a substantial thickness of overburden does not necessarily ensure safety from subsidence (Bruhn and others, 1981).

With total extraction, subsidence is concurrent and the overlying strata stabilize in a few months to a few years. Total extraction mining has been practiced for many years; however, the long, narrow pillar system was not as successful in recovering coal pillars as present total extraction methods using the current block system (Paul and Plein, 1935).

Figure 23 depicts a portion of an early total extraction mine in western Pennsylvania (Pittsburgh Coal) operated just after the turn of the century. Pillar extraction was conducted between 1909 and 1910. In the portion of the mine shown, about 78 percent of the coal was extracted, including some pillars; and about 22 percent was lost. Roof falls, squeezes, or other complications brought about by factors no longer evident may have been responsible for the inability to extract the pillars. As a result, the potential for subsidence remains a real possibility above this mine many years after its abandonment. More than 30 subsidence incidents have occurred above this mine in the past 10 years as the long, narrow pillars fail.

Some time after mining, complete collapse of abandoned entries and rooms is to be expected as a result of natural causes and activities of man. Until that point is reached, the ground surface overlying a mine may experience a variable frequency of subsidence incidents; perhaps only a few incidents immediately after mining, while the rock surrounding the openings is still relatively sound; increasing numbers of incidents for an extended period of time as progressive deterioration and failure of the rock surrounding the openings becomes more pronounced; and later, a diminishing number of incidents as the void spaces at mine level become fewer and fewer. This sequence is shown conceptually in Figure 24.

This implies that the possibility of future subsidence at a site

Figure 17. Interior damage to office building shown in Figure 16.

cannot be ruled out merely because subsidence has not been recognized in the first 50 or 100 years after mining. If abandoned mine openings beneath a site have not been designed for long-term stability, the potential for subsidence remains until the openings collapse, or until they are stabilized by backfilling, grout columns, or some other means (Gray and others, 1974). Precisely when collapse might take place in the absence of stabilization is not predictable. Even after subsidence has taken place at a particular site, the possibility of future subsidence may remain. Figure 25 shows a community that has experienced subsidence on at least four occasions over the past 50 years. Subsidence appears first to have taken place in the 1920's while mining was in progress. After abandonment of the mine, a long period of quiescence followed until 1964, when three homes were damaged by subsidence in an area measuring 300 by 400 feet that overlapped part of the 1921 subsidence area. This was followed in mid-1972 by subsidence of seven homes 400 feet east of the 1964 area, and in late 1972 by subsidence of six more homes in an area that partially coincided with the southerly part of the 1964 area. Subsidence stopped for the next few months but resumed in April 1973, affecting an area 450 by 800 feet and damaging 14 more homes. At nearly the same time, two other homes 1,100 feet east were similarly damaged by subsidence (Bruhn and others, 1981).

Multiple episodes of subsidence have been documented at 33 other sites in the Pittsburgh region, each episode representing a report of damage to a structure or the appearance of a hazardous condition. Figure 26 shows data from 19 of the sites. Data from the remainder are nearly identical. Multiple subsidence episodes of lesser significance may have taken place at these or other sites and not been recorded. Stephenson and Aughenbaugh (1978) report on multiple subsidence episodes at Johnston City, Illinois.

Inasmuch as multiple episodes of subsidence above abandoned mines are not uncommon, a subsidence occurrence at a site is perhaps prudently interpreted as one of a continuing series

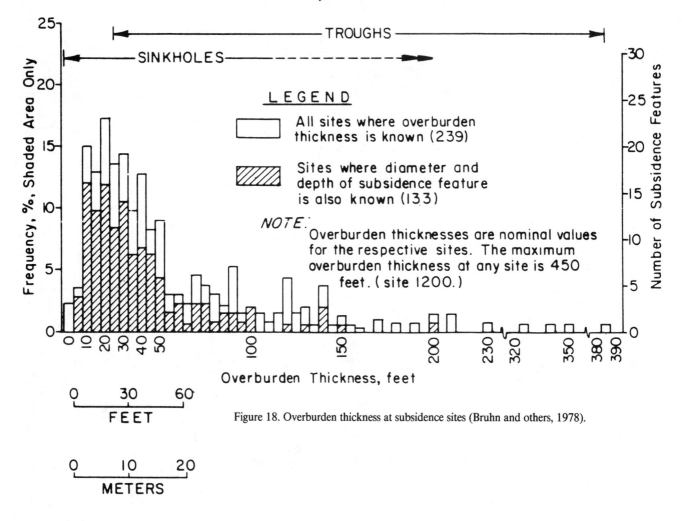

Figure 18. Overburden thickness at subsidence sites (Bruhn and others, 1978).

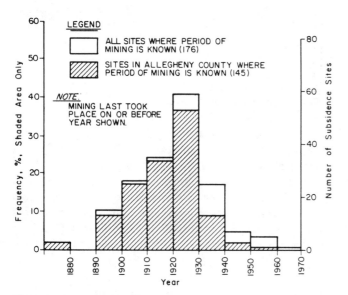

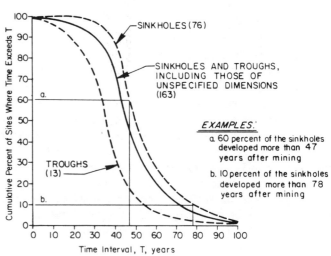

Figure 19. Periods of undermining at sites where subsidence has taken place (Bruhn and others, 1981).

Figure 20. Minimum time interval between mining and subsidence (Bruhn and others, 1978).

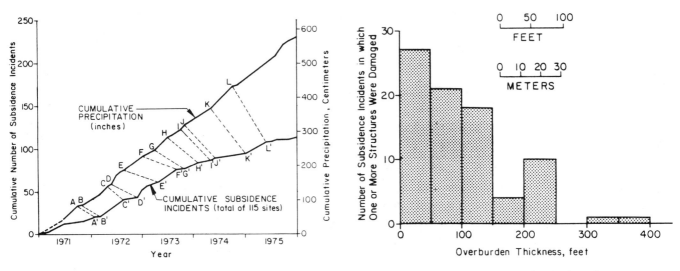

Figure 21. Relationship between cumulative precipitation and sinkhole development (Bruhn and others, 1978).

Figure 22. Overburden thickness at subsidence sites where one or more structures have been damaged (Bruhn and others, 1981).

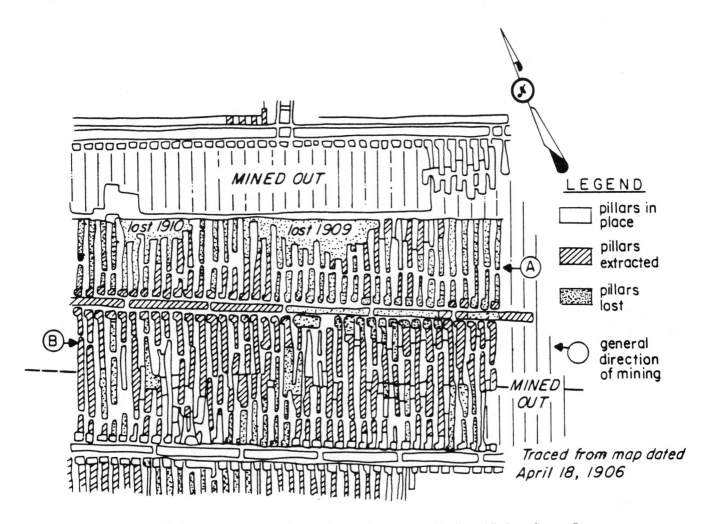

Figure 23. An early total extraction mine developed using long pillar lines, Allegheny County, Pa. (Bruhn and others, 1978).

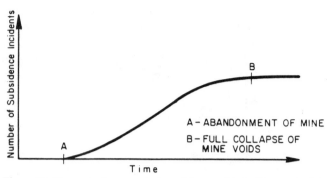

Figure 24. Conceptual representation of the possible incidence of subsidence above an abandoned mine containing open voids (Bruhn and others, 1981).

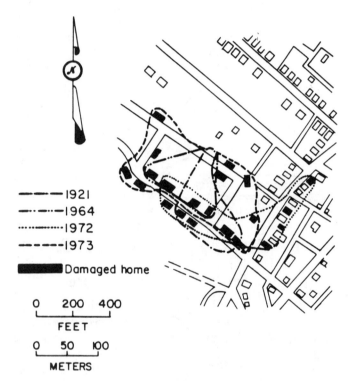

Figure 25. Community that has experienced multiple incidents of abandoned mine-related subsidence (Bruhn and others, 1981).

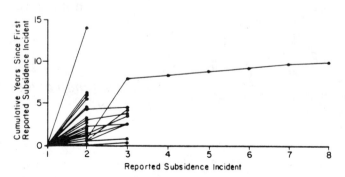

Figure 26. Intervals of recurrent subsidence at 19 sites overlying abandoned mines in the Pittsburgh Coal Region (Bruhn and others, 1981).

of subsidence episodes that may have begun sometime earlier and will continue sporadically until such time in the future as collapse of underground mine openings is complete. Pillar failure can be delayed, progressive, or sporadic. It should be evident that site surveillance programs of a few months' duration or, in fact, indefinite duration, cannot provide definitive evidence that a site overlying a mine with open voids will experience no future subsidence (Bruhn and others, 1981).

Based on the writers' experience with subsidence over abandoned mines, the following concepts are apparent:

1. Unless total extraction has occurred, there is no interval above an abandoned mine that is safe from subsidence, nor necessarily a reduction in severity of damage. Increased intervals above mine level, however, exhibit a reduced frequency of subsidence.
2. Unless total extraction has been achieved, subsidence may occur long after mining and may not be limited to a single episode.
3. In some cases, activities by man have hastened the onset of subsidence, if not initiated it.

Above abandoned mines, there are no means available as yet to predict exactly when or where subsidence might take place; and so subsidence must generally be expected anywhere unless it can be proved that the area has not been mined, that long-term pillar support has been provided, or that the mine voids are fully collapsed. Since methods of exploration to evaluate conditions for undermined sites and the various stabilization techniques that may be applied are extensively discussed by Gray and others (1974), Bell (1975, 1978), Abandoned Mined Lands (1981), Littlejohn (1979), Institution of Civil Engineers (1977), they will not be considered in this review.

CONSEQUENCES OF SUBSIDENCE

Damage to Buildings

Permanent reserves of unmined coal protect the central sectors of many old towns in coal mining districts. Most of these protected areas are probably secure against mine subsidence. With continued growth, however, many business and residential districts have grown far beyond the reserves to areas that have been undermined sometime in the past, or are now being undermined. About 30 buildings are currently reported damaged each year in western Pennsylvania as a result of mine subsidence. Commonly, two-thirds of these cases are associated with active mines and one-third with abandoned mines. Documentation of subsidence damage is extremely limited or nonexistent in most states, so the following discussion is restricted primarily to the Commonwealth of Pennsylvania, where records are available from the mid-1960's forward.

Subsidence damage to buildings associated with coal mining is of two types—that which takes place as mining progresses beneath the site, and that which takes place sometime after mining has been completed.

Damage during total extraction mining is most often associated with the traveling ground wave that accompanies the mine face as it passes from one end of the panel to the other. Where partial extraction is currently practiced in western Pennsylvania, the intention is to prevent ground movements altogether. The frequency of occurrence of subsidence under these conditions is far less than 1 percent. But where the subsidence has taken place, it can most often be attributed to crushing of mine pillars or to yielding of the mine roof or mine floor in response either to inadequate mining practices or to unforeseen geological or hydrological circumstances at mine level. As in cases of total extraction, damage to structures associated with partial extraction mining is linked to passage of a traveling wave and to the downwarping of the strata over the affected area of the mine. But here the traveling wave front is often irregular, and its ultimate position is unpredictable due to the uncertain nature of the failure process at mine level. In some cases, overburden load transfer (arching) from adjacent mining may induce such ground movements over areas that may have been stable for long periods of time. Generally, subsidence associated with partial extraction mining is unpredictable in both magnitude and extent, whereas that associated with total extraction can be predicted fairly reliably with regard to both areal extent and downward displacement of the ground surface, using angles of draw and other parameters reviewed earlier in this paper. Unfortunately, current knowledge does not permit prediction of damage to buildings, which is usually attempted by estimating the ground strains at a site and comparing them with allowable strains established empirically from observations of similar buildings. For lack of a better procedure, some persons in the United States have attempted to utilize the United Kingdom's National Coal Board (NCB) predictive techniques to assess the likelihood of subsidence damage at particular sites in the United States. This practice is to be discouraged in that it almost always underestimates the damage severity. The apparent reasons for this seem to be associated with differences in mining, geology, and building construction between the two areas. First, the NCB method strictly applies to longwall mining. About 90 percent of mining in the United States is by room and pillar methods, which may yield a pattern of ground strains significantly different from that of longwall. This has yet to be confirmed. Second, rock strata in the Appalachian coal fields typically contain a greater portion of sandstone, limestone, and other stiff rock units than do the primarily shale strata in the United Kingdom. Moreover, most rock strata in the United States have not been disturbed by previous mining, whereas in the United Kingdom it is not uncommon for several overlying seams to have been mined earlier. The resultant softening of the inherently more compliant rock strata in the United Kingdom leads to more modest ground curvatures with passage of the subsidence traveling wave which, in turn, leads to lower levels of damage to buildings. Third, differences in building construction between the two countries probably have an effect on damage severity equally as profound as the first two factors. Dwellings in the eastern United States typically have basements; those in the United Kingdom typically do not. The coupling between structure and ground surface is, therefore, significantly different. Basement repairs, almost without exception, represent the greatest portion of total repair costs in the United States. In a study of 250 homes in western Pennsylvania for the U.S. Bureau of Mines, the writers found that basement repair costs exceeded 50 percent of the total repair cost in 90 percent of the cases and exceeded 90 percent of the total repair cost in 50 percent of the cases (Bruhn and others, 1982).

These differences in mining, geology, and building construction between the United Kingdom and the United States preclude use of the NCB Handbook in this country, and emphasize the need for development of a comparable handbook for the United States.

Government agencies are now sponsoring subsidence projects in several locations in the Appalachian and Interior Coal Basins. One of the most comprehensive studies has been conducted at the Old Ben No. 12 Mine in Illinois. Other studies are currently underway in West Virginia, Pennsylvania, and Ohio. The studies are directed towards developing a better understanding of the mechanics of subsidence from mine level to ground surface, which will lead ultimately to development of an analytical model capable of predicting ground surface deformations given the type of mining, the height of extraction, the mechanical properties of the overburden and ground surface relief. Owing to the length of time required to mine a panel of coal, the results of these studies and development of the subsidence model will probably not be available until the late 1980's.

While development of a model is praiseworthy, there appears to be a preoccupation, at present, with studying the mechanics of subsidence at the expense of examining what seems to be the real problem—the response of buildings and the groundwater regime to deep mining. The current direction in research seems to stem largely from the mining and rock mechanics backgrounds of both the sponsoring agencies and the persons conducting the research. Damage to surface structures and modification of the groundwater regime are problems that historically have been solved by the civil engineering profession, notably with regard to earthquakes, consolidation settlements, etc. A pragmatic solution to the subsidence problem seems more likely from the civil engineering quarter than anywhere else. Building deformations due to foundation settlement, open excavations, and tunneling have been examined in considerable detail and provide a valuable source of information for subsidence engineers (Burland and others, 1977; Boscardin, 1980). Clearly, there is a need to observe the response of simple structures to subsidence while these comprehensive monitoring programs are underway. Such observations will lead to improvements in the design and construction of new homes and other buildings over areas about to be deep mined and, equally as important, put into perspective the level of accuracy of ground strain prediction that must be achieved for application to normal practice. It is significant in the writers' view that the level of damage observed in 250 undermined homes is virtually independent of overburden thickness.

This suggests that ground strain, which is expected to diminish with increase in overburden thickness, may not have as significant an effect on the level of damage as does structural design (Bruhn and others, 1982). The possibility that the level of damage severity may be more greatly influenced by the type of construction than by ground strain should be properly appreciated in establishing the level of sophistication that needs to be introduced into a predictive model to derive usable results.

Hydrogeologic Aspects of Subsidence

Although direct damage to buildings, highways, and utilities is the most costly consequence of subsidence in terms of total dollar value, the impact of subsidence on surface drainage and groundwater aquifers above underground mines may cause profound problems of a nonreparable nature. Young and Stoek (1916) reported these problems in Illinois, which were elaborated on by Herbert and Rutledge (1927). In topographically flat areas, such as Illinois, even rather modest ground surface subsidence of a foot or two can promote flooding of valuable farmlands, necessitating the installation of extensive drain tile systems. In central Illinois, permanent coal pillar support is left beneath extensive farming areas to avoid the potential flooding problems. In the process, this sterilizes much valuable coal.

In the eastern United States, where the ground surface is much more hilly, the problem with flooding of farmland is not nearly so prominent. However, flexure and fracturing of the rock units during subsidence has sometimes seriously affected the groundwater regime in one or more of the following ways:

1) Dewatering or draining an aquifer into the deep mine or into a lower permeable rock unit;
2) Reducing the areal extent of an aquifer by truncating it against a less permeable rock unit; and
3) Providing avenues through aquicludes for the mixing of waters from originally separate aquifers.

These effects, like damage to structures, appear to be highly dependent upon local geology as well as on the nature and extent of subsidence. Their possible consequences are reduction in yield from wells and springs and the degradation of groundwater quality. Although it is not uncommon to hear of wells or springs going dry during mining, there is little systematic documentation of this effect. Preliminary information reported by Rausch and Ahnell (1979) for locations above the Pittsburgh Coal near Morgantown, West Virginia, suggests that when the distance between mine level and the bottom of the well is less than 250 feet, the well is typically dewatered, and generally water level will not recover. Where the interval between mine level and the bottom of the well exceeds this amount, the well performance is generally not affected.

From a study of water inflow to 17 underground mines in western Pennsylvania, Roebuck (1980) reported that the following variables were important:

1. Proportion of the mine in a valley bottom setting;
2. Proportion of the mine that has caved;
3. The weighted perimeter of the mine;
4. The ratio of sandstone to nonsandstone in the first 80 feet above the mine roof.

Singh and Kendorski (1981) in a review of the impact of mining beneath bodies of water present recommendations for the minimum cover required to minimize flow into the mine with total extraction.

A recent study on the effects of underground coal mining on groundwater in the eastern United States (Sgambat and others, 1980) concludes that effects are highly dependent on the locations of the mine with respect to the natural flow system. Although contamination of groundwater exists in the immediate vicinity of coal mines on a regional basis, there is little evidence of gross groundwater contamination. Future deep mining in the Eastern Interior Basin and the southern Appalachians is likely to result in adverse groundwater effects in only very limited areas. The central Appalachians have a greater potential for such impacts. Pre-mine planning will help to minimize the adverse effects on groundwater.

A coupled finite element geomechanical-hydrology model is currently under development to predict groundwater disturbances associated with mine subsidence (Girrens and others, 1981).

CONCLUSIONS

Although a small number of technical people have done significant work on subsidence in the United States, many uncertainties remain in subsidence engineering. Progress has been made within the past few years in collecting and synthesizing available subsidence data in coal fields across the United States. By far, most data are found in the Northern Appalachian Coal Field and Illinois Basin.

The data demonstrate very clearly that subsidence above active mines is distinctly different than over abandoned mines in terms of the plan and profile geometry, the time and place of occurrence, and, to a certain extent, the governing mechanism. An understanding of one type of subsidence does not guarantee an understanding of the other. Conversely, it seems clear that the manner in which subsidence occurs above active or abandoned mines is much the same, whether it takes place in Illinois or in the Northern Appalachian Coal Field.

Distinct geologic differences have been recognized between and within the Appalachian and Illinois regions. To an extent not yet fully determined, these differences are reflected in variations in angles of draw, as well as in profiles of the subsidence development curves and other parameters. Unfortunately, the data are not sufficiently refined or sufficiently extensive to establish any reliable relationship between subsidence parameters and basic mine site stratigraphy as given by one or more borehole geologic logs. The inability to relate reported subsidence parameters to basic geologic constraints makes impossible a separate identification of the influence of soil and near surface rock on these parameters.

Moreover, very few measurements exist of ground strain, slope, and curvature, which would be of use in relating mining to severity of damage to buildings at ground surface. Data plots that have been synthesized for longwall, room and pillar, and pillar retreat mines in Illinois and the analytical expressions developed for longwall profiles in the Northern Appalachian Coal Field are certainly noteworthy achievements. However, relatively few cases of subsidence damage to homes have been documented above longwall panels in the Northern Appalachian Coal Field. In Illinois, the effects of longwall mining on two school buildings have been reported (Herbert and Rutledge, 1927). Similarly, the number of well-documented cases of home damage above abandoned mines in Illinois is rather small at this time, although several cases are currently under investigation (Mahar and Marino, 1982). More important, a vast majority of home damage cases have been documented above abandoned and active room and pillar and pillar retreat mines in the Northern Appalachian Coal Field for which almost no systematic subsidence profile data have been collected and synthesized. As a consequence, there is no method available to identify the ground strains, slope, and curvature prevailing at the time of damage for the majority of damage areas that are available for study, except for those few still visible. The lack of linkage between ground movement and structural damage makes the identification of soil and near surface rock effects on damage severity nearly impossible in a formal sense.

Sufficient data on subsidence over active mines have not been collected to permit accurate prediction of ground movements, which would allow controlled mining to balance surface strains or the rational design of structures to resist subsidence damage. An all-encompassing subsidence prediction model is of doubtful utility. Even in the United Kingdom, subsidence specialists use modifications of the NCB model based on local conditions. Thus, it is far better to develop detailed information on a regional basis to define angle of draw, critical area, strain profiles, and proper pillar dimensions for permanent support. The regulations associated with the Surface Mining Control and Reclamation Act require a subsidence control plan as part of a mining permit application (Von Schonfeldt and others, 1979; Riddle, 1980). This need to reliably predict subsidence and its impact on surface structures should result in the collection of field data required to develop adequate subsidence prediction models for U.S. coal mining regions.

Although abandoned coal mines containing voids are much older in Europe and the United Kingdom than the United States, the applied technologies are similar (Gray and others, 1974; Bell, 1978; Littlejohn, 1979). Prediction of subsidence over abandoned mines is likely to remain an educated guess even with improved exploration techniques, such as borehole cameras. Equipment of this type can greatly improve the control and effectiveness of stabilization programs. Site stabilization and special foundations for particular structures based on detailed engineering evaluations appear reasonable. Large-scale urban stabilization programs for residential areas such as suggested by Table 1 should be carefully considered. Insurance programs to provide assistance, if and when subsidence occurs, may be a better utilization of public funds (Abandoned Mined Lands, 1981; DuMontelle and others, 1981). Land use controls such as zoning of areas subject to subsidence should also be considered (U.S. Government Accounting Office, 1979).

REFERENCES CITED

Abandoned Mined Lands Reclamation Control Technology Handbook; Chapter No. 2, Mine Subsidence Control, 1981: Report prepared for U.S. Department of the Interior by GAI Consultants, Inc., Monroeville, Pennsylvania, 37 p., Contract No. J5101109.

Abel, J. F., Jr., and Lee, F. T., 1980, Lithologic controls on subsidence: Society of Mining Engineers of American Institute of Mining, Metallurgical and Petroleum Engineers (Preprint No. 80-314), 16 p.

—— 1980, Subsidence potential in shale and crystalline rocks: U.S. Geological Survey Open File Report 80-1072, 49 p.

Adamek, V., and Jeran, P. W., 1981, Evaluation of existing predictive methods for mine subsidence in the U.S.; in Proceedings, 1st Annual Conference on Ground Control in Mining, Morgantown, West Virginia University, p. 209–219.

Adler, L., and Sun, M., 1968, Ground control in bedded formations, Virginia Polytechnic Institute, Research Division, Bulletin 28, 266 p.

Alexander, T., 1979, Personal communication.

Anonymous, 1965, Longwall mining: Coal Age, New York, McGraw-Hill.

Ashley, G. H., 1928, Bituminous coal fields of Pennsylvania, Part I, Bulletin M6, Pennsylvania Topographic and Geologic Survey, 241 p.

Averitt, P., 1970, Stripping-coal resources of the United States—January 1, 1970: U.S. Geological Survey Bulletin 1322, 34 p.

Barla, G. B., and Boshkov, S., 1978, Investigations of differential strata movements and water table fluctuations during longwall operations at the Somerset Mine No. 60. Final Technical Report from the Henry Krumb School of Mines, Columbia University, New York, New York, under contract to U.S. Department of Energy, 149 p. (Available from National Technical Information Service, SE 9041-1).

Bauer, A., and Hunt, S. R., 1981 Profile, Strain and time characteristics of subsidence from coal mining in Illinois: Preprint, Workshop on Surface Subsidence to Underground Mining, Morgantown, West Virginia University, p. 207–219.

Bell, F. G., 1975, Site investigations in areas of mining subsidence: London, Newnes-Butterworths, 168 p.

—— 1978, Subsidence due to mining operations, in Bell, F. G., Foundation engineering in difficult ground, London, Newnes-Butterworths, p. 322–362.

Bieniawski, Z. T., 1981, Improved design of coal pillars for U.S. mining conditions, in Proceedings, 1st Annual Conference on Ground Control in Mining, Morgantown, West Virginia University, p. 13–34.

Bise, C. J., 1980, Pennsylvania's subsidence—Control guidelines; Should they be adopted by other states: Society of Mining Engineers of AIME (Preprint No. 80-5), 9 p.

Boscardin, M. D., 1980, Building Response to Excavation-Induced Ground Movements [Ph.D. Thesis]: University of Illinois at Urbana, 279 p.

Bräuner, G., 1973, Subsidence due to underground mining; 1. Theory and practices in predicting surface deformation; 2. Ground movements and mining damage: U.S. Bureau of Mines Information Circulars 8571 and 8572.

Britton, S. G., 1980, Mining multiple seams: Coal Mining and Processing, v. 17, no. 12, p. 64–70.

Bruhn, R. W., and others, 1978, Subsidence over the mined-out coal: American Society of Civil Engineers Spring Convention, Pittsburgh (ASCE Preprint 3293), p. 26–55.

—— 1981, Subsidence over abandoned mines in the Pittsburgh Coalbed, in Proceedings, 2nd, Ground Movements and Structures, J. Geddes, ed., Cardiff, Wales, 1980, London, Pentech Press, p. 142–156.

—— 1982, Survey of ground surface conditions affecting structural response to subsidence: Report prepared for U.S. Bureau of Mines by GAI Consultants, Inc., Monroeville, Pennsylvania under contract no. J0295014.

Burland, J. B., and others, 1977, Behavior of Foundations and Structures, State of the Art Report, Vol. II, 9th International Conference on Soil Mechanics and Foundation Engineering, Tokyo, p. 495–546.

Cassidy, S. M., 1973, Elements of practical coal mining: New York, Society of Mining Engineers of American Institute of Mining, Metallurgical and Petroleum Engineers, 614 p.

Cheng, D., and Peng, S. S., 1981, Analysis of surface subsidence parameters due to underground longwall mining in the Northern Appalachian Coalfield: Morgantown, West Virginia University, Department of Mining Engineering Report TR-81-1, 51 p.

Coates, D. F., 1965-66, Pillar loading: Mines Branch Research Reports, 168, 170, 180, Department of Mines and Technical Surveys, Ottawa, Canada, 84 p., 101 p., 71 p.

Conroy, P. J., 1978, Subsidence above a longwall panel in the Illinois No. 6 coal: American Society of Civil Engineers Spring Convention, Pittsburgh (ASCE Preprint 3293), p. 77–92.

—— 1980, Geotechnical analysis for longwall design: Coal Mining and Processing, v. 17, no. 12, p. 50–54.

Conroy, P. J., and Gyarmaty, J. H., 1981, Subsidence monitoring—Case History: in Proceedings, 1st Annual Conference on Ground Control in Mining, Morgantown, West Virginia University, p. 148–153.

Cortis, S. E., 1969, Coal mining and protection of surface structures are compatible: American Mining Congress Coal Convention, Pittsburgh, Pennsylvania, Mining Congress Journal, v. 55, n. 6, p. 84–89.

Dahl, H. D., and Choi, D. S., 1975, Some case studies of mine subsidence and its mathematical modeling, in Applications of Rock Mechanics, 15th Symposium on Rock Mechanics, 1973, New York, American Society of Civil Engineers, p. 1–21.

Department of Energy, 1982, Weekly Coal Production, Report of Energy Information Administration, 5/1/82.

DuMontelle, P. B., and others, 1981, Mine subsidence in Illinois: Facts for the homeowner considering insurance; Champaign, Illinois State Geological Survey, Environmental Geology Notes 99, 24 p.

Eavenson, H. N., 1942, The first century and a quarter of American coal industry: Baltimore, Waverly Press, Inc., 701 p.

Ganow, H. C., 1975, A geotechnical study of the squeeze problem associated with the underground mining of coal [Unpublished Ph.D. thesis]: Champaign, University of Illinois, 233 p.

Girrens, S. P., and others, 1981, Numerical prediction of subsidence with coupled geomechanical-hydrological modeling, in Proceedings, Workshop on Surface Subsidence Due to Underground Mining, Morgantown, West Virginia University, 8 p.

Gray, R. E., and Meyers, J. F., 1970, Mine subsidence and support methods in the Pittsburgh area: American Society of Civil Engineers, Proceedings, Journal of the Soil Mechanics and Foundations Division, v. 96, no. SM4, p. 1267–1287.

Gray, R. E., and others, 1974, State of the art of subsidence control, Report ARC-73-111-2550, prepared by General Analytics, Inc., for the Appalachian Regional Commission, Washington, D.C. (Available from National Technical Information Service, Springfield, Virginia as report no. PB 242465.)

—— 1977, Study and analysis of surface subsidence over the mined Pittsburgh Coalbed: Report prepared for U.S. Department of the Interior, Bureau of Mines. (Available from National Technical Information Service, Springfield, Virginia as report no. PB 281522.)

Great Britain, National Coal Board, 1975, Subsidence engineers handbook (second edition): London, National Coal Board, 111 p.

Greenwald, H. P., and others, 1937, Studies of roof movement in coal mines; Montour 10 Mine of the Pittsburgh Coal Company, U.S. Bureau of Mines Report of Investigations 3355, 41 p.

Gregory, C. E., 1980, A concise history of mining: New York, Pergamon Press, 259 p.

Haimson, B., and Fairhurst, C., 1970, In-situ stress determination at great depth by means of hydraulic fracturing, in Proceedings, Symposium on Rock Mechanics, 11th, Berkeley, California, 1969, published as Rock mechanics-theory and practice, New York, American Institute of Mining, Metallurgical, and Petroleum Engineers, p. 559–584.

Herbert, C. A., and Rutledge, J. J., 1927, Subsidence due to coal mining in Illinois: U.S. Bureau of Mines Bulletin 238, 59 p.

Holland, C. T., 1962, Design of pillars for overburden support, Parts I and II: Mining Congress Journal, v. 38, no. 3, p. 24–28; v. 38, no. 4, p. 66–71.

HRB-Singer, Inc., 1977, Nature and distribution of subsidence problems affecting HUD and urban areas: U.S. Department Housing and Urban Development, 113 p. (Available from National Technical Information Service, Springfield, Virginia as report no. PB 80172778.)

—— 1980, Technical and economic evaluation of underground disposal of coal mining wastes: Report prepared for U.S. Department of the Interior, Bureau of Mines. Contract No. JO285008.

Hunt, S., 1979, Characterization of subsidence profiles over room and pillar coal mines in Illinois, in Proceedings, Illinois Mining Institute, v. 86, p. 50–65.

Hunter, D. W., 1972, Bridgewall mining: A new concept: Coal Age, v. 77, p. 84–89.

Hustrulid, W. A., 1976, A review of coal pillar strength formulas: Rock Mechanics, v. 8, p. 115–145.

Institution of Civil Engineers, 1977, Ground subsidence: London, Institution of Civil Engineers, 99 p.

Jeremic, M. L., 1981, Coal mine roadway stability in relation to lateral tectonic stress—Western Canada: Mining Engineering, v. 33, p. 704–709.

Johnson, W., and Miller, G. C., 1979, Abandoned coal-mined lands; Nature, extent, and cost of reclamation: U.S. Department of the Interior, Bureau of Mines, 20 p.

Jones, D. C., and Hunt, J. W., 1954, Coal mining (third edition): University Park, Pennsylvania, The Pennsylvania State University, 3 v.

Karmis, M., and others, 1981, A study of longwall subsidence in the Appalachian Coal Region using field measurements and computer modeling techniques, in Proceedings, 1st Annual Conference on Ground Control in Mining, Morgantown, West Virginia University, p. 220–229.

Kauffman, and others, 1981, Room and pillar retreat mining; a manual for the coal industry, U.S. Bureau of Mines Information Circular 8849, 228 p.

Keystone Coal Industry Manual, 1981, New York, McGraw-Hill, 1419 p.

Kohli, K. K., and others, 1980, Surface subsidence due to underground longwall mining in the Northern Appalachian coal field: Society of Mining Engineers of AIME (Preprint No. 80-53), 8 p.

Littlejohn, G. S., 1979, Consolidation of old coal workings: Ground Engineering, v. 12, no. 4, p. 15–21.

Mahar, J. W., and Marino, G. G., 1981, Building response and mitigation measures for building damages in Illinois: in Proceedings, 1st Annual Conference on Ground Control in Mining, Morgantown, West Virginia University, p. 238–252.

Maize, E. R., and Greenwald, H. P., 1939, Studies of roof movement in coal mines, 2. Crucible Mine of the Crucible Fuel Company: U.S. Bureau of Mines Report of Investigations 3452, 19 p.

Maize, E. R., and others, 1941, Studies of roof movement in coal mines; 4. Study of subsidence of a highway caused by mining coal beneath: U.S. Bureau of Mines Report of Investigations 3562, 11 p.

McCulloch, C. M., and others, 1975, Selected geologic factors affecting mining of the Pittsburgh Coalbed: U.S. Bureau of Mines Report of Investigations 8093, 72 p.

Munson, D. E., and Eichfeld, W. F., 1980, Evaluation of European empirical methods for subsidence in U.S. coal fields: Albuquerque, New Mexico, Sandia Laboratories Report SAND80-0537, 25 p.

Newhall, F. N., and Plein, L. N., 1936, Subsidence at Merrittstown air shaft near Brownsville, Pennsylvania: American Institute of Mining, Metallurgical, and Petroleum Engineers, Transactions, v. 119, p. 58–94.

O'Rourke, T. D., and Turner, S. M., 1979, A critical evaluation of coal mining subsidence patterns, in Proceedings, 10th Ohio River Valley Soils Seminar on the Geotechnics of Mining, Lexington, Kentucky, p. 1–8.

Paul, W. J., and Plein, L. N., 1935, Methods of development and pillar extraction in mining the Pittsburgh Coalbed in Pennsylvania, West Virginia and Ohio: U.S. Bureau of Mines Information Circular 6872, 31 p.

Peng, S. S., and Cheng, S. L., 1981, Predicting surface subsidence for drainage prevention: Coal Mining and Processing, v. 18, no. 5, p. 84–85.

Poad, M. E., 1977, Single entry development for longwall mining; Research approach and results at Sunnyside No. 2 mine, Carbon County, Utah: U.S. Bureau of Mines Report of Investigations 8252, 29 p.

Rausch, H. W., and Ahnell, G., 1979, Ground-water drainage associated with coal mining in Monongalia County, West Virginia [abs.]: Conference of the American Water Resources Association, Las Vegas, Nevada.

Riddle, J. M., 1980, Dealing with subsidence and SMCRA: Mining Engineering, v. 32, no. 12, p. 1702–1704.

Roebuck, S. J., 1980, Predicting groundwater flows to underground coal mines in Western Pennsylvania [M.S. thesis]: University Park, Pennsylvania State University, 102 p.

Sgambat, J. P., and others, 1980, Effects of underground coal mining on ground water in the Eastern United States: Geraghty & Miller, Inc., Annapolis, Maryland, 182 p. (Available from National Technical Information Service, Springfield, Virginia as report no. PB 80-216 757.)

Singh, M. M., and Kendorski, F. S., 1981, Strata disturbances prediction for mining beneath surface water and waste impoundments, in Proceedings, 1st Annual Conference on Ground Control in Mining, Morgantown, West Virginia University, p. 76–89.

Sopworth, A., 1898, Discussions on subsidence due to coal workings: Institution of Civil Engineers Minutes of Proceedings, v. 135, p. 165–167.

Speck, R. C., 1979, A comparative evaluation of geologic factors influencing floor stability in two Illinois coal mines [Ph.D. thesis]: Rolla, University of Missouri-Rolla, 265 p.

Stemple, D. T., 1956, A study of problems encountered in multiple-seam coal mining in the Eastern United States: Bulletin of the Virginia Polytechnic Institute, Engineering Experiment Station Series No. 107, 64 p.

Stephenson, R. W., and Aughenbaugh, N. B., 1978, Analysis and prediction of ground subsidence due to coal mine entry collapse; in Proceedings, 1st, Large Ground Movements and Structures, J. Geddes, ed., Cardiff, Wales, 1977, New York, John Wiley & Sons, p. 100–118.

Tandanand, S., and Powell, L., 1981, Consideration of overburden lithology for subsidence prediction: in Proceedings, 1st Annual Conference on Ground Control in Mining, Morgantown, West Virginia University, p. 17–29.

U.S. Bureau of Mines, 1976, Minerals Yearbook.

U.S. Government Accounting Office, Report by Comptroller General, 1979, Alternatives to protect property owners from damages caused by mine subsidence, Report CED-79-25, 41 p. (Available from National Technical Information Service, Springfield, Virginia as report no. PB290869.)

Vesic, A. C., 1975, Bearing capacity of shallow foundations; in Winterkorn, H. F., and others, eds., Foundation engineering handbook: New York, Van Nostrand Reinhold Company, p. 121–147.

Von Schonfeldt, H., and others, 1979, Subsidence and its effect on longwall mine design: Paper presented at Annual American Mining Congress Coal Convention, St. Louis, May 20-23.

Wade, L. V., and Conroy, P. J., 1980, Rock mechanics study of a longwall panel: Mining Engineering, v. 32, no. 12 p. 1728–1735.

Wardell, K., 1953, Some observations on the relationship between time and mining subsidence: Institution of Mining Engineers Transactions, v. 113, 1953-1954, p. 471–483.

—— 1969, Ground subsidence and control: Mining Congress Journal, v. 55, no. 1, p. 36–42.

Wildanger, E. G., and others, 1980, Sinkhole type subsidence over abandoned coal mines in St. David, Illinois: Abandoned Mined Lands Reclamation Council, Springfield, Illinois, 88 p.

Young, L. E., and Stoek, H. H., 1916, Subsidence resulting from mining: University of Illinois Bulletin No. 91, Urbana, Illinois, 205 p.

Zwartendyk, J., 1971, Economic aspects of surface subsidence resulting from underground mineral exploitation [Ph.D. thesis]: University Park, Pennsylvania State University. (Available from National Technical Information Service, Springfield, Virginia, as report no. PB 207512.)

Manuscript Accepted by the Society April 18, 1984

Printed in U.S.A.

Coal mine subsidence—western United States

C. Richard Dunrud
U.S. Geological Survey
Box 25046, MS 972
Denver Federal Center
Denver, Colorado 80225

ABSTRACT

Coal mine subsidence is the local lowering of the ground surface caused by coal mining. Subsidence processes above underground mines consist of a gradual downwarping of the overburden into coal extraction panels, which causes depressions, or a sudden collapse into individual mine openings, which causes pits. Subsidence type (depressions, pits), amount, areal extent, rate, and duration are controlled by (1) thickness of coal mined, (2) mine geometry and mining methods, and (3) thickness, lithology, structure, and hydrology of the bedrock and surficial material in the mining area. Subsidence in surface mining areas is caused by compaction of rehandled overburden material (spoil), dewatering of aquifers near surface mines, and (or) stress and strain readjustments. Depressions (troughs), cracks, and pits (sinkholes) can occur under either underground or surface mining procedures.

Maximum vertical displacement in depressions above extraction areas large enough to cause maximum subsidence in the western United States ranges from about 45 to 90 percent of the thickness of coal mined. However, pits may be deeper than the mining thickness, where the caved material can move laterally into adjacent mine openings. In surface mining areas, although little detailed information is available in the western United States, subsidence may range from a few percent to perhaps as much as 10 percent or more of the thickness of the reclaimed spoil. The limit angle, which determines the area of subsidence above underground mines, ranges from about 5° to 30° (from a vertical reference) above underground mines in the western United States. In surface mining areas, however, the subsidence area can range from nearly equal to the mining area, where compaction of rehandled spoil occurs, to many times the mining area, where aquifers are dewatered and undergo compaction.

The time or duration between mining and complete subsidence above underground mines commonly ranges from a few months to a few years, where downwarping occurs above extraction panels, to many years or decades, where pillars are not mined. The duration between mining and the occurrence of pits (sinkholes), however, can vary from a few decades to many decades or even centuries. The time necessary for depressions and pits in surface mining areas is not well known but apparently depends on factors such as methods of emplacing and grading and rate of wetting the rehandled material, rate of dewatering of aquifers near the mine, and stress-strain readjustments.

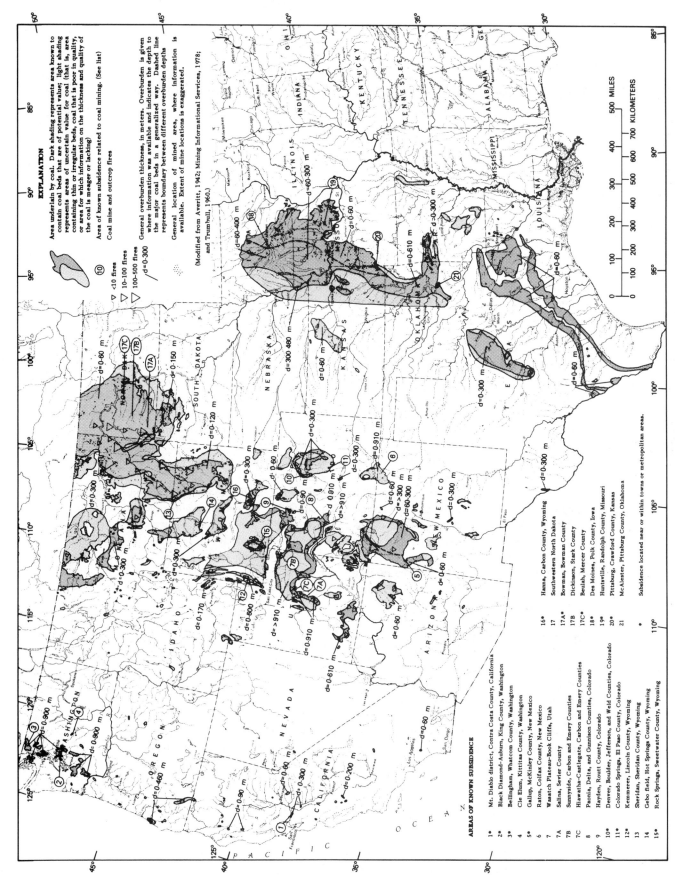

Figure 1. Map showing coal mining, subsidence, and fires reported, or known to the author as of 1980, in the conterminous states west of the Mississippi River. Compiled by Jill M. Jappe and C. Richard Dunrud.

TABLE 1. AREA UNDERLAIN BY COAL DEPOSITS AND AREA AFFECTED BY COAL MINING IN THE WESTERN UNITED STATES*

State	Area underlain by coal deposits	Area of surface mining	Area of underground mining	Total area mined	Abandoned mine area
Arizona	8,300	25	†	25	†
Arkansas	12,300	40	105	145	105
California	600	<5	30	35	35
Colorado	76,700	35	180+	215+	180+
Idaho	1,300	§	§	§	§
Iowa	51,800	60	285	345	315
Kansas	48,700	195	245	440	410
Louisiana	3,500+	§	§	§	§
Missouri	59,600	320	270	590	270
Montana	133,000	50	10	60	20+
Nebraska	800	0	§	§	§
Nevada	100+	0	§	§	§
New Mexico	64,800+	25	§	25+	§
North Dakota	72,500	50	65	115	65+
Oklahoma	37,600	135	235	370	235
Oregon	1,500	<1	§	§	§
South Dakota	18,000	5	†	5	5
Texas	40,500	40	10	50	10
Utah	38,900	5	100+	105+	§
Washington	4,000+	10	95+	105	§
Wyoming	103,600	290	145+	435	155+
Totals	778,100+	1,290	1,775+	3,065+	1,805+

*Areas are in square kilometers (1 square kilometer = 0.386 square miles = 247 acres); areas are only approximate because of disagreement among sources, lack of recent data, and the procedures used in estimating these figures.

+Plus symbol indicates that the actual area may be more, or even double, these figures in some cases.

†Only scattered local underground mines are located in the State; information on the extent and location of these mines is not available as of 1980.

§No information available as of 1980.

SOURCES

Adams, J. R., and Vanston, J. H., 1975, Coal and lignite in Texas; a brief review: The University of Texas at Austin Center for Energy Studies Public Information Report 1, p. 9-32.

Arnold, E. C., and Hill, J. M., 1980, New Mexico's energy resources '79: New Mexico Bureau of Mines and Mineral Resources Circular 172, p. 28-38.

Averitt, Paul, 1942, Coal fields of the United States: U.S. Geological Survey, Mineral Resources of the United States Map, scale 1:2,500,000, 2 sheets.

_____ 1969, Coal resources of the United States, January 1, 1967: U.S. Geological Survey Bulletin 1275, p. 32.

Bateman, A. F., Jr., 1964, Montana's coal resources: U.S. Geological Survey Open-File Report, 20 p.

Beikman, H. M., Gower, H. D., and Dana, T. A. M., 1961, Coal reserves of Washington: Washington Division of Mines and Geology Bulletin 47, 115 p.

Berge Exploration, 1978, Coal resources of the U.S.: Denver, Colorado, Berge Exploration, Inc., scale 1:2,500,000.

Boreck, D. L., Jones, D. C., Murray, D. K., Schultz, J. E., and Suek, D. C., 1977, Colorado coal analyses, 1975: Colorado Geological Survey Information Series 7, 112 p.

Brownfield, M. E., 1981, Oregon's coal and it's economic future, in Oregon Geology: Oregon Department of Geology and Mineral Industries, v. 43, no. 5, p. 59-67.

Bush, W. V., and Gilbreath, L. B., 1978, Inventory of surface and underground coal mines in the Arkansas Valley coal field: Arkansas Geological Commission Information Circular 20-L, 15 p., and map.

Cole, G. A., Daniel, J. A., Heald, B. P., Fuller, D., and Matson, R. E., 1981, 1980 oil and gas drilling and coal production summary for Montana: Montana Bureau of Mines and Geology Open-File Report MBMG-59, 6 p., and map.

Cole, G. A., 1981, Fossil fuels map of Montana: Montana Bureau of Mines and Geology Map MBMG-60, scale 1 in.=1 ft.

Doelling, H. H., 1972, Central Utah coal fields; Sevier-Sanpete, Wasatch Plateau, Book Cliffs and Emery: Utah Geological and Mineralogical Survey Monograph Series 3, 571 p.

Doelling, H. H., and Graham, R. L., 1972a, Southwestern Utah coal fields: Alton, Kaiparowits Plateau and Kolob-Harmony: Utah Geological and Mineralogical Survey Monograph Series 1, 333 p.

_____ 1972b, Eastern and northern Utah coal fields: Vernal, Henry Mountains, Sego, La Sal-San Juan, Tabby Mountain, Coalville, Henrys Fork, Goose Creek, and Lost Creek: Utah Geological and Mineralogical Survey Monograph Series 2, 409 p.

Evans, A. K., Uhleman, E. W., and Eby, P. A., 1978, Atlas of western surface-mined lands; coal, uranium, and phosphate: U.S. Fish and Wildlife Service, p. 10-11, 24-29, 36-61, 68-77, 82-131, 140, 156-188, 250, and D-2 - D-8.

Evans, T. J., 1974, Bituminous coal in Texas: Texas University at Austin, Bureau of Economic Geology Handbook 4, 65 p.

Friedman, S. A., 1974, An investigation of the coal reserves in the Ozarks section of Oklahoma and their potential uses: Oklahoma Geological Survey Report, 117 p.

Glass, G. B., 1978, Wyoming coal fields, 1978: Geological Survey of Wyoming Public Information Circular 9, 91 p.

Harl, N. E., Achterhof, J. B., Anderson, P. F., and Wiese, Karen, 1977, Coal mine maps for eight Iowa counties: Iowa State University Miscellaneous Bulletin 12, p. 17, 45, 53, 67, 98-99, 133, 155, and 181.

Hershey, H. G., 1947, Mineral resources of Iowa: Iowa Geological Survey map, scale 1:1,000,000.

Hinds, Henry, 1951, The coal deposits of Missouri: Missouri Bureau of Geology and Mines, v. 11, 2d series, p. 3, 8, 18, and maps.

Johnson, Wilton, and Miller, G. C., 1979, Abandoned coal mine lands—Nature,extent, and cost of reclamation: U.S. Bureau of Mines, 29 p.

Jones, D. C., Schultz, J. E., and Murray, D. K., 1978, Coal resources and development map of Colorado: Colorado Geological Survey Map Series 9, scale 1:500,000.

Kiilsgaard, T. H., 1964, Coal, in Mineral and water resources of Idaho: U.S. 88th Congress, 2d session, Senate Committee on Interior and Insular Affairs, Committee Print, p. 58-66.

Kirkham, R. M., and Ladwig, L. R., 1979, Coal resources of the Denver and Cheyenne Basins, Colorado: Colorado Geological Survey Resource Series 5, 70 p., 5 pls.

Landis, E. R., 1964, Coal, in Mineral and Water resources of South Dakota: South Dakota State Geological Survey Bulletin No. 16, p. 147-151.

_____ 1965, Coal resources of Iowa: Iowa Geological Survey Technical Paper 4, 142 p.

Lockwood, Helen, 1947, Coal fields of northwestern South Dakota: South Dakota State Geological Survey Map, scale 1 in.=5.7 mi.

Meagher, D. P., and Aycock, L. C., 1942, Louisiana lignite: Louisiana Geological Survey Pamphlet 3, 56 p.

Mining Informational Services, 1977, 1977 Keystone coal industry manual: New York, McGraw-Hill, p. 570-593, 608-613, 622-644, 648-653, 670-678, 684-690, 707-719, and map.

_____ 1978, 1978 Keystone coal industry manual: New York, McGraw-Hill p. 456-477, 494-500, 509-531, 535-540, 557-565, 571-578, and 595-619.

Myers, A. R., Hansen, J. B., Lindvall, R. A., Ivey, J. A., and Hynes, J. L., 1975, Coal mine subsidence and land use in the Boulder-Weld coal field, Boulder and Weld Counties, Colorado: Colorado Geological Survey, EG-9, pls. 1 and 2 (prepared by Amuedo and Ivey).

Robertson, C. E., 1971, Evaluation of Missouri's coal resources: Missouri Geological Survey and Water Resources Report of Investigations 48, p. 1-22, and maps.

Stroup, R. K., and Falvey, A. E., 1969, Coal reserves for steam-electric generation in Kansas and Missouri areas of Missouri River Basin: U.S. Bureau of Mines Preliminary Report 174, p. 1, 6, and 11, figs. 1 and 2.

Trumbull, James, 1960, Coal fields of the United States: U.S. Geological Survey Map, Sheet 1, scale 1:5,000,000.

U.S. Department of Agriculture, 1979, The status of land disturbed by surface mining in the United States; basic statistics by state and county as of July 1, 1977: Soil Conservation Service SCS-TP-158, 124 p.

U.S. Geological Survey, 1974, Stripping coal deposits of the northern Great Plains, Montana, Wyoming, North Dakota, and South Dakota: U.S. Geological Survey Map, scale 1:1,000,000.

Additional sources: written and (or) oral communications with State officials in the State Geological surveys of Arizona, California, Colorado, North and South Dakota, Idaho, Iowa, Kansas, Oregon, Wyoming, Iowa, Missouri, Montana, New Mexico, Texas, Utah, and Washington; and government officials at the Office of Surface Mining and the U.S. Bureau of Mines. Information also was obtained from unpublished data on mine fires and subsidence from the U.S. Bureau of Mines.

INTRODUCTION

Lands underlain by abandoned coal mines are being developed for residential and industrial use at an increasing rate to accommodate population increases. At the same time, coal mining activities are increasing rapidly in the western United States in response to increasing demands for energy. Conflicts are therefore increasing about how the land should be best used in the public interest.

The effects of subsidence on surface development and on the environment must be known in order to make decisions on the best use of lands underlain by abandoned mines or by economic coal deposits. Subsidence processes also must be understood in order to analyze and predict subsidence type, amount, rate, and duration above abandoned and active mines to help establish land use priorities where conflicts exist; reduce hazards to life and property; and reduce financial loss to individuals, counties, states, and the nation.

The nature and extent of coal mining and mine subsidence must be known if hazards and potential financial losses are to be known accurately. Coal deposits underlie about 784,000 km^2 (303,000 mi^2), or about 15 percent of the land surface in the western United States (Fig. 1). Coal mines are located in 18 of the 21 states west of the Mississippi in approximately 2,800 km^2 (area known as of 1980), or 0.4 percent of the land underlain by coal deposits. Of this total, about 2,000 km^2, or 70 percent of the mined lands, were mined by underground methods. Some form of subsidence has occurred locally above many of these mining areas.

Purpose and Scope

The purpose of this report is to discuss the nature and extent of subsidence in the western United States, subsidence processes in underground and surface coal mines, and factors controlling subsidence in order to provide a basis for further study and analysis of subsidence. Results of case studies of subsidence and fires caused by underground mining, subsidence hazards, subsidence prediction and control, and land use also are discussed. The report also compares results of subsidence in the western United States with those of other countries.

Definition of Terms and Symbols

Coal mine subsidence is defined as the local lowering of the ground surface into underground mine openings or by the compaction, or decrease in void space, of rehandled earth materials in surface mining areas. It includes all vertical and horizontal deformations within the overburden that are caused by the movement of coal, rock, and surficial material into underground coal mine openings and includes vertical and horizontal movements of the ground surface caused by surface mining, exclusive of slope failures.

Overburden includes all rock and surficial material that overlies the coal bed or coal mine openings. Mine openings are all underground cavities created during the mining process. They include entries, rooms, and crosscuts. Mining faces are coal or rock faces being mined to produce the mine openings or to extract coal. The goaf (also called gob) is a mined-out area or abandoned mine area. Critical extraction area (also called critical area) is the mined area necessary to cause maximum surface subsidence; subcritical extraction areas are less than critical areas; and supercritical extraction areas are larger than critical areas. The minimum critical extraction area is a circular area, whose diameter equals the length of a square critical extraction area and is also called area of influence (Fig. 7-A).

The limit line is defined as a straight line from the edge of the mine area to the limit of measurable subsidence at the surface. The break line is a straight line from the edge of the mine area to the point of maximum extension at the surface. The limit angle (ϕ) (also called draw angle) is the angle that the limit line makes with a vertical reference. The break angle (β), which is less than the limit angle, is the angle that the break line makes with a vertical reference (Fig. 7).

Maximum vertical subsidence (S) is the maximum vertical displacement measured above a given mine area or mine opening for a given thickness of coal mined; S_h is the maximum horizontal displacement caused by the local lowering of the ground surface. Slope (M) is the change of vertical displacement with respect to unit horizontal distance of the original ground surface. Curvature (C) is the change in slope with respect to unit length of surface in radians/m. Horizontal strain (e) is the ratio of the change in length of the ground surface to its original length that is caused by curvature; extension is defined as positive and shortening as negative (Fig. 7).

NATURE AND EXTENT OF COAL MINING AND SUBSIDENCE

Coal is mined in increasing quantities in many states west of the Mississippi River; thick and extensive coal deposits are being produced for use in electric power stations and for other uses. The coal commonly is mined by underground methods where the overburden is thicker than about 60 m (200 ft) thick and by surface methods where the overburden thickness is less than about 60 m (200 ft) (Fig. 1). Coal mining by underground methods commonly is considered economic where the overburden thickness is less than about 915 m (3000 ft). Coal deposits locally occur at depths in excess of 1830 m (6000 ft) in states such as Utah and Colorado. On a state-by-state basis, the area underlain by coal deposits ranges from about 100 km^2 in Nevada to as much as 133,000 km^2 in Montana.

The area affected by past and present coal mining ranges from a few square kilometers in South Dakota and Idaho to more than about 350 km^2 in Iowa, Kansas, Missouri, Oklahoma, and Wyoming (Fig. 1). The aggregate thickness of coal mined by underground and surface methods, in the past and at the present time, ranges from a few decimeters to a meter or so, in a single bed in Iowa, to as much as 92 m (300 ft) from 12 different beds in the Kemmerer, Wyo., area (Sprouls, 1982, p. 42-45). The overburden thickness above underground mines ranges from a few meters to about 795 m (2600 ft).

The total area underlain by underground mines, known as of 1980, ranges from a few square kilometers or less in Idaho, Louisiana, Nebraska, Nevada, and South Dakota, to about 150 to 300 square kilometers in Colorado, Iowa, Kansas, Missouri, Oklahoma, and Wyoming (Table 1 in Fig. 1). Information on the extent of abandoned underground coal mines is difficult to obtain because compilation is only now beginning in many of the states as of 1980. However, some of the larger areas underlain by abandoned underground mines, such as in the metropolitan areas of Denver and Colorado Springs, Colo.; Rock Springs, Wyo.; Gallup, N.M.; Black Diamond-Auburn, Wash.; Pittsburg, Kansas; and Des Moines, Iowa, are located near or within major population centers.

Mining Methods

The room-and-pillar (R&P) and longwall (LW) mining methods, which commonly were used in the past and still are used today in many mines of the western United States, are outgrowths of mining methods that were adapted from those used in the 13th to 19th centuries in the United Kingdom (Fig. 2). The longwall method also was used in some areas, such as in the Des Moines, Iowa, area. The room-and-pillar method of mining was developed in the 17th century in various areas of the United Kingdom, where the overburden was too thick to mine by surface methods or from individual shafts. Haulageways were driven outward from shafts; pillars were left to support the haulageways, some of which were arched to support the roof for the life of the mine (Piggot and Eynon, 1978, p. 750). During the 16th and 17th centuries, more regular mining patterns were used where rail haulage was implemented and where mechanized pumps were used to dewater the mines. Access to the mines was by shaft, where no coal outcrop existed in the area, or by portal, where the coal cropped out.

During the 17th century, various room-and-pillar geometries were used in the United Kingdom, depending on the background of miners and site conditions. Examples include the bord-and-pillar (England), stoop-and-room (Scotland), and post-and-stall (South Wales) (Piggot and Eynon, 1978, p. 752-754; Fig. 2A, B, C). The coal commonly was partially extracted in order to support the roof.

The Shropshire, or longwall, and "short" longwall, or shortwall, methods were developed in the 17th century in England and Scotland, respectively (Piggot and Eynon, 1978, p. 753; Littlejohn, 1979, p. 22). With these methods, a face of coal, considerably longer than mining faces developed in room-and-pillar areas, was mined. The mine roofs near the working face and haulageways were supported with mine refuse, timbers, quarry rocks, and cribs (Fig. 2D).

The panel and barrier mining geometry was developed in

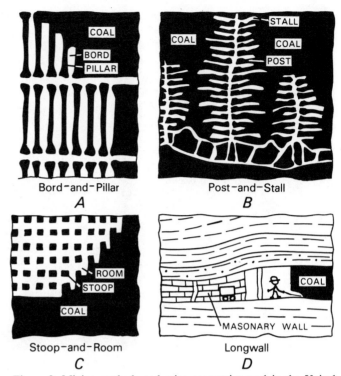

Figure 2. Mining methods and mine geometries used in the United Kingdom during the 17th century. A through C are plan views; D is a profile. A. Bord-and-pillar (England). B. Post-and-stall (South Wales). C. Stoop-and-room (Scotland). D. Longwall. Modified from Littlejohn (1979, p. 25).

northern England in the early 1800's (Piggot and Eynon, 1978, p. 754). Haulageways, supported by barrier pillars, were driven to the extraction areas (panels). The coal commonly was extracted completely in the mining panels where no subsidence restrictions existed. This panel-barrier geometry provided haulageways that were protected by adjacent barrier pillars and therefore allowed more extensive areas to be mined from common shafts or portals. Pillars commonly were extracted shortly after development, in order to prevent premature closing of the mine openings (squeezes). Various combinations and modifications of the stoop-and-room, bord-and-pillar, or post-and-stall geometries were used in many mines of the western United States during the 1800's and early to mid-1900's (Fig. 3A, B).

In currently active room-and-pillar coal mines in the western United States, the pillars commonly are extracted shortly after room development to minimize hazardous mine deformations. The longwall and shortwall methods are essentially extensions of the Shropshire and shortwall methods of the 17th century, which incorporate 20th century state-of-the-art mechanization for cutting and hauling coal. Coal faces across a panel are cut with track-mounted cutting machines (longwall) or continuous mining machines (shortwall), which use self-advancing roof-support systems and ventilation systems that are designed to comply with modern mine health-and-safety standards (see, for example, Peng, 1978, p. 14-25, 208-280 (LW, R&P); Kauffman and others,

Figure 3. Aerial oblique photographs of subsidence depressions and pits in the Sheridan, Wyo., area. *A*. Depressions and local pits occur above areas mined in a geometry similar to the bord-and-pillar method (Fig. 2*A*) to right of reclaimed land, which was mined by surface methods; overburden is about 15 to 20 m thick; large, half-moon-shaped pit in right center is about three times deeper than thickness of coal mined, where surface drainage was diverted to mine workings and subsurface erosion occurred (Nov. 1981). *B*. Irregular pattern of pits apparently occurs above bord-and-pillar and post-and-stall mine workings; overburden ranges in thickness from about 5 to 45 m (Nov. 1981).

1981 (R&P); Olson and Tandanand, 1977, 198 p. (LW); and Kuti, 1979, p. 1593–1602 (LW) for details on mine layout and mining procedures). The panel-pillar mine layout facilitates mine haulage, increases development and extraction efficiency, mine stability, and the health and safety of the miners.

Subsidence Features

Subsidence features above underground mines in the western United States consist of depressions (troughs), cracks, and pits (also called cave holes, plump holes, or crown holes) (Fig. 3A, B). In plan view, depressions or troughs commonly are rectangular where the overburden is less than about 15 m (50 ft) thick but become more elliptical as the overburden thickens (Dunrud and Osterwald, 1980, p. 17). In cross section, the depressions are bowl-shaped with smooth, continuous margins that locally are disrupted by cracks. They form above mine panels where the coal remaining cannot support the weight of the overburden. Maximum depth of the depressions is less than the thickness of coal mined.

Subsidence pits commonly are circular in plan and typically form steeply dipping to vertical rims. They usually form above individual mine openings where the overburden thickness is less than about 10 to 15 times the thickness of coal mined. Pits can be deeper than the thickness of coal mined (Dunrud and Osterwald, 1980, p. 19; Fig. 3A). Pits also can occur within depressions, where collapse occurs above individual mine openings and where pillars, adjacent to these openings, yield from the weight of the overburden (Fig. 3A).

Subsidence features in surface mining areas also consist of depressions, cracks, and local pits. Although few studies and measurements have been made in the western United States, the amount of subsidence apparently is only a small percentage of the thickness of coal mined. Subsidence depressions, cracks, and local pits may occur where (1) differential compaction occurs because of unequal compaction of spoil during rehandling, lithologic changes in earth materials, or local increased saturation of rehandled earth materials; (2) coal or rocks that occur near or within the mine area are dewatered; or (3) differential loading of the ground causes stress-strain readjustments.

SUBSIDENCE PROCESSES ABOVE UNDERGROUND MINES

Subsidence is caused by the lowering or collapse of the overburden or roof into the underground mine openings, by the yielding of coal pillars, and (or) by the yielding of rocks above or below the pillars. Knowledge of in-situ stresses, stress changes, and deformations around underground mine openings is important to the analysis and prediction of subsidence. Knowledge of the behavior of the bedrock and surficial material also is important to the analysis of subsidence effects on other resources and on natural and manmade surface structures.

In-Situ Stresses

In-situ stresses in sedimentary rocks are caused primarily by (1) the weight of the overburden, (2) sedimentation processes, (3) tectonic stresses, and (4) topography. The state of stress present during mining reflects the total loading history of the rocks and therefore may be significantly different from that expected from gravitational loading of present overburden thicknesses. Stresses in rocks in deep mines, for example, may be less than stresses produced by the weight of the overburden, whereas the stresses in shallow mines can be greater than those expected due to overburden loading (Corlett and Emery, 1959, p. 377; Jaeger and Cook, 1979, p. 374–375).

If gravitational loading is assumed, the vertical component of stress (σ_v) produced by the weight of the overburden can be estimated by:

$$\sigma_v = \rho g d \quad (1),$$

where ρ is the density of the overburden, g is gravitational acceleration, and d is the overburden thickness. Under this assumption, the vertical stress gradient is about 24 MPa/km (1 psi/ft, 24 kPa/m where $\rho = 2400$ kg/m^3.

The horizontal component of stress (σ_h) in horizontally stratified rocks and in tectonically stable areas commonly is estimated by:

$$\sigma_h = \left(\frac{\nu}{1-\nu}\right)\sigma_v \quad \text{or} \quad \frac{\sigma_h}{\sigma_v} = \frac{\nu}{1-\nu} \quad (2),$$

where ν equals Poisson's ratio (for example, Hoek and Brown, 1980, p. 95). The vertical gradient of horizontal stress therefore is 0.43 σ_v, or about 10 MPa/km (0.45 psi/ft or 10 kPa/m), where $\rho = 2400$ kg/m^3 and $\nu = 0.3$.

In the past decade, however, horizontal stresses ranging from about 0.5 to 5.5 times the vertical stress have been measured at depths ranging from 100 to 1,000 m below the surface (Hoek and Brown, 1980, p. 95–101). Horizontal stresses can be greater than the vertical stress due to tectonic activity, processes of sedimentation and compaction, erosion and deposition of reworked material, and time-dependent strain readjustments of the rock mass at depth.

Intergranular adjustments of grains during compaction and loading, for example, may cause the horizontal stresses to be 0.4 to 0.75 times the vertical stress, according to Capper and Cassie (1963, p. 99–100). Horizontal stresses may be greater than the vertical stress in areas where rocks are folded and cut be reverse faults. Conversely, horizontal stresses may be less than stresses estimated by (2) in areas where normal faulting is present. Stresses produced by folding can alternately increase and decrease horizontal stresses in anticlines and synclines (Jeremic, 1981b, p. 704–708).

Horizontal and vertical stresses may become about equal at depth by undergoing time-dependent deformation. Subsequent erosion could cause a decrease in the vertical stress, as material is eroded, whereas horizontal stresses might decrease only slightly because the rocks commonly are laterally restrained.

Topographic effects from cliffs, canyons, and ridges also alter vertical and horizontal stresses. Horizontal stresses may be reduced relative to the vertical stresses near ridges and cliffs; vertical stresses may be reduced relative to the horizontal stresses beneath canyons.

Mine-Induced Stresses and Deformations

Excavation of mine openings changes the in-situ or premining stresses. Overburden previously supported by the material excavated now must be supported by the coal and rock adjacent to the mining openings. The in-situ stresses commonly are transferred to unmined coal and rock by the formation of compression arches above rocks of the immediate roof (for example, Grond, 1957, p. 157–158, April; p. 197–205, May; Whittaker and Pye, 1977, p. 305–307; Mohr, 1956, p. 140–152; Schoemaker, 1948, p. 6–7) and by the immediate roof rocks behaving as plates (for example, Stephansson, 1971, p. 5–71). Large-scale deformations—such as downwarping in the roof, bulging and heaving of mine rib and floor, and related fracturing and flowing—can occur in the coal and rock where the stress increase near the openings exceeds the strength of the material. The strength of the coal and rock also is reduced, compared with its premining state, because the confining stresses are reduced.

Mining in the panel areas commonly consists of either extracting a maximum amount of coal or mining part of the coal bed. Where complete extraction procedures are used, a maximum amount of coal is mined by room-and-pillar, longwall, or shortwall methods; the mine openings are temporarily supported until extraction is complete; the mine roofs are allowed to cave; and the remaining pillars yield until the remaining coal and the caved rocks can support the overburden stress. Where partial extraction procedures are used, the coal is mined by room-and-pillar methods and pillars are left for support.

The roofs of the mine openings break and cave as the mine openings are widened beyond the limits of stable roof spans. Results of analytical and physical model studies (for example, Stephansson, 1971, p. 5–71) and underground mapping studies by the writer in the western United States indicate that the ratio of the thickness of the rock units in the roof to the width of mine opening often controls the type of failure likely to occur. The roof rocks commonly break along steeply dipping to vertical fractures where the rocks are thinner than about one-fifth of the width of the mine openings and where flexural tensile stresses exceed their tensile strength. En-masse failure often occurs along inclined shear fractures where the rocks are thicker than one-fifth to one-half of the width of the mine openings and where shear stresses exceed the shear strength of the rocks. Blocky to slabby fragments a few centimeters to a meter or so in maximum dimension commonly are produced where thinly layered rock units fail, whereas masses of rock as large as the width of the mine opening can fail en-masse along shear fractures in thick rock units (Fig. 4).

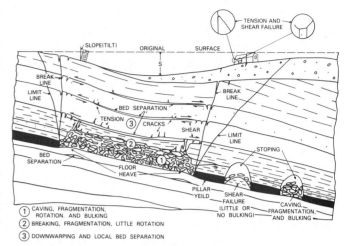

Figure 4. Conceptual diagram showing coal mine subsidence processes and deformations.

Convergence of mine openings. The roof and floor of the mine converge until the overburden load is supported by (1) the coal remaining after mining, (2) the rocks above and below the coal, (3) the caved rocks in the extraction areas, and (4) any material that is placed in the mined-out areas (goaf). The load-carrying capacity of the coal pillars reportedly decreases with increasing convergence of the mine roof and floor, whereas the load-carrying capacity of the caved and backfilled material increases with increasing convergence of the mine roof and floor (Jaeger and Cook, 1979, p. 491–492). The ratio of subsidence to extraction thickness is controlled by the load-carrying capacity of the remaining coal, roof and floor rocks, and the caved rock. This ratio may be very small where pillars are designed to support the roof, or it may approach 1 where nearly all the coal is mined and the caved rocks eventually compact to nearly their original volume.

Stress, deformation, and strength of pillars. Coal pillars commonly consist of a yielded and fractured zone near their perimeter and a less fractured core zone. Much or most of the load-carrying capacity of pillars occurs within this core zone. Studies show that the more important factors to consider in design of stable pillars are (1) average vertical stress on pillars, (2) stress distribution within the pillars, (3) in-situ pillar strength, (4) shape and width-to-height ratio of pillars, and (5) factor of safety (ratio of pillar strength to stress) (for details see, for example, Bieniawski, 1981, p. 13–22; Babcock and others, 1981, p. 23–34; Pariseau, 1980, p. 57–72).

Pillar strength is controlled by the lithology, structure, and lateral confining stress provided by the roof and floor rocks, in addition to the strength of the coal. Pillar strength and confining stress levels may be high where strong roof and floor rocks with large coefficients of friction, such as sandstones containing ripple marks or casts of burrowing animals are present. Conversely, pillar strength and confining stress may be low where weak rocks with low coefficients of friction, such as claystones and soft shales, occur above and below the coal. Pillar strength and con-

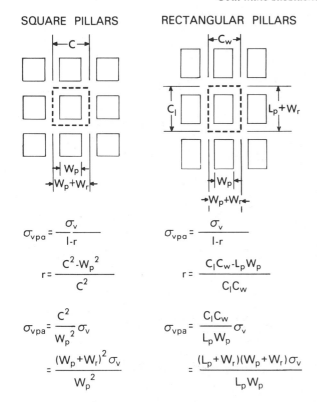

Figure 5. Relations among extraction ratio (r), mining dimensions, and average vertical stress (σ_{vpa}) on square and rectangular pillars in room-and-pillar mining area, where the length and width of the extraction panel is greater than the overburden thickness. Modified from Peng (1978, 175–176).

fining stress may be high to intermediate where strong rocks, such as limestones and siltstones, occur above and below the coal. Care should be used, for example, in evaluating confinement where the roof rocks are strong, but are overlain by weak material, such as shale or thin coal beds (rider coals). Lateral movement can occur in the weaker rocks or in the underlying pillars, which may cause buckling or breaking in the stronger rocks.

Partial extraction procedures may be required where subsidence is to be reduced to a minimum. Stresses and strength of pillars must be determined in these instances.

The average vertical stress on coal pillars of uniform size can be estimated by the tributary method, where the lateral extent of the mining area is greater than the overburden thickness (for example, Peng, 1978, p. 175–177; Cook and Hood, 1978, p. 136). Each pillar is assumed to support an equal amount of in-situ vertical stress (σ_v) and one-half of the four adjacent rooms. The average vertical stress on each square pillar (σ_{vpa}) (Fig. 5) is related to the vertical stress (σ_v), the centerline spacing of the rooms (C), the width of the rooms (W_r), the pillar width (W_p), and the coal extraction ratio (r) by:

$$\sigma_{vpa} = \frac{\sigma_v}{1-r} = \frac{(W_p + W_r)^2}{W_p^2} \sigma_v \quad (3).$$

The average vertical stress on each rectangular pillar is:

$$\sigma_{vpa} = \frac{C_l C_w}{L_p W_p} \sigma_v = \frac{(L_p + W_r)(W_p + W_r)}{L_p W_p} \sigma_v \quad (4),$$

where L_p is pillar length, and C_l and C_w are the centerline spacings of the rooms (Fig. 5).

The strength of coal pillars is governed by height, in addition to length and width. Pillars become stronger as their width-to-height ratio increases. The strength of pillars (S_p) commonly is related to the thickness of coal mined (t) and pillar width (W_p) by:

$$S_p = \sigma_c [A + B(\frac{W_p^a}{t^b})] \quad (5),$$

where W_p for rectangular pillars equals the square root of the product of pillar length and width (Peng, 1978, p. 184–189), σ_c is the uniaxial compressive strength of a cube of coal with sides equal to, or greater than the critical size (which commonly is about 1.5 m), A ranges from 0 to 1000, B ranges from 0.2 to 2800, and the exponents a and b range from about −0.3 to 2 and 0 to 2, respectively. Pillars reportedly are stable where A = $1000/\sqrt{t}$, B = 20, a = 2, b = 2 (the value for A apparently is in English units; see, for example, Babcock and others, 1981, p. 26–28; Bieniawski, 1981, p. 16–20; Peng, 1978, p. 184–193 for details).

Extraction ratios decrease with increasing overburden thickness or with decreasing strength of the coal in order for the pillars to provide adequate support. Extraction ratios for pillar stability for various depths, assuming that each pillar supports an equal amount of the overburden load, can be estimated by rearranging (3) to give:

$$r = 1 - \frac{\sigma_v}{\sigma_{vpa}} \quad (6).$$

Using a safety factor of 1.6 and assuming an average overburden density of 2500 kg/m³, Cook and Hood (1978, p. 137) report that the extraction ratio for various overburden thicknesses and strengths of coal pillars can be estimated by:

$$r = 1 - \frac{1.6\sigma_v}{S_p}, \quad r = 1 - \frac{(25)(1.6)}{S_p} = 1 - \frac{40\,d}{S_p} \quad (7).$$

However, the extraction ratio is 0 at a depth of 0.5 km if S_p = 20 MPa, using (7).

Deformation of roof and floor rocks. As many as five different types of deformation can occur around mine openings, depending on such factors as the mine geometry, mining method, and goaf treatment; thickness and strength of the overburden; and lithology and structure of the rocks above and below the mine openings. The types of deformation include (1) downwarping with tensile failure and local buckling of roof rocks (buckling is most common in thin-bedded rock units), (2) shear failure of coal, roof rocks, and floor rocks, (3) tensile failure of rock units overhanging the mine openings, (4) uplift or heave of floor rocks, and (5) punching of pillars into roof and floor rocks, where the strength of the coal exceeds the strength of the rocks.

In mine openings with large width-to-height ratios, such as in coal extraction panels, downwarping, flexure, and tensile failure are common, as mine roof span areas are increased; shear failure also may occur locally (Fig. 4). Breaking, fragmentation, and rotation of fragments commonly occur in the mine roofs because large differential vertical displacements occur over short lateral distances. En-masse failure also can occur along joints, bedding, and other preexisting fractures. Higher in the roof, downwarping, with local breaking and only minor rotation of fragments, occurs because vertical deflections are reduced by the underlying caved rock that has increased in volume. Higher in the roof, where vertical deflection relative to horizontal distance is further reduced, downwarping of rock units as multiple plates causes local bed separations and local tension fractures and compression features.

Studies by the writer and others (for example, Kanlybayeva, 1964, p. 12; Ropski and Lama, 1973, p. 118) indicate that three rather distinct zones of deformation in the roof above extraction panels can be distinguished. These include, from the mine roof upward, a zone of (1) caving, fragmentation, and rotation of fragments for a vertical distance equal to about two to three thicknesses of coal mined; (2) breaking, fragmentation without much rotation, and separation along bedding planes for a distance of another mining thickness or so; and (3) downwarping and flexure of strata as multiple plates, with local bed separations and local fracturing in the remaining overburden. Large shear stresses can occur across lithologic contacts and at the neutral surfaces of rock units subjected to downwarping and flexure (Fig. 4).

Bulking. Bulking, or volumetric increase, of the caved rock fragments relative to their in-place volume, which is a major factor controlling subsidence, is controlled by the type of deformation around and within the mine openings (Fig. 4). Both temporary and permanent bulking may occur. Temporary bulking occurs where rocks separate along bedding planes during downwarping of strata, but where they eventually close. Permanent bulking commonly occurs where permanent increase in void space occurs in the caved rock or in downwarped rocks with different strengths. The bulking factor therefore is controlled by the size and shape of the broken rocks, the geometry of the cave zones, the contact stresses among rock fragments within the rubble pile, and the relative strengths of the affected rocks.

The bulking factor commonly is greatest where rock fragments fall to the floor of the mine openings in a random fashion and where strong, massive rocks occur in the downwarped zone. The bulking factor for the caved rocks decreases as the mine openings fill up because the fall distance is reduced and the blocks are more likely to assume their bedding attitude. Although the bulking factor of the caved rubble may decrease with increased loading, presence of water, and time, rock fragments usually occupy a significantly larger volume than rock in place. En-masse failure, however, causes very little bulking where little or no rotation occurs (Fig. 4).

The bulking factor (B) for caved fragments is defined as (Fig. 6):

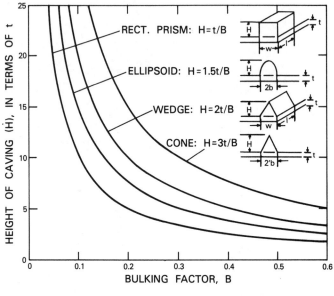

Figure 6. Maximum height of stoping for various caving geometries. Graph shows the relation between height of caving (H), in terms of thickness of coal mined (t), and the bulking factor (B) for various caving geometries. Sample calculation for one-half the volume of an ellipsoid of revolution is shown.

$$B = \frac{V_c - V_o}{V_o} \qquad (8),$$

where V_c is the volume of the caved rock and V_o is the volume of the rock in place. The maximum possible height of caving or stoping is related to the thickness of coal mined (t) and the cavity geometry, assuming uniform bulking throughout the caved area (for example, Piggot and Eynon, 1978, p. 761–762). For bulking factors of 50 percent and 20 percent (0.5 and 0.2), the maximum caving heights (H) for a rectangular prism, ellipsoid, wedge, and cone cavity geometry are:

rectangular prism (also cylinder)—H = t/B; H = 2t and 5t (8-a),
ellipsoid of revolution (half volume)—H = 1.5t/B; (8-b),
 H = 3t and 7.5t

wedge—H = 2t/B; H = 4t and 10t (8-c),
cone—H = 3t/B; H = 6t and 15t (8-d).

The most common cavity geometry observed by the writer are geometries approximating an ellipsoid, although prism, cylinder, wedge, and cone geometries also were observed locally in the Sheridan, Hanna, and Kemmerer, Wyo., areas.

Bulking factors commonly range between 0.3 and 0.5, according to Piggot and Eynon (1978, p. 761–762). However, they may range from 0.2 to 0.5, where the overburden consists primarily of soft claystones and shales, such as in the Sheridan, Wyo., area (Dunrud and Osterwald, 1980, p. 12–20). The bulking factors could be lower than 0.2 and caving could be higher than predicted by even the conical goemetry, where the caved rocks fall into water-filled mine openings and are totally or partially disaggregated and (or) transported laterally by the water.

Surface Subsidence

Studies and observations by the author in Utah and western Colorado (Dunrud, 1976, 39 p.) and in northeastern Wyoming (Dunrud and Osterwald, 1980, 49 p.) indicate that subsidence is caused by downwarping of the overburden into mining panels, by successive caving (stoping) of the roofs of individual mine openings, and by other processes such as shaft failure. Both downwarping and stoping can occur above mine openings, depending on (1) width-to-height ratio of the mine openings, (2) mining procedures, and (3) overburden thickness and strength (Fig. 4). Summaries of the results of various analytical and finite element model studies of subsiding overburden, as single or multiple plates, are discussed elsewhere (for example, Voight and Pariseau, 1970, p. 730–735; Berry, 1978, p. 781–811; Savage, 1981, p. 195–198; Singh, 1979, p. 97–99).

Downwarping. Results of subsidence studies by many investigators (for example, Shadbolt, 1978, p. 724–726; Zwartendyk, 1971, p. 14–33, 123–144) show that vertical lowering, or downwarping, of the surface and overburden rocks causes (1) vertical displacement (s), (2) horizontal displacement (s_h), (3) slope (tilt) (M), (4) curvature (C), and (5) horizontal strain (e). According to the National Coal Board of the United Kingdom (NCB, 1975, p. 28) and Wardell (1971, p. 209), slope and strain are proportional to the ratio of maximum vertical displacement to overburden thickness (S/d). According to Wardell (1971, p. 209), curvature is proportional to S/d^2. Slope and strain, which are the two most important subsidence parameters used in determining potential effects on surface structures, therefore, increase as the thickness of coal mined increases and as the overburden thickness decreases. For example, slope and horizontal strain double as the mining thickness increases two fold or the overburden thickness decreases twofold.

Analysis of profiles of subsidence depressions show that vertical displacement increases inward from the margin of the depressions, reaches about one-half maximum subsidence at the point of inflection, and attains a maximum value in the middle of the depression (Fig. 7). Maximum subsidence commonly occurs in the middle of the subsidence depression (trough) above critical mining areas (areas which cause maximum vertical displacement) or in the flat parts of the subsidence depression above supercritical extraction areas (Fig. 7A, B). Horizontal displacement and slope increase inward from the margins of the depressions to a maximum at the point of inflection, or transition point, and decrease again to zero at the point of maximum subsidence.

Tensile strain and convex bending of the ground surface increase inward from the limit of subsidence to a maximum about midway between the limit of subsidence and the point of inflection, and they decrease again to zero at the point of inflection. Compressive strain and concave bending of the ground surface increase inward from the point of inflection to a maximum about midway between the point of inflection and the point of maximum subsidence and decrease to zero where the point of maximum subsidence occurs. Mining two adjacent panels, which are separated by a barrier pillar, can cause curvature and strain above the barrier of about twice the amount measured above a single panel because the two subsidence profiles are additive where they overlap above the barrier (Fig. 7C).

Affected area. The area affected by subsidence depressions increases with overburden thickness at constant limit angle. The limit lines also tend to steepen somewhat (limit angles decrease relative to vertical reference) with decreasing overburden thickness. For example, in the Sheridan, Wyo., area, the subsidence areas are only slightly greater than the mining panels, where the overburden is less than about 25 m thick, but they commonly increase at a greater rate than can be accounted for at a constant limit angle as the overburden becomes thicker than about 25 m. Analytical model studies of a subsiding elastic layer by Savage (1981, p. 195–198) also indicated that the limit angle decreases with decreasing overburden thickness. The shapes of subsidence depressions approximate the shape of the mining panels, where the overburden is less than about 25 m thick, but become more circular or elliptical as the overburden becomes thicker than about 25 m (Dunrud and Osterwald, 1980, frontispiece A, p. 17, 24–25).

The critical extraction area, or area of influence, is the area that must be mined in order to cause maximum surface subsidence. The diameter of the critical area (D_c) observed in the United Kingdom, for example, is about 1.4 to 1.5 d (Fig. 8). The critical extraction area (A_c) can be calculated, once the critical extraction angle (γ) is known, from the equation:

$$A_c = \pi R_c^2 = \frac{\pi}{4} D_c^2 \qquad (9),$$

where:

$$R_c = d \tan\gamma \, ; \, D_c = 2 \, d \tan\gamma \qquad (9\text{-a}).$$

The critical extraction angle can be defined as the angle that a straight line makes from the solid coal boundary of the extraction panel to the point at which maximum subsidence occurs (Fig. 7A). It is often assumed that the critical extraction angle equals the limit angle (for example, Shadbolt, 1975, p. 115; Pöttgens, 1979, p. 269); however, it commonly is greater than the limit angle in the western United States. For example, the critical extraction angle in the Salina, Utah, and Somerset, Colo., areas averages about 25° and 35°, whereas the limit angles range from 5° to 20° and 15° to 25°, respectively.

Vertical displacement. Maximum vertical displacement (S) is controlled by the thickness of coal mined (t), the width (W)

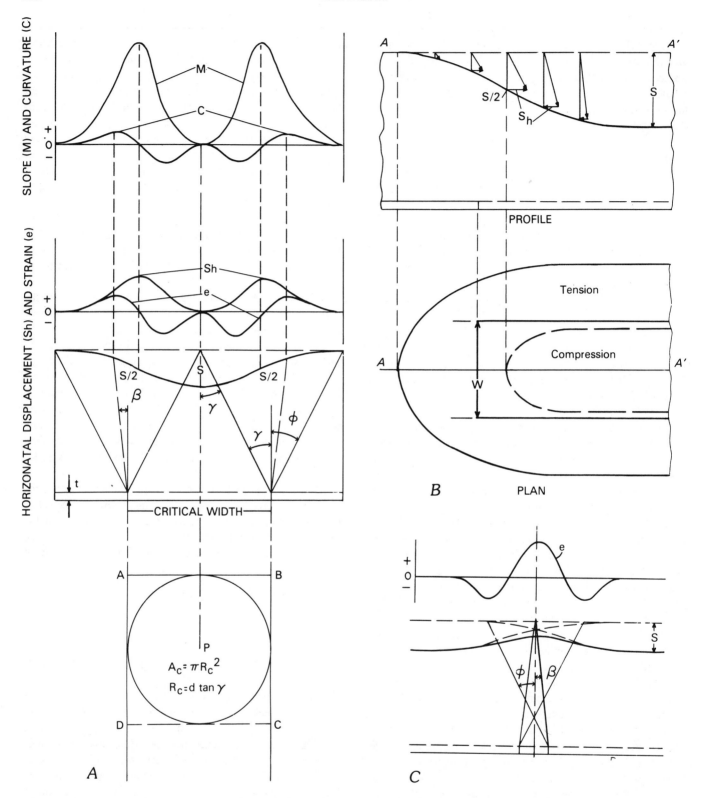

Figure 7. Conceptual subsidence maps and profiles. *A.* Subsidence profile, slope or tilt (M), curvature (C), horizontal displacement (S_h), and strain (e) above a critical extraction panel; circular area, A_c, or rectangular area, ABCD (area of influence, critical extraction area), must be extracted before maximum subsidence S occurs at point P. *B.* Map and profile of a subsidence depression (trough), showing vertical and horizontal displacement and zone of tension and compression above an extraction panel of critical width. Modified from Shadbolt (1978, p. 725–728), National Coal Board (1975, p. 3–5), and Singh (1979, p. 94–96). *C.* Superposition of subsidence profiles above a barrier pillar between two adjacent extraction panels; tensile strain is about twice that for a single panel.

and length (L) of the mine panels, and overburden thickness (d). The length of the panel must equal its width, or else the length will be the dominant dimension controlling the amount of displacement (Fig. 8). Plots of subsidence ratio (a = S/t) versus mining width to overburden thickness ratio (W/d) show that subsidence ratio commonly increases markedly where W/d ranges from about 0.3 to 1.0 and attains a constant value where W/d ranges from about 1.2 to 2.2 in the United Kingdom and about 1.0 to 1.4 in the western United States (Orchard *in* Shadbolt, 1978, p. 727–730; Wardell, 1971, p. 206).

Maximum vertical displacement can be estimated by:

$$S = a\,t \qquad (10),$$

where the subsidence ratio (a) is the ratio of maximum subsidence to the average thickness of coal mined (Fig. 8). Subsidence above vertically superimposed coal mines in n coal beds can equal the sum of the subsidence caused by each mine or:

$$S = a_1 t_1 + a_2 t_2 + \ldots + a_n t_n = \sum_{i=1}^{n} a_i t_i \qquad (11).$$

The values $a_1, a_2, \ldots, a_n$ may be about equal where mining and geologic conditions are uniform, or they may vary significantly where mining and geologic conditions are variable.

Rate and duration of subsidence. Subsidence development curves (Wardell, 1954, p. 471–482) are useful to correlate subsidence rate and duration in areas where different geologic conditions occur and where different extraction panel geometries and goaf treatments are used (Fig. 9A; NCB, 1975, p. 42–43). The ratio of vertical displacement at a point (P in Fig. 9) of measurement to maximum vertical displacement (s/S) is plotted in terms of the ratio of the distance from the mining faces beneath this point to the overburden thickness (X/d), or critical diameter (D_c). The curves show that (1) subsidence begins when longwall mining faces or pillar lines are within about 0.2 d of the point of measurement, (2) the rate of subsidence is greatest when longwall faces or pillar lines move from beneath the point of measurement to about 0.5 d beyond it, (3) subsidence is 90 percent to 95 percent complete when extraction is complete within the area of influence of the point of measurement, (4) the development curves vary only slightly above caved or mechanically backfilled longwall panels, and (5) the rate of subsidence is less where the workings are pneumatically backfilled than it is with other goaf treatments.

Results of studies above room-and-pillar and longwall panels in the western United States show that subsidence development curves above room-and-pillar mining panels are similar to longwall panels, where coal extraction ratios are similar to longwall panels and where the pillar lines are straight and perpendicular to the direction of mining, as in longwall mining. However, subsidence may continue for months, years, or even decades where many coal pillars or pillar remnants are left unmined and the coal, roof rocks, and (or) floor rocks continue to yield with time (Fig. 9B).

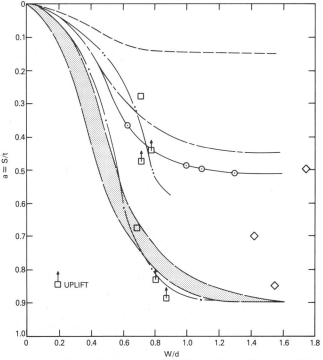

Figure 8. Graph showing maximum subsidence for the Somerset, Colo., and Salina, Utah, areas. Graph shows the ratio of maximum vertical subsidence to thickness of coal mined (a = S/t) versus the ratio of the width of the mining panel to overburden thickness (W/d) for the Somerset, Colo., area (circles), the Salina, Utah, area (squares), and York Canyon, N. Mex., area (diamonds; upper point is in draw; middle is on ridge; lower point is maximum for area; Gentry and Abel, 1978, p. 191–220). Shaded area represents typical curves for longwall extraction areas in the United Kingdom; long and short dashed curve represents average subsidence above mechanically backfilled mines; short dashed curve is average for hydraulically backfilled mines (modified from NCB, 1975, p. 9; Wardell, 1971, p. 207). Curve with dash and single dot is from measurements in overburden composed of interbedded shales, mudstones, and lenticular sandstones in the Salina, Utah, area; curve with dash and two dots is from measurements above a massive sandstone about 30 m thick in the Salina, Utah, area that overlies the interbedded shales, mudstones, and lenticular sandstones. Uplift may be caused by expansion of swelling clays, downwarping in adjacent panels, and (or) buoyant effects of water.

Subsidence development curves also show that the duration of subsidence at a point commonly depends on the time necessary to extract the coal within the critical extraction area beneath the point (Wardell, 1954, p. 475–480; NCB, 1975, p. 42–43). The curves indicate that, although the rate of subsidence is nonlinear with respect to the movement of longwall faces or pillar lines within the critical extraction area, the duration of subsidence is inversely proportional to the rate of mining (R_m) and directly proportional to the overburden thickness (d) and the tangent of the critical angle of extraction (γ) (Fig. 9; NCB, 1975, p. 43) when only the beginning and completion of mining within the critical extraction area are considered. The duration (T), in relation to the diameter of the critical extraction area (D_c), may be estimated by:

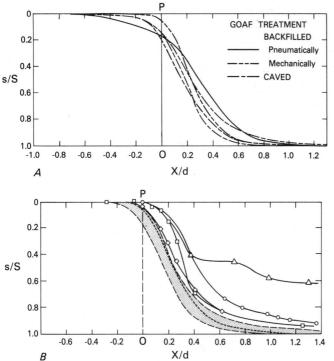

Figure 9. Subsidence development curves for backfilled and caved longwall and room-and-pillar mines in the United Kingdom and western United States. A. Graph shows typical curves of the ratio of vertical subsidence at point P to maximum vertical subsidence (s/S) in relation to the ratio of the horizontal distance of longwall mining faces from point P to the overburden thickness (X/d) in the United Kingdom. B. Subsidence development curves for the Somerset, Colo. (circles and triangles), Salina, Utah (squares and diamonds), and York Canyon, N. Mex. (dotted line; Gentry and Abel, 1978, p. 205–206), areas. Shaded area represents curves for caved and mechanically backfilled longwall mines in the United Kingdom. Curve connecting circles is for a critical (W/d = 1.25) room-and-pillar extraction panel and curve connecting triangles is for a subcritical (W/d = 1.0) panel in the Somerset area. Curves connecting squares and diamonds are above the centerline and chain pillars, respectively, of a subcritical room-and-pillar extraction panel (W/d = 0.75) in the Salina area where a massive sandstone about 30 m thick occurs at the surface. Residual subsidence of about 20 percent above subcritical panel (triangles) in Somerset, Colo., area may be caused by yielding of unmined pillars after mining was completed.

$$T = \frac{D_c}{R_m} = \frac{2d\tan\gamma}{R_m} \quad (12).$$

According to (12), the duration of subsidence above a longwall panel is 0.56, 1.12, and 2.24 years for mining rates of 1,000, 500, and 250 m/yr, where the overburden is 400 m thick and the critical extraction angle is 35°. The duration of subsidence measured above a subcritical room-and-pillar extraction panel in the Salina, Utah, area was equal to the duration calculated by (12). The duration of subsidence measured in a critical room-and-pillar extraction panel in the Somerset, Colo., area, however, was about 20 percent greater than the duration calculated by (12) (0.85 vs 0.72 years).

Slope, curvature, horizontal displacement, and strain. Slope (tilt) and curvature can be estimated from profiles of subsidence depressions from the first and second derivatives or finite differences of the profile with respect to horizontal distance (for example, Sherwood and Taylor, 1957, p. 207; Kunz, 1957, p. 38–166). Horizontal displacement and horizontal strain can be estimated from slope and curvature by assuming that the near-surface material in downwarped areas behaves as plates, that a modified simple beam analogy of plate deformation provides a means of estimating horizontal displacement and strain of the plates, that planes at right angles to the neutral surface before bending remain planes at right angles to the neutral surface after bending, and that shear strain can be neglected. Horizontal strain commonly is predominant where bending occurs in plates that are less than about one-fifth of the width of the mining area, based on model studies by various investigators (for example, W. Z. Savage, oral communication, 1982; Stephansson, 1971, p. 9–11) and on field studies by the writer.

Slope or tilt (M) between two points of measurement along a subsidence profile (x_{i+1}, x_i, Fig. 10) can be determined by:

$$M_i = \frac{s_{i+1} - s_i}{x_{i+1} - x_i} = \frac{\Delta s_i}{l} \quad (13),$$

where Δ is the first descending finite difference of tabulated vertical displacements at points x_i and x_{i+1} and l is the distance between the two points, which is a constant that is small enough such that the curved surface over the interval can be approximated by a straight line. Curvature of the profile at point x_i (C_i) is defined as the infinitesimal change of slope with respect to arc length (Sherwood and Taylor, 1957, p. 207). Curvature can be estimated by:

$$C_i = \frac{M_{i+1} - M_i}{x_{i+1} - x_i} = \frac{\Delta M_i}{l^2} = \frac{\Delta^2 s_i}{l^2} \quad (14),$$

where the tangent of the slope angle is about equal to the angle in radians.

Assuming that the beam concept is valid for material near the surface, horizontal displacement, s_{hi+1}, at point x_i can be determined from the slope of the beam, M_i, measured between points x_{i+1} and x_i by (Fig. 10):

$$s_{hi+1} = k\, h\, M_i = k\, h\, \frac{\Delta s_i}{l} \quad (15),$$

where k is a correction factor that varies from one-half for elastic beams with equal tensile and compressive strength (for example, Singer, 1951, p. 124–127) to about two-thirds for beams of cohesionless sand (Lee and Shen, 1969, p. 139–166), h is the thickness of the beam and l equals about 0.05 d. However, profiles can be made by drawing smooth curves through the data points further apart than 0.05 d and then measuring from the profile at intervals of about 0.05 d (NCB, 1975, p. 34–36).

Horizontal strain at point x_{i+1} (e_i) in Figure 10 is equal to the ratio of the change in length of the ground surface to the

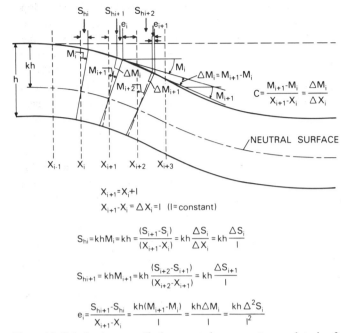

Figure 10. Relation between displacement, slope, curvature, and strain of a beam where the thickness of beam (h) is less than about one-fifth of the span and shear strain may be neglected. Spacing between points of measurement (x_i, x_{i+1}) is small enough that the first and second order finite differences of vertical displacement with respect to horizontal distance ($\Delta s/\Delta x$ and $\Delta^2 s/\Delta x^2$) closely approximate the first and second derivatives (ds/dx and d^2s/dx^2) (see, for example, Kunz, 1957, p. 125–132 for details).

initial length. It therefore equals the change in horizontal displacement with respect to horizontal distance or:

$$e_i = k h \frac{M_{i+1} - M_i}{l^2} = k h \frac{\Delta M_i}{l^2} = k h C_i \quad (16),$$

where C_i is the curvature (reciprocal of the radius of curvature) at point x_{i+1}. Horizontal strain due to bending, using the beam concept in very gently curved surfaces, such as subsidence profiles, can therefore be determined by measuring either the change in horizontal displacement or the change in slope with respect to horizontal distance. In subsidence studies, lengthening is positive and shortening is negative.

Horizontal strain (e) caused by curvature (C) is estimated from subsidence profiles in the United Kingdom (NCB, 1975, p. 36–37) by:

$$e = 0.16(C)^{\frac{1}{2}} \quad (17).$$

This formula, although not dimensionally homogeneous (C is in rad/m), is based on observed changes in lengths between benchmarks and the curvature measured in many longwall mining areas in the United Kingdom. It might also be used to estimate the strain due to curvature in the western United States in subsidence areas where the thickness of near-surface material, which may behave as a plate, cannot be determined, although strains estimated using (17) may often be lower than the actual values.

Horizontal displacement and strain in bedrock and surficial material, using the plate and beam analogy, may be estimated only approximately in some areas. Strain may be continuous in certain areas and discontinuous (tension cracks or compression features) where the stress exceeds the strength of the material. The bedrock commonly also is jointed and locally faulted, so that the profiles may not be definable as continuous functions. The neutral surface also may be difficult to determine because of marked variations in lithology and in tension and compression moduli.

Cracks and compression features (bulges or ridges) occur where tensile and compressive stresses exceed the tensile and compressive strength of the material, respectively. The ground surface above mining areas, therefore, may be subjected to tension, compression, and decompression causing cracks to open and close and bulges or ridges to form and then collapse again as mining faces or pillar lines move beneath it. For example, DeGraff and Romesburg (1981, p. 123–127) found that tension cracks closed by as much as 31 percent to 100 percent in the Salina, Utah, area, as pillar lines moved out of the area of influence of the cracks. Cracks and bulges commonly remained unchanged above barrier pillars or mine boundaries, however, until either the pillars or solid coal adjacent to the mine areas were mined or until they were altered or erased by mass-gravity movements, erosion, and deposition.

Where mining faces or pillar lines are actively mined, the ground surface, surficial material, and bedrock are subjected to a subsidence wave that travels at the rate of mining above the mining faces or pillar lines and subjects them to vertical and horizontal displacements, tilt, curvature, and strain. According to Wardell (1954, p. 480–482), the strain produced by traveling subsidence waves above active longwall mining faces in the Yorkshire coal field of northern England is less than the strain at the margins of stable depressions; but its position, in plan view with respect to the mining face, is about the same for either stable or enlarging depressions. Peng (1978, p. 340), however, reports that the maximum tensile and compressive strains of enlarging depressions are less than those for stable depressions.

Subsidence parameter graphs for the Somerset, Colo., and Salina, Utah, areas (Fig. 11A, B) and the Sheridan, Wyo., area (Fig. 11C), which were derived in part from profile and slope curves of NCB (1975, p. 17) and in part by the writer, relate the subsidence profiles to solid coal boundaries (ribside) and to slope (M), horizontal displacement (s_h), curvature (C), horizontal strain (e), limit angle (ϕ), and critical extraction angle (γ). The profiles are plotted in terms of the ratio of vertical subsidence to total subsidence (s/S) versus the ratio of horizontal distance to ribside to overburden thickness (X_r/d). They also show the limit angle and critical extraction angle. The graphs can be used to determine subsidence area, displacement, slope, and curvature with respect to distance from ribside in terms of S/d and X_r/d where subsidence profiles have been measured. They may also be used to estimate displacement, slope, curvature, and horizontal strain

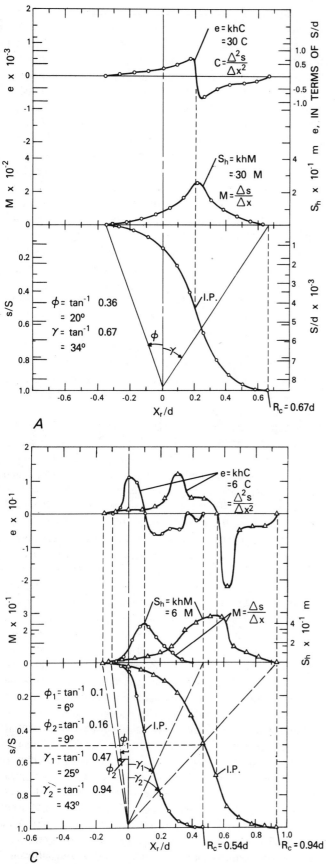

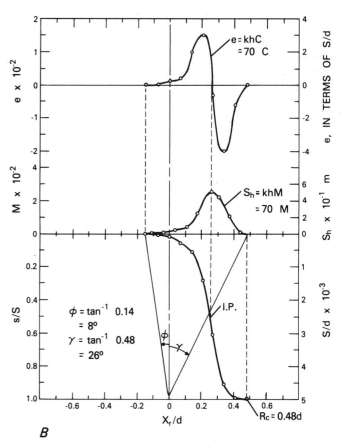

Figure 11. Subsidence parameter graph for the Somerset, Colo., Salina, Utah, and Sheridan, Wyo., areas. Vertical displacement ratio (s/S), slope or tilt (M), horizontal displacement (S_h), and horizontal strain (e) are derived from a transverse profile in the Somerset, Colo., area and from longitudinal profiles in the Salina, Utah, and Sheridan, Wyo., areas. Limit angles (ϕ) for the Somerset, Colo., Salina, Utah, and Sheridan, Wyo., areas are 20°, 8°, and 6° to 9°, respectively; critical extraction angles (γ) are 34°, 26°, and 25° to 43°, respectively. For the Somerset, Colo. (A), Salina, Utah (B), and Sheridan, Wyo. (C), areas, maximum slope (tilt) and strain are: Somerset—M_m = 3 S/d (0.025 m/m), Salina—M_m = 5 S/d (0.025 m/m), Sheridan—M_m = 4 S/d (0.22 m/m), and M_m = 2.5 S/d (0.28); Somerset—E = −0.7 S/d (−0.006 m/m), Salina—E = −4 S/d (−0.020), Sheridan—E = 2.3 S/d (0.02 m/m) and E = −2 S/d (−0.04 m/m).

above extraction panels in advance of mining, once these parameters have been determined for similar geologic and topographic settings in the mining area.

Summary of subsidence parameters. Maximum vertical and horizontal displacement, slope (tilt), strain, and limit angle depend on mine geometry, overburden thickness and strength, and goaf (gob) treatment. Representative values, which are based on observations above many mines in other countries and selected mines in the western United States (Table 2), are summarized below:

1. The ratio of maximum vertical displacement to mined thickness above caved, critical, or supercritical longwall and room-and-pillar extraction panels commonly ranges from about 0.45 to 0.9 in the western United States, 0.7 to 0.9 in the United Kingdom, 0.85 to 0.9 in the Ruhr coal field of Germany and France, 0.7 in the Silesian coal field of Czechoslovakia and Poland, 0.6 to 0.9 in the U.S.S.R., and as much as 0.96 in the Limburg coal field of the Netherlands. As much as 0.05 uplift was measured above mine panels in the Somerset, Colo., and Salina, Utah, areas of the western United States.

2. The ratio of maximum horizontal displacement to vertical displacement is not as well documented as vertical displacement, but is as much as 0.8 in the Netherlands (Pöttgens, 1979, p. 269), 0.15 in the coal fields in the United Kingdom, 0.35 to 0.45 in the Ruhr and northern French coal fields, to 0.3 to 0.35 in the coal fields of the U.S.S.R, and 0 to 1.2 in the western United States.

3. Maximum change of slope or tilt is controlled by mining depth, panel width and length, and site factors. In longwall mine panels, in flat-lying coal beds in the United Kingdom, for example, it reportedly ranges from a maximum of about 3.5 S/d where W/d = 0.45 to 2.75 S/d where W/d exceeds 1. However, in the western United States, slopes of about 3 and 5 S/d (0.025 m/m) were measured above room-and-pillar panels in the Somerset, Colo., and Salina, Utah, areas where W/d equaled 1.25 and 0.8, respectively. In the Sheridan, Wyo., area, maximum slope equaled about 2 to 4 S/d (0.2–0.3 m/m) above supercritical extraction areas (Figs. 11, 12).

4. Horizontal strain depends on the shape of the subsidence profile, which in turn varies with area mined and the overburden thickness and strength (Fig. 12). Maximum compressive strain, according to NCB (1975, p. 29), equals about 2 S/d where W/d = 0.3, but decreases to about 0.5 S/d where W/d equals or exceeds 1.0. Maximum tensile strain equals about 0.8 S/d, where W/d = 0.45, but decreases to about 0.7 S/d, where W/d exceeds 0.8. Maximum tensile and compressive strains of about 0.9 S/d (0.020–0.022 m/m) were measured by Gentry and Abel (1978, p. 210) above a supercritical longwall panel in the York Canyon, N. Mex., area. Maximum strain of −0.7 to −4 S/d (−0.006 to −0.02 m/m) was estimated by the writer from profiles using (16) in the Somerset, Colo., and Salina, Utah, areas, respectively, and −1 to 2.3 S/d (−0.05 to 0.12 m/m) for supercritical extraction areas in the Sheridan, Wyo., area (Figs. 11A, B, C; 12).

5. The position of the point of inflection, or transition point,

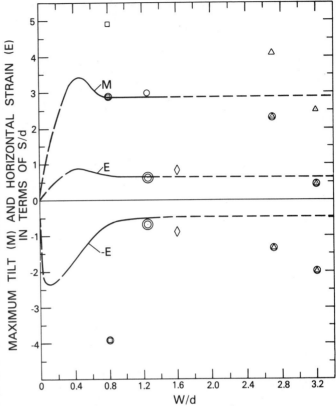

Figure 12. Comparison of maximum slope and strain caused by subsidence in the western United States and in the United Kingdom. Maximum slope and compressive and tensile strain are plotted in terms of the ratio of maximum vertical subsidence to overburden thickness (S/d), for various ratios of mining widths to overburden thicknesses (W/d) in the United Kingdom (solid and dashed lines from NCB, 1975, p. 29) and in the western United States (data points). Circles, squares, and triangles are maximum slope (tilt) derived from subsidence profiles for the Somerset, Colo. (M_m = 3.0 S/d = 0.025 m/m; W/d = 1.25), Salina, Utah (M_m = 5.0 S/d = 0.025 m/m; W/d = 0.8), and Sheridan, Wyo. (M_m = 2.5 S/d = 0.3 m/m and 4 S/d = 0.2 m/m), areas, respectively (Fig. 11). Circled circles, squares, and triangles are maximum horizontal compressive and tensile strain derived from curvature of the profiles, using equation (16) for the Somerset, Colo. (E = −0.7 S/d = −0.006 m/m, E = 0.6 S/d = 0.005 m/m), Salina, Utah (E = −4 S/d = −0.02 m/m, E = 3 S/d = 0.015 m/m), and Sheridan, Wyo. (E = 2.3 S/d = 0.12 m/m, E = −1.4 S/d = −0.07 m/m, and E = −2 S/d = 0.22 m/m, E = 0.45 S/d = 0.05 m/m), areas, respectively. Diamonds are maximum compressive and tensile strain measured in the Raton, N. Mex. (E = −0.9 S/d = −0.022 m/m, E = 0.8 S/d = 0.020 m/m) (Gentry and Abel, 1978, p. 216), area.

between positive and negative curvature and tensile and compressive strain, commonly ranges from directly above the panel boundary (ribside), where W/d = 0.4 to about 0.15 d above the extraction area from ribside, where W/d is greater than about 1.0 (NCB, 1975, p. 14). The inflection point measured by the writer in the Somerset, Colo., and Salina, Utah, areas is 0.2 to 0.25 d above the extraction area from ribside and reportedly also is 0.25 d in the Limburg area of the Netherlands (Pöttgens, 1979, p. 278). In the Sheridan, Wyo., area, where the overburden is 20

TABLE 2. SUMMARY OF SUBSIDENCE PARAMETERS IN SELECTED COUNTRIES AND IN AREAS OF THE WESTERN UNITED STATES

Coal field or area	Mining and site factors						Subsidence parameters*											
	Rock strength†			Thickness and width§														
	Medium (%)	Low (%)	Very low (%)	t (m)	d (m)	W/d	° (ϕ)	° (γ)	a (S/t)	S_h/S	s_h/s	M_m m/m (×10⁻²)	M_m S/d	C_m Rad/m (×10⁻⁴)	C_m S/d²	−E m/m (×10⁻³)	−E S/d	+E m/m (×10⁻³)	+E S/d

Extraction panels in selected mining areas of the world**

Netherlands (Limburg) Caving	..	..	..	..	..	††	35–45	..	0.96	0.8	..	..	..	..	..	..	..	..	..
France (Pas de Calais)	..	..	..	..	..	††	35	..	..	..	..	..	..	..	..	..	..	..	..
§§	..	..	..	..	..	††	..	..	0.25–0.35	..	..	..	..	..	..	..	..	..	..
***	..	..	..	..	..	††	..	..	0.45–0.55	..	..	..	..	..	..	..	..	..	..
Caving	..	..	..	..	..	††	..	..	0.85–0.90	0.4	..	..	..	..	..	..	..	..	..
Germany (Ruhr)	..	..	..	..	..	††	30–45	..	0.45–0.90	..	..	..	..	..	..	..	..	..	..
***	..	..	..	..	..	††	..	..	0.45–0.50	..	..	..	..	..	..	..	..	..	..
Caving	..	..	..	..	..	††	..	..	0.90	0.35–0.45	..	..	..	..	..	..	..	..	..
Poland-Czechoslavakia (Silesia) §§	..	..	..	..	..	††	..	..	0.12	..	..	..	..	..	..	..	..	..	..
Caving	..	..	..	..	..	††	..	..	0.70	..	..	..	..	..	..	..	..	..	..
United Kingdom	..	..	..	..	..	††	25–35	..	0.15–0.90	..	..	..	..	..	..	..	..	..	..
§§ and ***	..	..	..	..	..	††	..	..	0.15–0.50	..	..	..	..	..	..	..	..	..	..
Caving, strip packing.	..	..	..	..	..	††	..	..	0.70–0.90	0.16	..	..	..	..	..	..	..	..	..
United States (caving)																			
Eastern	..	..	..	..	..		15–27	..	0.10–0.60	..	..	..	..	..	..	..	..	..	..
Central (LW)	..	..	..	..	..	††	0– 8.5	..	0.50–0.60	..	..	..	..	..	..	..	..	..	..
Western (R&P, LW)	..	..	..	..	..	††	5–25	..	0.45–0.90	0–1.2	..	2–30	2–5	2–350	9–45	6–220	0.7–4	2–30	0.25–0.7
U.S.S.R. (caving)	..	..	..	..	..	††	30	..	..	..	..	..	..	..	..	..	..	..	..
Donets-Kizelov	..	..	..	..	..	††	..	..	0.60	0.30	..	..	..	..	..	..	..	..	..
Kuznetsk-Karaganda	..	..	..	..	..	††	..	..	0.70	0.3 –0.35	..	..	..	..	..	..	..	..	..
Chelyabinsk	..	..	..	..	..	††	..	..	0.90	0.30	..	..	..	..	..	..	..	..	..
Donbass-Lvov-Volyn	..	..	..	..	..	††	..	..	0.8 –0.9	0.3 –0.35	..	..	..	..	..	..	..	..	..

Western United States (caving)

Somerset, Delta, and Gunnison Counties, Colo. †††	..	..	..	3	..	..	15–25	..	0.36–0.52	..	..	..	..	..	..	..	..	..	..
1.	25	70	5	3	177	1.25	20	34	0.52	0.12x	0.4x	2.5	3	−2	−14	−6	−0.7	5	0.6
2.	25	70	5	3	171	1.00	..	..	0.49	..	..	..	..	..	..	..	..	..	..
3.	25	70	5	3	241	0.64	..	..	..	..	..	..	..	..	..	..	..	..	..
Salina, Sevier County, Utah §§§	..	..	..	3	..	..	8–20	..	0.45–0.88	0.3x	0.4x	..	..	..	..	..	..	..	..
1.	35	60	5	3	220	0.86	20	..	0.88	..	..	..	..	..	..	..	..	..	..
2.	35	60	5	3	255	0.68	20	..	0.68	..	..	..	..	..	..	..	..	..	..
3. ****	45	50	5	3	260	0.71	12	..	0.48	..	..	..	..	..	..	..	..	..	..
4. ****	45	50	5	3	267	0.78	8	26	0.45	0.4	1.1	2.5	5	−3	−45	−20	−4	15	3
York Canyon, Colfax County, N. Mex. ††††	65	35	0	3	104	††	5–15	..	0.5 –0.85	0–1.2x	..	..	..	..	..	−22x	−0.9x	20x	0.8
Sheridan, Sheridan County, Wyo. §§§§									0.5 –0.85										
1. *****	0	10	90	3.6	30	††	6	25	0.50	0.25	1.3	22	4	200	36	72	−1.4	120	2.3
2.	5	5	90	3.6	20	††	9	40	0.65	0.2	1.2	28	2.5	−360	−20	−220	−2	50	0.45

* Subsidence parameters include limit angle (ϕ), vertical and horizontal subsidence (s, s_h), maximum vertical and horizontal subsidence (S, S_h), maximum slope (M_m), maximum curvature (C_m), maximum horizontal compressive strain (−E) and tensile strain (+E) E=khC_m (16); k is estimated to be 0.6 for medium and low strength rocks and 0.7 for very low strength rocks; x after number indicates measured value for horizontal displacement (s_h) and horizontal strain (e); .., means no data.

† Rock strength: strength of rocks in overburden--medium strength rocks include sandstones and siltstones with carbonate or siliceous cement and with unconfined compressive strength values ranging from 50 to 100 MPa; low strength rocks include poorly cemented sandstones, siltstones, shales, coal, and mudstones with unconfined compressive strength ranging from 25 to 50 MPa; very low strength rocks include poorly consolidated to unconsolidated, soft claystones and mudstones with unconfined compressive strengths ranging from 1 to 25 MPa (strength classification modified from Deere and Miller's, strength classification of intact rock in Hoek and Brown, 1980, p. 25; see p. 14-37 for details on rock strength).

§ Thickness and width include average thickness of coal mined (t), average overburden thickness (d), and width of extraction panel (W).

** Data from Peng (1978, p. 283-284), Wardell (1971, p. 207), and Pöttgens (1979, p. 267-271).

†† Critical or supercritical extraction panels in Western United States and elsewhere in most cases.

§§ Hydraulic backfilling in extraction panels.

*** Pneumatic or mechanical backfilling in extraction panels.

††† Measurements made by R. E. McKinley, RLS, and by C. Richard Dunrud.

§§§ Measurements by Dall Dimick and Leonard Barney (Southern Utah Fuel Co., written and oral communs., 1980, 1981, 1982).

**** Massive sandstone about 30 m thick at ground surface.

†††† Data from Gentry and Abel (1978, p. 191-220); overburden lithology from C. L. Pillmore (oral commun., 1982).

§§§§ Abandoned mines; extraction ratios in extraction areas are estimated to range between 65% and 80%; subsidence data determined by the author by reconstructing original alluvial surfaces, by measuring existing surfaces with computer-linked, first-order photogrammetric plotter, and by taking the difference between the two surfaces at 1.5 m intervals.

***** Measurements are above barrier pillar between two extraction areas.

to 30 m thick, the point of inflection ranges from 0.1 to 0.55 d on the extraction side of the panel boundary (Fig. 11).

6. Limit angles range from about 35° in the coal fields of France, 35° to 45° in the Netherlands, 30° to 45° in Germany, 25° to 35° in the United Kingdom, 5° to 30° in the United States, and 30° in the U.S.S.R. (Table 2).

Stoping. Stoping (caving) commonly occurs where mine roofs become unstable and fail intermittently, where the mine support deteriorates or is removed, and (or) where the roof rocks are alternately wetted and dried (Fig. 13). Stoping is common above individual mine openings, where the width-to-height ratios of the openings are less than about 4 to 1, where the strength of the rocks is exceeded, and where large vertical deflections of rock

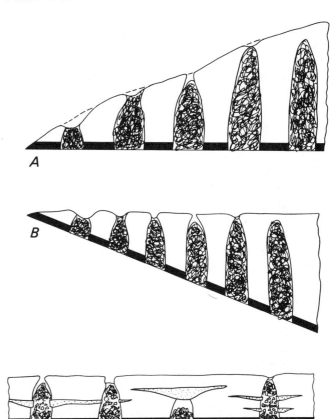

Figure 13. Profiles and photographs showing subsidence pits in various stages of development. *A* and *B*. Subsidence pits of various ages caused by variable overburden thickness; pits range from bowl-shaped depressions, that are a few decades old, to newly formed pits with overhanging rims and initial bowl-shaped depressions due to topographic relief (*A*) or dipping coal beds (*B*). *C*. Subsidence pits of varying ages due to varying lithology; sandstones reduce the rate of stoping and locally may halt upward movement of mine cavities. Pits with overhanging rims change to vertical rims and then to bowl-shaped depressions by mass-wasting and mass-gravity movements. Initial bowl-shaped pits may form by yielding and flowing into small cavities near the surface, rather than by collapse and subsequent mass movements. Modified from Dunrud and Osterwald (1980, p. 17–25). *D*. Aerial oblique photographs showing pits with overhanging to vertical rims and underground cavities. Flow of intermittent stream is diverted underground by the pits.

layers occur in short lateral distances. This cavity geometry commonly causes the rocks to break along bedding planes and joints rather than to undergo beamlike deflection, as is common a few mining thicknesses above extraction panels with larger width-to-height ratios. Shear failure also can cause the roof to fail en-masse or in individual layers, where the rocks weaken with the reduction in confining stress (Fig. 4), with wetting and drying, or with time. Tensile failure also can occur where bending occurs in rock units that are not laterally restrained.

The mine openings begin to move upward where the roof rocks break and fall to the mine floor. The volume of the openings decreases as the cavities move upward due to bulking. Stoping continues until the cavities reach the surface, the cavities become filled with rock fragments (Fig. 13A, B), or strong rocks (sandstones, limestones, etc.) are present and further caving is delayed or possibly prevented (Fig. 13C, Dunrud, 1976, p. 17–38).

Chimneying is caving of rocks above mine openings where the areas of the roof in cross section are small (as small as 3 m in diameter), where the cross sectional area of the caved zone remains nearly constant, and where the caving can progress up through hundreds of meters in as little as a few days (Obert and Duvall, 1967, p. 578–579). The process, which may be similar to stoping, apparently occurs where rocks are weakened by faults and joints or where such factors as water or steam (from underground coal fires, for example) promote rapid caving.

Subsidence pits occur where the overburden thickness is less than or equal to the possible caving height (Eq. 8, 8-a to d; Figs. 6, 13). The near-surface material can suddenly fail en-masse (plug failure of Obert and Duvall, 1967, p. 579–580) with little or no warning where the cavities have moved up to within a few meters or less of the surface and where the shear strength of the near-surface material is exceeded. Plug failure commonly occurs when the material becomes saturated.

Affected area. The aggregate area of newly formed pits usually is less than the aggregate mine area, in contrast to the area affected by downwarping. Rims of the initial pits commonly dip outward (are overhanging) or dip vertically (Fig. 13A, B, C, D). The areas become larger as the rims of the pits fail by mass-gravity movements (toppling, falling, slumping, sliding) and mass wasting, and are reduced to bowl-shaped depressions in a few years or decades, depending on the geotechnical properties of the near-surface material. The areas of the bowl-shaped depressions commonly are about equal to or slightly greater than the areas of the mine openings. The depression rims are asymmetrical on steep slopes; the downslope rim is steeper than the upslope rim (Fig. 13A).

Bowl-shaped pits also form initially where small cavities are located near the surface by flowage of the near-surface materials rather than by collapse en-masse (Fig. 13A, B, C). Bowl-shaped pits have been observed by the writer in the Sheridan, Wyo., area, where cavities above the mine openings were nearly filled with caved material and where the near-surface material consists of soft claystones, colluvium, or soil.

Vertical displacement. Subsidence pits can be deeper than the thickness of coal mined where the caved material is transported laterally into adjacent mine cavities. Some of the deepest pits in the Sheridan, Wyo., area are located where the pits intercept surface drainage and (or) ground water (Fig. 3A). Flows can be established from the surface or underground aquifers to the underlying mine openings through pits. Bulking of the caved debris can be reduced where material is transported to underground mine openings adjacent to the caved opening. Lateral transport of material due to gravity also increases as the dip of the coal beds increases. In the Sheridan and Kemmerer, Wyo., mining areas, for example, pits about three times the thickness of coal mined were observed by the writer where pits intercepted streamflows.

Pits also occur in unconsolidated material above tension cracks in the underlying bedrock. Presumably, most surficial material can extend more without rupturing than can most bedrock. For example, cracks as much as 25 cm (10 in) wide were observed in the Sheridan, Wyo., area beneath soil and colluvium that showed no signs of failure, except where local pits occurred and the underlying crack was exposed (Dunrud and Osterwald, 1980, p. 17, 20).

Rate and duration. The rate and duration of stoping and of subsidence pit formation is much more variable than it is where downwarping occurs. Roof collapse may not begin for years or decades after the mines are abandoned. For example, the time between mining and subsidence pit formation ranged from many years to many decades in the Sheridan and Kemmerer, Wyo., areas, the Denver-Boulder Metropolitan area, and the Des Moines, Iowa, area. According to Littlejohn (1979, p. 22), subsidence is still occurring in Great Britain above mines that were abandoned centuries ago. The rate of pit formation, once the cavities have moved to within a meter or so of the surface, however, can be very rapid. Pits can then form within a few minutes to a few hours.

The duration of pit development commonly increases as the overburden becomes thicker or stronger. In the Sheridan, Wyo., area, for example, pits generally become younger as the overburden thickens (Fig. 13A, B) or where well-indurated sandstones in the overburden reduce the rate of stoping (Fig. 13C). Pits occur only in a narrow belt above mine openings near the outcrop in the Cumberland mining area, near Kemmerer, Wyo., because the overburden thickens in short lateral distances above the steeply dipping coal and because the stoping rate is reduced by the presence of a well-cemented sandstone in the mine roof (Fig. 13C). Where the overburden thickness is variable because of uneven topography or dipping coal beds, pit development is limited to areas where the mines are located at depths less than the maximum height of stoping.

Ground failure can occur suddenly above cavities that have migrated to within about a meter or so of the surface. Pit formation in these instances commonly is not preceded by significant depressions, although rumbling sounds sometimes can be heard a few minutes to a few days before failure when near-surface caving occurs, particularly where large, underground cavities are located near the surface (Dunrud and Nevins, 1981, sheet 1). Earth tremors can be caused by the breaking and caving of material into the cavities and by movement of the resulting pressure waves in them.

Downwarping and stoping. Both downwarping and stoping are observed where the remaining pillars or pillar remnants (stumps) eventually yield or the coal and rocks yield and flow into adjacent mine openings. Depressions and pits are the final result of downwarping and stoping where the overburden is thinner than the stoping height (Fig. 3A). The chronology of occurrence depends on coal extraction percentage, mine geometry, and the strength of the coal and rocks above and below the remaining pillars or stumps.

Pits commonly precede depressions where coal extraction ratios are small and the remaining coal pillars are strong enough to support the weight of the overburden. Depressions may occur later, where the coal pillars or rocks above and below the pillars weaken with time or with deterioration by air or water. The remaining coal pillars also can be weakened or burned where air and water can enter the mine and cause either a reduction in strength of the pillars or spontaneous combustion.

Depressions often occur first, where coal extraction ratios are large enough to cause yielding of pillars and (or) yielding of the rocks above and below the pillars. Pits may occur later, perhaps many decades later, where cavities still exist after yielding of coal and rock ceases and where the overburden is thinner than the stoping height. Depressions only may occur where the underground cavities are filled, either by yielding and flowage of pillars or rock above and below them or by bulking of the caved fragments (Dunrud and Osterwald, 1980, p. 17–20).

Shaft Subsidence. Subsidence pits also have been caused by collapse of seals or staging above unfilled coal mine access shafts in the Sheridan, Wyo., area and in other areas of the western United States. The pits may be similar in general appearance to subsidence pits that occur above mine openings, but may be much deeper, particularly where large voids occur below fill material that eventually fails. Deep pits also can occur where the shaft linings eventually fail (see Taylor, 1975, p. 131–133; Littlejohn, 1979, p. 25–30; Down and Stocks, 1978, p. 238–239 for details).

FACTORS CONTROLLING UNDERGROUND SUBSIDENCE PROCESSES

The most important factors controlling the area, amount, rate, and duration of subsidence in the western United States—in addition to mine geometry and extraction ratio, mining method, and goaf treatment already mentioned—are geology and hydrology, geotechnical properties of the coal and rock above and below the coal, topography, and slope angle of the surface.

Mining Factors

Mine geometry and extraction ratio commonly are the two most important factors controlling subsidence. Mine geometry includes the size and configuration of the mining panels and individual openings within them, the height of the openings and pillars, and the spatial relation to any other vertically superposed mines. Extraction ratio is the ratio of coal extracted to the total amount of coal in the mined bed.

Mine geometry and extraction ratio. Downwarping is the most common subsidence process above extraction panels where the width-to-height ratios of the mining areas are greater than about 0.5 to 1.0 and where enough pillars in room-and-pillar areas are removed to cause yielding in either the pillars remaining after mining or in the rocks above or below them. Stoping is the most common process above haulageways, air courses, and rooms. Where downwarping occurs, the maximum amount, area, rate, strain, and duration of subsidence above critical or supercritical mining panels are controlled by the extraction ratio, height of caving, and fragmentation. Subsidence may be delayed or permanently reduced above subcritical mining panels because compression arches may span the panels such that much of the surface subsidence is caused by yielding of the barrier pillars or coal at the panel boundaries.

Subsidence can be reduced by leaving pillars for support. This procedure reduces extraction ratios, however, and only may delay subsidence, where pillars and roof and floor rocks eventually weaken and yield with time and with wetting and drying. Mine fires can also reduce overburden support, where air and water can enter the mine openings through mine shafts, portals, or subsidence cracks and pits and cause spontaneous ignition.

Mining methods. Subsidence may be greater above longwall panels than above room-and-pillar mining panels of similar width and length, where pillars or remnants of pillars (stumps) are left behind. However, complete extraction in longwall panels can promote breaking and fragmentation in a greater thickness of roof rocks than in room-and-pillar mining areas, where unmined pillars or stumps reduce caving in the roof. Surface subsidence commonly decreases as the height of caving (H) increases (Eq. 8-a to d; Fig. 6). The decrease depends on the ultimate height of caving, caving geometry, bulking factor, and compressibility of the caved fragments. Pillars or stumps remaining in room-and-pillar extraction panels can reduce caving height and bulking by locally supporting the roof and preventing caving.

Subsidence above room-and-pillar mining panels commonly is more variable than above longwall panels because the extraction ratios and heights of caving within the panels are more variable. In addition, subsidence can occur after the mines are abandoned (residual subsidence), where remaining pillars or stumps yield or deteriorate with time (Fig. 9B). Maximum subsidence can be greater than the amount of subsidence that occurs above longwall panels.

Rate of mining can affect surface strain and duration of subsidence. Surface strain has been locally reduced by mining with the longwall method at a constant and rapid rate (for example, Briggs, 1929, p. 80–81) because the amount of strain commonly decreases as the rate of mining increases. Goodwin (*in* Briggs, 1929, p. 101), for example, reported that strain caused by longwall mining at a rapid and constant rate in the Hyde, Manchester area of England was reduced to the point that three-story houses were not damaged, even where as much as 3 m of subsidence occurred.

Goaf treatment. Results of studies above longwall mines in Great Britain (Shadbolt, 1978, p. 729; NCB, 1975, p. 22; Fig. 8) indicate that subsidence can often be reduced from 70 percent to 90 percent of the mining thickness to 45 percent to 50 percent by mechanically backfilling (stowing) the abandoned mine areas. It might be reduced to 5 percent to 30 percent of the mining thickness by hydraulically backfilling sand into the abandoned mine areas (Wardell, 1971, p. 207; Hudspeth *in* Briggs, 1929, p. 85). Subsidence was reduced by as much as 50 percent, where hydraulic backfilling was used instead of mechanical backfilling in the Liege coal field, Belgium (Briggs, 1929, p. 85). Fayol (*in* Briggs, 1929, p. 56) reported that subsidence was reduced in the Commentry mines of France by backfilling with strong shale instead of weak claystone.

The compressibility of hydraulically backfilled material varies with the type of material used and confining pressure (Table 3). It reportedly is lowest for sand and highest for refuse from the mine and wash plant (Briggs, 1929, p. 84).

Site Factors

The major site factors controlling subsidence displacement, tilt, horizontal strain, area, rate, and duration are (1) thickness of coal beds, (2) lithology, texture, structure, and hydrologic conditions of the bedrock above and below the coal bed, (3) geotechnical properties of the bedrock and surficial material, and (4) topography and slope angle. Lithology consists of the composition of grains and cement (or binder) that make up the bedrock. Texture is the size, shape, and arrangement of grains or fragments. Primary structures consist of bedding, laminations, and stratification; secondary structures are dip of bedrock, faults, joints, and cleat. Important geotechnical properties of the rocks are density, strength, deformation moduli, and slake durability.

Thickness of coal beds. The thickness of the coal bed and height of the mine workings in the bed control maximum subsidence. Subsidence caused by mining more than one superimposed coal bed can equal the sum of the subsidence caused by mining each bed (Eqs. 10, 11; Briggs, 1929, p. 103–106). The ratio of subsidence to thickness of coal mined (subsidence ratio), however, commonly decreases as the mining thickness increases, for a given overburden thickness. For example, Goldreich (*in* Briggs, 1929, p. 47–52) reported that subsidence was nearly equal to the thickness of coal mined in the Ostrau-Karwin region of Czechoslovakia where the mining thickness was less than 1.5 m but that it decreased, in relation to mining thickness, where the thickness of coal mined was greater than 1.5 m. The decrease may occur because the height of caving and bulking often increases with increased mining thickness.

Lithology and structure. Downwarping, stoping, caving and fragmentation, and bulking are governed by strength and structure of the overburden rocks and their behavior in response to wetting and drying. Sandstones and siltstones with a tight packing framework, cemented with silica or carbonate material or limestones, relatively free of joints or faults, for example, may be strong enough to support mine roofs in extraction panels without caving until extensive areas are mined and also reduce the height of caving after caving occurs. They can halt the stoping process above entries and rooms for many decades or even centuries. Strong rocks near the mine roofs also may cause high abutment stresses during mining. This can cause bumps or squeezes near the mining faces or pillar lines, force miners to abandon the area before all the coal is mined, and cause irregularities in the overlying subsidence profile, which in turn can locally increase tilt and surface strain.

Strong overburden rocks that occur near the surface can reduce subsidence from what it would be if the overburden were composed of weak, jointed rocks. In the Salina, Utah, area, for example, subsidence amounting to about 45 percent of the thickness of coal mined was measured above a strong sandstone about 45 m thick, whereas subsidence of as much as 90 percent of the thickness of coal mined was measured above a sequence of shales, mudstones, and lenticular sandstones. Maximum subsidence can vary greatly, even beneath overburden of similar lithology, however, where different mining methods and goaf treatments are used. Horizontal displacement and horizontal strain can be greater, in relation to vertical displacement, where thick rock units occur at the surface and behave as plates.

The bulking factor commonly is less for soft, thinly bedded rocks, such as shales, than it is for hard, thickly bedded to massive rocks, such as limestones or quartzites. Soft, tabular fragments commonly undergo less initial volumetric increase during caving, and they also may recompact more under load than blocky fragments. For example, the volume of caved limestones or sandstones and siltstones, which are cemented with silica or carbonate, may be virtually unaffected by wetting or cyclic wetting and drying, whereas the volume of caved mudstones, evaporite rocks, or sandstones and siltstones, with a clay binder, may be greatly reduced where the rock breaks into increasingly smaller fragments or disaggregate where subjected to wetting and drying.

TABLE 3. COMPRESSIBILITY OF VARIOUS MATERIALS
USED IN HYDRAULIC BACKFILLING*

Material	Compressibility (%)	(%)
	Confining Pressure (MPa)	
	7.5	15
Sand	1.75	2.5
Sand and gravel	3	5
Sand, ashes, and wash-plant refuse	5	9
Granulated slag from blast furnace	9	14
Refuse from mine and wash plant	18	22

*Modified from Briggs (1929, p.84).

Expansive clays can cause volume and stress increases where they become wet. Post (1981, p. 197–202) measured volume increases ranging from about 30 percent to 95 percent, where smectite clays (saponite, nontronite, montmorillonite, etc.) became saturated and were held at a constant pressure. Swelling pressures ranged from 0.5 to 3 Mpa (70–415 psi) as the clays became saturated and were held at constant volume.

The dip of the coal bed also can affect subsidence amount. Studies in 1897 by Dortmund Board of Mines in the Westphalian coal region of Germany (Briggs, 1929, p. 58–59; Bräuner, 1973a, p. 41) showed that the maximum vertical displacement (S) was related to the thickness of coal mined (t) and the dip of bedrock (θ) by:

$$S = a\ t\ \cos\theta\ (\theta \leqslant 65°) \quad (18),$$

where the subsidence ratio (a) equaled about 0.8 for caved workings, but varied from about 0.4 for beds dipping less than 10°, 0.3 for beds dipping from 10° to 35°, 0.25 for beds dipping from 35° to 65° for mechanically backfilled mine openings. Maximum subsidence increases again where the dip is greater than about 65° (Briggs, 1929, p. 99; NCB, 1975, p. 18). The mining thickness (t) may approach the *width* of the mining panel, rather than the *thickness* of the coal bed, where the dip is nearly vertical.

The dip of the coal bed also affects the limit angle, break angle, and the geometry of the subsidence profile in relation to mining panels (Fig. 14). Jicinsky and Goldreich (*in* Briggs, 1929, p. 45, 110) observed that the limit lines are steeper above the updip side of the mine in steeply dipping beds than they are on the downdip side. Fayol (*in* Briggs, 1929, p. 55–57) reported that, in the Commentry mining area of France, the center of the subsidence depression shifts progressively updip relative to the center of the mining panels with increasing dip and that the subsided area also increases. Kaneshige (1971, p. 173–176) observed in models that maximum subsidence occurred above the updip boundary of the mine, where the overburden is composed of thick rock sequences, but occurs above the downdip boundary of the mine, where the overburden consists of thin rocks.

A reasonably consistent relation between dip of bedrock, limit angle, and break angle has been measured in coal fields of many different countries, provided that the limit and break angles for gently dipping beds are adjusted for that particular region (Fig. 14). This relation may apply to the western United States, although little is known about subsidence in dipping terrane. As the dip of bedrock increases from 0° to 45°, (1) the limit and break line above the updip side of the mine steepen by 20° to 30°, (2) the limit and break line above the downdip side of the mine flatten by 20° to 30°, and (3) the center of the subsidence depression is offset downdip 0° to 30° relative to the center of the mining panel. As the dip becomes greater than about 45°, the subsidence area again decreases and becomes more symmetrical with respect to the mining area.

Tensile strain has been observed to increase on the downdip side of the mining panel and to decrease on the updip side of the

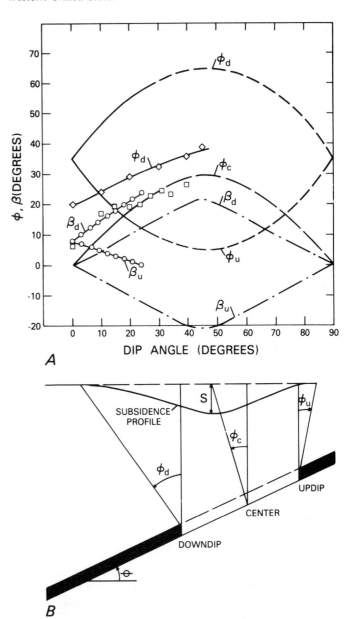

Figure 14. Relation between dip of bedrock and subsidence. *A*. Relation between limit and break angles on updip side (ϕ_u, β_u) and downdip side (ϕ_d, β_d) of a mining panel in relation to the dip of bedrock (θ); relation between the panel centerline and the point of maximum subsidence at the surface (ϕ_c) with the dip of bedrock. Circles are from O'Donahue (*in* Briggs, 1929, p. 112); squares are from Goodwin (*in* Briggs, 1929, p. 113); diamonds are from Hausse (*in* Briggs, 1929, p. 60–61). Solid and dashed lines are from NCB (1975, p. 18) and Marr (1959, p. 702); dotted and dashed lines are from Jicinsky (*in* Briggs, 1929, p. 45–46); modified from NCB (1975, p. 18). *B*. Typical subsidence profile in relation to mining panel and dip of bedrock (θ).

mine as the dip of bedrock increases. For example, tensile strains above longwall panels in beds dipping 16° reportedly are about 50 percent greater on the downdip side and 50 percent less on the updip side of the mine panel than they are above mines in horizontal beds, whereas tensile strains above panels in beds dipping about 35° are about 70 percent greater on the downdip side and 70 percent less on the updip side of the mine panel than they are above a mine in horizontal beds (NCB, 1975, p. 32–33). Tension cracks were observed by Goodwin (*in* Briggs, 1929, p. 101–102) to be more common in the Broadstairs Colliery, England, as the dip of the coal beds increased, but no mention is made of the position of the cracks in relation to the mining panels.

Limit angles and critical extraction angles commonly are less where overburden is composed of thick, jointed sandstones, siltstones, or limestones than they are in overburden consisting of shales, mudstones, claystones, or thin sandstones and limestones. According to observations by Mohr (1956, p. 141), the limit lines and break lines steepen in strong, jointed rocks and flatten in weak rocks. Statistical correlations of the percentage of limestone and sandstone in the overburden with limit angle by Abel and Lee (1980, p. 11) indicate that a 20 percent increase in the percentage of limestone in the overburden steepens the limit line by about 12°, but that a 55 percent increase in the percentage of sandstone is required to increase the limit line by 12 percent.

In the Somerset, Colo., area, where overburden rocks typically consist of about 70 percent shale and mudstone, 25 percent sandstone and siltstone, and 5 percent calcareous siltstone, the limit angle ranges from 15° to 25°. In the Salina, Utah, area, however, the limit angle ranges from about 5° to 10°, where a massive sandstone about 30 m thick underlies the surface; but it is about the same as the Somerset area, where interbedded shales, mudstones, and sandstones occur at the surface. The critical extraction angle is about 25° in the Salina area, but is about 35° in the Somerset, Colo., area (Fig. 11A).

Subsidence rate and duration also are controlled by lithology, in addition to mine geometry, mining methods, and goaf treatment. Subsidence in relation to position of pillar line above a thick, massive sandstone in the Salina, Utah, area, for example, was delayed about 0.1 d, compared with subsidence above shales and thin sandstones in the Somerset, Colo., area (Fig. 9B). However, once subsidence began, the rate of vertical settlement, in relation to mining rate, is greater in the Salina area than it is in the Somerset area.

Faults within the subsidence area commonly affect the limit angle by providing planes of weakness that can localize subsidence. Briggs (1929, p. 121–123) observed that subsidence is localized along faults where they are oriented subparallel to the limit lines (limit planes in three dimensions). He observed that the subsidence area commonly is less than normal where coal faults are present in the mine area and where they dip more steeply than the limit planes (Fig. 15A). but this area is greater than normal where the faults dip less steeply than the limit planes (Fig. 15B). Bell (1975, p. 13–15) observed that sudden movements along faults are the greatest and most common where mine panels terminate on the footwall sides of faults. As much as 2.5 m (8 ft) of renewed displacement occurred along fault projections about 700 m above room-and-pillar mines in the Sunnyside, Utah, area (Fig. 16A) and as much as 35 cm of new displacement along faults in the Salina, Utah, area (Fig. 16B).

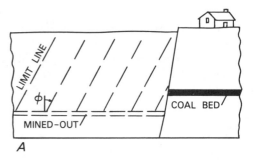

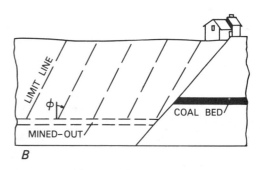

Figure 15. Influence of faults on subsidence amount and area. *A.* Building may be protected from subsidence, where fault plane is steeper than limit angle. *B.* House may be damaged, where located on fault, even though it is located outside the normal limit of subsidence because fault plane is flatter than attitude of limit angle. Modified from Briggs (1929, p. 122).

Hydrology. Surface water can affect the rate of subsidence and the rate of erosion and deposition in subsidence areas. Bruhn and others (1978, p. 36) report that the rate and frequency of occurrence of pits increase with increasing precipitation during a three- to eight-month period preceding pit development. Once formed, cracks and pits can widen and change geometry rapidly during periods of high precipitation, where located on slopes underlain by readily erodible material, such as soil, colluvium, and claystone. For example, the pits, which diverted surface water underground above mine workings in the Kemmerer and Sheridan, Wyo., areas, enlarged in area by twofold to fourfold, compared with the areas of common pits, and deepened twofold to threefold. However, cracks and pits with no underground connection can fill rapidly, where inundated by sediment-laden water.

Ground water and surface water also can be diverted through faults, cracks, and separations along bedding planes to other formations or to the mine workings (Fig. 4). For example, in the Salina, Utah, area, the predominant inflows of water occur where faults, such as the one shown in Figure 16A, transect the mine workings. In the Sunnyside mining area of Utah, the author observed that vegetation was dying near offset subsidence cracks (Fig. 16A), apparently as a result of dewatering of the root zones. F. W. Osterwald (U.S.G.S., oral communication, 1979) also observed that spring flow in the area had decreased, compared with the late 1950's and early 1960's, when he mapped the area.

Figure 16. Photographs showing fault-controlled subsidence. *A.* Graben as much as 2.5 m deep and 50 m wide is located about 700 m above mine workings in Sunnyside, Utah, area. *B.* Renewed faulting displaces ground surface as much as 35 cm for about 150 m along a preexisting fault, which is located about 270 m above mine workings in the Salina, Utah, area.

However, the decrease could be related to different levels of precipitation during these periods, and precipitation was not measured. Large amounts of water also have been observed to flow suddenly through cracks and pits to underlying mine workings in sufficient volume to fill them with water in a few hours, particularly where faults provide extensive planes or weakness (for example, Young and Stoek, 1916, p. 19–20, 56–57).

The presence of ground water also may affect subsidence. For example, maximum subsidence ranges from 90 percent to 96 percent of the thickness of coal mined in the United Kingdom and the Netherlands, respectively, where large amounts of ground water and surface water are present. By contrast, maximum subsidence commonly amounts to about 50 percent to 70 percent of the thickness of coal mined in the Somerset, Colo., and Raton, N. Mex., areas, where the mines are dry (Dunrud and Osterwald, 1980, p 26). In the Salina, Utah, area, however, where the mine is wet and the caved rocks commonly are saturated, subsidence amounted to as much as 89 percent of the thickness of coal mined (Fig. 8).

Geotechnical properties. Important geotechnical properties of rocks controlling subsidence are density, compressive, tensile, cohesive, and shear strengths, bulk modulus, Young's modulus, shear modulus, Poisson's ratio, and slake durability (resistance to disaggregation during cyclic wetting and drying; see Franklin and Chandra, 1972, p. 325–338 for details). These properties are controlled primarily by the composition, texture, cement, and structure of the bedrock and surficial material. Rock masses derive much of their strength from confining stresses and attendant increases in the frictional resistance across bedding, stratification, joints, and (or) faults (for example, Handin and Haeger, 1958, p. 2892–2934; Dunrud and Osterwald, 1980, p. 12). Strength of rock masses therefore are often reduced near mine openings or in caved and downwarped zones.

Compressive, tensile, shear, and cohesive strength of intact rocks, such as drill core, can be estimated from the point load index strength test (Broch and Franklin, 1972, p. 669–693; Hassani and others, 1980, p. 543–565; and Stimpson and Ross-Brown, 1979, p. 182–188). However, the strength of the rock mass often cannot be determined from tests on rocks at the outcrop because rocks, such as claystones and shales, are weakened rapidly by weathering, whereas rocks such as sandstones and siltstones can be strengthened by case hardening. The strength of coal and rocks in place can be orders of magnitude less than the strength of coal and rock determined from drill cores. For example, the uniaxial compressive strength of coal and rock reportedly decreases exponentially with increasing size of cubic or core samples tested (Peng, 1978, p. 184–185; Hoek and Brown, 1980, p. 155–156).

Downwarping and stoping is controlled in part by the strength, deformation moduli, and Poisson's ratio of the rocks. For example, sandstones, limestones, and siltstones with carbonate cement may support the roofs of mine openings and prevent or greatly delay stoping. They also may support the roof above mining panels over large areas, prevent caving, and cause high abutment stresses. The orientation of mine-induced fractures can be controlled by the angle of internal friction of the rock and bedding planes or joints. The tensile strength of most rocks, however, is low and they fail readily where subjected to tensile stresses. Rocks with low compressive and shear strengths, such as claystones, shales, and mudstones, may deform plastically in response to flexural stresses caused by downwarping. Shear stresses generated across bedding and stratification planes of rocks undergoing flexure can exceed the shear strength of the rock or the frictional resistance across them.

The strength of the bedrock also can control the amount and rate of compaction of the caved fragments. Compaction of the bulked fragments is governed by their compressive, tensile, and shear strengths over long periods of time and their response to wetting and drying. The bulking factor may be significantly reduced where the fragments yield with time or with increased loading or with wetting and drying. For example, linear strain of about 0.35 (35 percent) was measured on loose piles of crushed rock, where subjected to about 1 MPa vertical stress, whereas strain of only about 0.16 (16 percent) was measured on sand and crushed rock, contained in a cylinder, where subjected to 15 MPa (Peng, 1978, p. 223). Volumes and pressures also may increase where expansive clay is present in the caved rock and it becomes saturated. Vertical expansion of as much as 5 percent of the thickness of coal mined was measured two years after mining was completed in extraction panels in the Salina, Utah, area (Fig. 8). The uplift may have been caused in part by expansion of clay, when it is exposed to water.

The strength of rocks above and below coal pillars controls deformation, where stresses increase on pillars. Weak claystones may deform plastically and move outward from beneath the pillars and upward into the mine openings, where stresses are greater than their yield strength, particularly where these rocks become saturated and pore water reduces the strength (Hoek and Brown, 1980, p. 153–154) and the effective confining stress of the rock (for example, Handin and others, 1963, p. 717, 724–755; Rockaway and Stephenson, 1982, p. 7–14; Jeremic, 1981a, p. 39–46; Jeremic, 1981b, p. 706–708). Similarly, the strength of bedded, jointed, and faulted rocks also can be reduced where water is present in these discontinuities and where the pressure of the water reduces the normal and shear stress across them (for example, Jaeger and Cook, 1979, p. 219–225).

The presence of water and (or) alternately humid and dry air can weaken mine roof and floor rocks, such as soft shales and claystones (Young and Stoek, 1916, p. 71). According to Badger and others (*in* Bell, 1975, p. 38–39), weakening can be caused by the dispersion of clay colloids, dissociation of ions, and (or) by wetting and drying. Colloidal dispersion is caused by ion or cation exchange of clays and shales, where water is introduced into the rock. Wetting and drying can disintegrate rocks by alternating producing large positive and negative capillary pressures.

The deformational characteristics of weathered bedrock or surficial material and their response to wetting and drying and to time determine whether the material will crack or stretch in

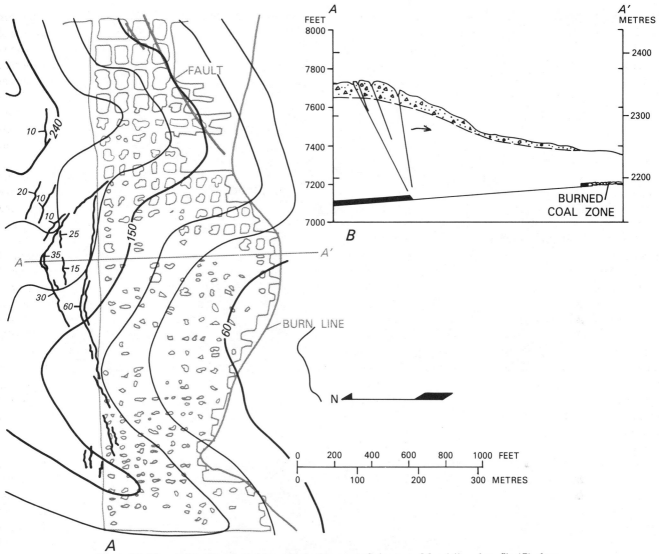

Figure 17. Map and profile of subsidence in the Somerset, Colo., area. Map (*A*) and profile (*B*) along cross section A-A' show relation between a room-and-pillar extraction panel and tension cracks that developed in the overburden (contour interval of overburden thickness = 30 m). Numbers by cracks indicate maximum width, in centimeters.

response to curvature and strain (for example, Jaeger and Cook, 1979, p. 308–325). For example, delays of one to two years were observed between the time cracks were believed to occur in bedrock, based on subsidence measurements, to the time they occurred in overlying surficial material in the Somerset, Colo., and Salina, Utah, areas. The surficial material apparently yielded without cracking until it was weakened by cyclic wetting and drying caused by cyclic precipitation.

Cracks commonly are roughly perpendicular to the direction of maximum extension in subsidence depressions, where surficial material occurs at the surface. However, extension can occur along preexisting joints and faults at angles of as much as a few tens of degrees to the normal of maximum extension. In the Salina, Utah, area, for example, the writer mapped cracks along two sets of joints and one set of faults that trended at angles of as much as 30° to the normal to the direction of maximum extension. However, the cracks commonly were roughly perpendicular to the direction of maximum extension, in surficial material greater than about a meter thick.

Topography and slope angle. Topography and surface slope angle commonly affect maximum subsidence, strain, and subsidence profile geometry by altering overburden stresses and changing the subsidence profile geometry. Mining panels located near cliff faces or steep canyon walls, for example, can cause the overburden to behave in a manner more analogous to cantilevered beams, or block rotation, rather than to laterally constrained beams, particularly where coal is mined near the outcrop.

The writer measured cracks many tens of meters long and as much as 35 cm wide in overburden above a massive barrier pillar

Figure 18 (this and facing page). Aerial photographs of surface mines. *A*. Aerial oblique view of surface mine near Decker, Mont. (May 1978). Topsoil is removed by scrapers ahead of overburden stripping and stockpiled; next overburden is drilled, blasted, removed by dragline, and cast by dragline into the cut where coal has been removed; spoil is then graded and revegetated (gray strips). *B*. Coal is blasted and removed by shovel and trucks. *C*. Ridges of dragline-dumped spoil were graded many years after emplacement near Colstrip, Mont. Elongate depressions, which apparently are caused by differential compaction, are located in areas between graded ridges in foreground.

C

that separated two mining panels near a cliff face in the Geneva mine area in east-central Utah. Maximum average horizontal strain across the cracked area was about 0.02 m/m; maximum cumulative extension equaled about 0.40 t (Dunrud, 1976, p. 9). Cracks ranging from a few millimeters to as much as 60 cm wide and equaling about 1 m in cumulative horizontal displacement were mapped in 1981 by the writer above the boundary of a room-and-pillar mining panel in the Somerset, Colo., area (Fig. 17). The coal was extracted in a panel oriented approximately parallel to the contour of a slope averaging 25°. The coal was mined to the burned coal zone near the outcrop. The combination of steep slope, lack of lateral constraint, and the absence of a coal barrier between the burn zone and the mining panel apparently caused cantilever-type (block rotation) failure, judging from the amount of extension and concentration of cracks in this area.

The magnitude of subsidence, slope, and strain often varies with direction of mining in relation to direction of slope, where coal is mined beneath ridges and valleys. Gentry and Abel (1978, p. 203-204), for example, reported that maximum subsidence was 25 percent to 30 percent greater on a ridge than it was in an adjacent draw above a supercritical longwall panel in the Raton coal field of New Mexico (Fig. 8). Tension cracks were reported to be wider, more extensive, and more abundant on slopes facing in the direction of mining than they were on slopes that faced opposite to the direction of mining. Cracks were observed by the writer to be wider and more abundant in the Somerset, Colo., area on slopes that faced in the direction of mining than on slopes that faced opposite to the direction of mining. Tension cracks also were more abundant in the Salina and Sunnyside, Utah, areas near canyon rims, where extension occurred due to lack of lateral support.

SUBSIDENCE PROCESSES IN SURFACE MINES

In the western United States, surface mining operations commonly consist of (1) removing and stockpiling the topsoil, (2) removing the remaining overburden above the coal, (3) mining the coal, (4) replacing, grading, and sometimes compacting the overburden material, (5) replacing the topsoil, and (6) revegetating the area (Fig. 18). Mining begins by excavating an initial cut to expose the coal so that it can be mined. The overburden material removed commonly is stockpiled for later use. After mining the coal from the initial cut (box cut), the overburden from the next cut is placed in the initial cut (Fig. 18A). This sequence continues for the life of the mine. The rocks in the overburden often are broken and mixed during the rehandling process. The coal commonly is removed in benches, where more than one coal bed occurs or where the overburden is too thick to remove in one cut or is too thick to be stable in a single highwall during mining. The overburden and coal often are drilled and blasted to facilitate excavation.

Compaction of Spoil

Subsidence in surface mining areas commonly is related to thickness of rehandled earth material (spoil), the method of removing and replacing the bedrock and surficial material, and the geologic, geotechnical, hydrologic, and climatic conditions at the site, rather than the thickness of coal mined. The aggregate thickness of the coal mined and the bulking factor of the overburden material removed and replaced, however, control the difference in ground elevation before and after the coal is removed. Subsidence, which may be a small percentage of the overburden thickness, also can be affected by the behavior of the unmined bedrock and surficial material near the mines and the highwalls.

Studies in the United States and abroad (for example, Brawner, 1978, p. 13–28; Paul, 1978, p. 29–34; Hutchings and others, 1978, p. 136–161; Charles and others, 1978, p. 229–251; and Smith, 1980, p. 1–102), studies by White and Farrow (1982, p. 385–388), and reconnaissance studies by the author in the Powder River basin of Wyoming and Montana indicate that subsidence and other types of mine deformation occur by (1) mass-gravity failures of highwalls and rehandled overburden (spoil), (2) compaction of spoil, (3) dewatering of aquifers near and within the mine, and (4) stress-strain readjustments in the material near and within the mine area caused by changes in crustal loading and (or) hydrostatic pressures. Mass-gravity failures (slabbing, toppling, slumping, falling, sliding, and flowing) are not considered subsidence processes and thus are not discussed in this report. Subsidence processes are discussed only briefly; further details can be found in the above references.

Subsidence caused by compaction or differential compaction of spoil is related to porosity, changes in porosity, and permeability of the spoil with loading, wetting, cyclic wetting and drying, and with time. It is controlled by the (1) thickness and strength of the overburden, (2) geologic and geotechnical properties of the bedrock and surficial material, (3) methods of handling the spoil material, (4) ground water and surface water, and (5) time since emplacement.

Subsidence commonly is proportional to the thickness of the replaced material for a given bulking and composition of the material. Subsidence can increase with increased bulking, unless the material retains its porosity during loading, wetting and drying, and with time. The bulking factor commonly is greater where large fragments of strong sandstones, siltstones, or limestones occur in the rehandled spoil than it is where small fragments of weak mudstone or shale are present after replacement. Subsidence could be greater, however, where these larger fragments are water soluble or where the grain matrix is held together by water-soluble material. The fragments could then crush, slake, and disintegrate with loading, wetting and drying, and time, and consolidate more than spoil where fragments are smaller and bulking is less to begin with.

For example, at the Horsley strip (open cast) mine in England, subsidence was measured with borehole extensometers emplaced in graded spoil that averaged about 50 m thick (Charles and others, 1978, p. 229–251). Most of the overburden was removed by power shovels and transported by dump trucks. The spoil consisted of about 10 percent boulders, with the remainder composed of smaller fragments, sands, and silts. The larger fragments commonly segregated from the finer material during spoil replacement. Tests of borehole samples revealed a porosity of 20 percent to 40 percent, a moisture content of 10 percent to 100 percent, and a permeability of more than 10^{-2} cm/s.

During an observation period of about three years, subsidence amounted to about 0.25 percent to as much as 1.2 percent of the thickness of replaced material. It was observed that (1) as much as 0.5 m of subsidence occurred during the period of measurement; (2) the rate of subsidence caused by self-loading decreased about linearly with the logarithm of time; (3) subsidence was about two to seven times greater, where the material became saturated, than it was where it remained dry and settlement was caused by self-loading; (4) subsidence was significantly less than the average, or reversed (uplift), where large piles of overburden first were placed on the graded spoil and were then removed; and (5) very little subsidence occurred in a water pond area, where the spoil was saturated shortly after emplacement. Cavities, as much as 0.5 m in diameter, were observed during drilling and sampling operations (Charles and others, 1978, p. 232).

The methods of handling and grading the spoil can control differential settlement. For example, considerable differential settlement occurred in the Colstrip, Mont., area in dragline-dumped spoil rows, where grading was delayed for many years after emplacement. Differential compaction apparently was caused by the spoil being more well compacted beneath the rows, because of self-loading, than it was between the rows (Fig. 18C). However, at another mine in the area—the Rosebud Mine—differential settlement of spoil over a 9-year period, which was graded 30 to 40 years after emplacement in rows by dragline, ranged from less than 10 cm to about 25 cm. This amount of subsidence reportedly had little or no effect on a railroad track that was located on the spoil (White and Farrow, 1982, p. 386–387).

The spoil at the Rosebud Mine consists of broken and mixed particles which range between 0.25 to 1 m^3 in volume and 1,500 to 1,800 kg/m^3 in density, averages 72 percent sand, 16 percent silt, 12 percent clay, and about 12 percent in moisture content. The mean annual rainfall at Colstrip was about 45 cm (18 in) from 1971 to 1976 (White and Farrow, 1982, p. 385). About 50 percent of the moisture occurred during intense spring storms of short duration. Subsidence in spoil that is saturated from intense storms of short duration may be less than subsidence in spoil saturated by rising water tables. Where this type of precipitation is dominant, only a small volume near the top of the spoil may become saturated, whereas a large volume of the spoil can be saturated by rising water tables.

Dewatering of Aquifers and Stress-Strain Readjustments

Subsidence can be caused by dewatering and compaction of aquifers and by stress-strain readjustments near and within surface mines. Hutchings and others (1978, p. 136–161), for example, observed subsidence within a 100 km^2 area around a large open pit mine in thick brown coal deposits of the Latrobe Valley, Victoria, Australia. Movements included (1) regional vertical and horizontal movements caused by declining pressure in confined aquifers and lowering of free aquifers near the pit during dewatering operations, (2) uplift at the bottom of the pit caused by pressure relief of confined aquifers, (3) vertical and horizontal movements caused by stress-strain readjustments due to crustal load and topographic changes, and (4) horizontal block movements of coal on saturated clays near the rims of the pit caused by water in joints and cracks near the rims of the pit.

The coal beds, which contain interbeds of clays, sands, and silts, are as much as 180 m thick. The overburden is only 10 to 20 m thick. The pits are as much as 180 m deep and 2.5 km wide. Studies indicate that the vertical and horizontal movements are caused by consolidation of coal and overburden, where confined aquifers near the pits are dewatered and pore water pressures are reduced, and by reduction of vertical and horizontal confining stresses near and within the excavations.

Results of subsidence measurements showed that the vertical displacement ranges from about 0.2 m to as much as 1.25 m. Subsidence averages about 0.5 m over a 100 km^2 area around the Morwell and Yallourn open pit mines. Horizontal displacements range from 0.2 to 2 m within a 2 km radius of the Morwell pit. Cracks are common along coal cleat and in the overburden within about 600 m of the rims of the pits, whereas vertical and horizontal movements without cracking occur at distances greater than about 600 m of the pit rims. Vertical and horizontal earth movements affected a larger area, as the area and depth of the pits increased. During the last 50 years of operation, subsidence first affected the immediate mining area, then the stability of the ground beneath the powerplant and support structures located about 1 km west of the pit. Finally, it affected water storage reservoirs, gas lines, four steam powerplants, a factory, roads, and communities within the 100 km^2 area.

COAL MINE FIRES

Coal mine fires are common and widespread in the western United States. Many hundreds of outcrop fires and many tens of underground fires have been reported in the states west of the Mississippi River (Fig. 1). Fires are burning in portions of five abandoned underground mines, covering an area of about 3 km^2, in the Sheridan, Wyo., area (Dunrud and Osterwald, 1980, p. 30). Local fires also occur in surface mines in the western United States. Many of the outcrop and underground fires are caused by subsidence and, in turn, cause more subsidence. Many, if not most, of the fires ignite spontaneously where oxygen, water, and fine coal occur in proportions conducive to ignition.

Ignition Processes and Combustion Products

Results of laboratory tests during the last 50 years indicate that factors contributing to spontaneous heating and ignition include (1) availability and flow of oxygen through fine coal, (2) size of the coal particles, (3) rank of the coal, (4) moisture content, (5) temperature, and (6) other factors, such as pyrite content, structure, and mining practices (Kim, 1977, p. 2–6). Once ignited, fires can continue and accelerate, where air is available.

Field studies indicate that fires ignite where air and water come in contact with fine coal fragments and dust, which have large surface area-to-volume ratios. Air and water enter underground mines through poorly sealed portals and shafts and (or) through subsidence cracks or pits. Once the coal ignites, the fires can support combustion and spread by intaking fresh air and exhausting the products of combustion through cracks, pits, and poorly sealed portals and shafts (Dunrud and Osterwald, 1980, p. 31). Voids produced by the fire can cause further subsidence and more cracks and pits, which in turn provide more air to the fire and increase the burn rate.

Subsidence depressions, cracks, and pits occur where fires consume coal in the mine and at nearby mine boundaries and soften the coal adjacent to the burn front (Shoemaker and others, 1979, p. 140–153). The heat and steam produced by the fires also may weaken the overburden, particularly where soft claystones or shale are present, and increase the stoping rate. The burn rate of the fires commonly is sporadic, where caving of the overburden temporarily reduces local ventilation to the fire until ventilation is reestablished again by the development of more pits and cracks.

Studies of fires that were started artificially by the U.S. Bureau of Mines in the underground mine at Bruceton, Pa., indicate that thermal propagation velocities are about 2.5 cm/hr in the coal bed (Kuchta and others, 1982, p. 13–24). The studies also showed that the fire temperature drops 200° to 600°C within about 100 hours after the fires are sealed; that the main fire gases are CO, CO_2, CH_4, and H_2; and that the concentration of particulates in the smoke is a sensitive indicator of the temperature in the fire area.

Noxious and toxic chemicals, steam, and smoke, which are exhausted into the atmosphere in varying quantities through pits and cracks, can pollute the air and water and be a hazard to life (Down and Stocks, 1978, p. 57–85; Fig. 19). Analysis of the gasses from underground coal mine fires in the Sheridan, Wyo., area, for example, revealed traces of carbon disulfide, carbon oxysulfide, organic sulfur compound(s), methane, and carbon monoxide. The exhausted gasses also commonly contained more carbon dioxide and helium than normal atmosphere. Combustion products also can pollute ground water or surface water where

Figure 19. Surface effects of a coal mine fire in the Sheridan, Wyo., area. *A.* Aerial oblique view showing plumes of steam and smoke emanating from subsidence pits and cracks above a fire in an abandoned underground mine and in a coal bed overlying the mine (Nov. 1981). Depths to the fires range from about 5 to 40 m. Ground temperatures in flame-filled cracks on leading edge of fire pit (middle) measured 925°C in April, 1976. *B.* Collapse above fire in an underground coal mine occurred without warning; snow had not even melted before the collapse occurred (Jan. 1979).

the water comes in contact with dissolved or transported products of combustion.

Surface and Underground Effects

Fires in shallow underground mines also have reached the surface locally or have come close enough to the surface to raise ground temperatures enough to melt the winter snows, cause early greening of vegetation, or keep the vegetation green all winter (Fig. 19A). Fires also can move upward into the overburden and eventually erupt at the surface, which can cause fires in overlying coal beds and in nearby vegetation at the surface (for example, Dunrud and Osterwald, 1980, p. 32). Collapse may occur suddenly and without warning, where (1) the steam and heat of the fires weaken overburden rocks and surficial material, (2) the overburden is less than about 10 to 15 times the thickness of the coal bed(s), and (3) fires have started recently and the resulting heat flow has not yet affected the surface.

Fires also occur locally in currently active, deep coal mines. Mine deformation studies by the author and others in Utah and Colorado, for example, revealed that local fires are common where air, fine coal, moisture content, and high temperatures promote spontaneous ignition. Elevated temperatures can be caused by spontaneous heating, geothermal heating, or nearby natural coal fires. A fire in a mine in the Somerset, Colo., area forced closure of the mine for about a year. Recovery operations involved mine personnel entering the mine a few hundred meters at a time in breathing apparatuses, erecting airtight seals, reestablishing ventilation to the seal, and sealing and flooding the fire areas until the entire mine was ventilated and the fire contained.

Fires also occur by spontaneous ignition in operating and abandoned surface mines of the western United States (for example, Dunrud and Osterwald, 1980, p. 37–41; Hertzberg, 1978, p. 47–49; Brawner, 1978, p. 24–25). The fires usually can be controlled with the equipment used for mining, whereas in underground mines, the extent and exact location of the fires are often unknown and the mines are inaccessible. Fire control procedures for extensive fires in underground mines consist of isolation trenches, smothering, and flooding (Johnson and Miller, 1979, p. 18–21). Fire control measures for outcrop fires commonly consist of loading out the burning coal, smothering, isolation trenches, and local flooding.

SEISMIC ACTIVITY

Stress concentrations and accumulated strain around mine openings can cause sudden, violent releases of strain energy (rockbursts, bumps) within the coal and rock or along faults, stratification planes, and joints (for example, Obert and Duvall, 1967, p. 582–611; Dunrud and Osterwald, 1965, p. 174–175; Osterwald and Dunrud, 1965, p. 168–174; Peng, 1978, p. 343–347). Together with rock creep and flow, rockbursts are a mechanism of adjustment of the rock mass toward equilibrium conditions which are disrupted by mining. The spontaneous releases of strain energy cause earth tremors which range from a level that can be detected only by sensitive seismic recording equipment, to as great as magnitude 3 or 4 on the Richter scale. The larger tremors locally have damaged brick chimneys and other massive structures.

Seismic studies by the author and others in coal mining areas in the Sunnyside, Utah, and Somerset, Colo., areas indicate that the sources of the larger magnitude tremors commonly occur where mining and subsidence processes have caused either stress concentrations, such as in the abutment zones adjacent to mining panels, or a reduction in vertical stresses, such as in the rock beneath the mined-out areas (Dunrud, 1976, p. 28–29). Many of the larger magnitude tremors in the Sunnyside, Utah, area during 1967–1970, for example, were located along faults near the mine workings in plan view, and were located near the mine workings to as much as 2,500 m beneath the mine workings in profile view (Dunrud and others, 1972, p. 32–33). Tremors caused by subsidence processes above abandoned coal mines also may cause earth tremors that can be severe enough to alarm local residents (Johnson and Miller, 1979, p. 6).

Tremors commonly are numerous near underground coal fires. In the Sheridan, Wyo., area, for example, small earth tremors were recorded above two mines, which were believed to be caused by breaking, caving, and sudden releases of steam and air in the mine and in rocks above the fire areas (Dunrud and Osterwald, 1980, p. 41). Tremor frequencies on the order of 10 to 1,000 Hz also were measured within about 5 to 15 m of a fire pit. The number of tremors detected increased by about an order of magnitude when a seismometer was moved about six times closer to the fire pit. Fire fronts therefore might be detected and located with seismic or seismo-acoustic equipment capable of detecting tremors in the frequency range of 10 to 10,000 Hz.

SUBSIDENCE HAZARDS

Subsidence depressions, cracks, or pits can endanger life and property and also can reduce the value of the land. Changes, diversion, and pollution of surface water and ground water and coal mine fires also can endanger the health and safety of people and animals and may further reduce the value of the land. The risk to life and property depends on the location of subsidence features relative to population centers. Subsidence and fires in uninhabited areas pose little risk to life and property, even though they may be a threat to the environment; however, they can pose a high risk to life and property in populated areas or where transportation routes are located.

Many of the currently active or proposed underground coal mines are located in areas remote from population centers. However, a significant number of the abandoned underground coal mines are located near or within population centers beneath overburden that ranges from a few meters to a hundred meters or so thick. For example, about 200 abandoned underground mines are located near the Denver Metropolitan area, Colo., where the overburden thickness ranges from a few meters to more than 120 m (400 ft). Depressions, cracks, and pits have locally damaged utility lines, highways, and buildings near or within the towns of Lafayette, Erie, Louisville, and Firestone (Myers and others, 1975, p. V-10 to V-13; J. L. Hynes, oral communication, 1981). Depressions and pits are locally a potential hazard in a residential subdivision in the southwest Denver area (Amuedo and Ivey, Inc., written communication, 1982). Underground coal fires are locally present in the Marshall and Erie, Colo., areas. Subsidence also has occurred in, within, or near such towns as Rock Springs, Glenrock, Hanna, and Kemmerer, Wyo.; Colorado Springs, Colo. (Foster, 1979, p. 1A); Gallup, N.M.; Black Diamond, Renton, Auburn, Wash.; Bowman, N.D.; and Huntsville, Mo. As many as 120 coal mines are located beneath about 20 percent of Des Moines, Iowa, and adjacent areas, at depths of 12 to 70 m (40 to 230 ft); subsidence has occurred recently in two areas of Des Moines (Avcin and others, 1979; Fig. 1).

Hazards to Life

Subsidence hazards to life range from very low to low, where cracks are a few millimeters to a few decimeters wide or where wider cracks can be seen and avoided. Moderate to high risks can exist where cracks are wider than a few decimeters, particularly where they are obscured by vegetation. Risks to life can be high where pits and large cracks are present near the ground surface and the added weight of a pedestrian or vehicle can trigger surface collapse. Risk to life also can be high above coal mines in steeply dipping coal beds, where pits can occur suddenly and be a significant percentage of the *width* of the mining area. In the Westphalian coal field of western Germany, a team of horses was lost when the ground being plowed above a coal mine in an inclined bed suddenly collapsed beneath them, and the rims of the pit failed and covered them up (Briggs, 1929, p. 100).

Subsidence hazards can be high above room-and-pillar mines, where the pillars were not removed and the overburden is less than the stoping height above the mine openings (Dunrud and Osterwald, 1980, p. 47). Pits may occur without warning in these areas because caving can occur to the surface without filling with bulked debris. A large pit suddenly formed in a street of Rock Springs, Wyo., in 1976. A four-year-old boy barely escaped as the pit began to form where he stood (*Denver Post,* Oct. 12, 1976).

Ths risk to life can be very high where transportation routes are constructed above unknown mine openings and where the overburden thickness is less than the caving height. For example, large cavities were encountered above a forgotten coal mine in the Sheridan, Wyo., area during construction of Interstate 90 in 1982 (Wyoming State Highway Dept., written communication, 1982). Pits could have occurred suddenly in the highway and posed a threat to travelers if the mine cavities had not been discovered and mitigated during the construction stage.

The greatest risk to life occurs where cavities and cracks are located near the surface above underground fires, where the stoping rate exceeds the rate of heat transfer and no signs of the fire below are present, and where the added weight of a pedestrian can trigger collapse (Fig. 19B). In Centralia, Pa., for example, a 12-year-old boy narrowly escaped death when the ground collapsed beneath him and he fell into a smoking, steaming pit that had suddenly formed in his grandmother's yard. Fortunately, the boy was able to grab a tree root and hang on until he could be rescued. The gasses in the pit, which was located about 60 m above a burning coal mine, contained carbon dioxide, carbon monoxide, and sulfur oxides (Current Science, 1981, p. 6–7).

Hazards to Property

Hazards to property range from low to very high, depending on the type of deformation, the type and size of structures, and the type of material used in construction (NCB, 1975, p. 45–58). Long, rigid structures with no flexibility, such as welded pipelines and brick buildings, can be damaged easily by subsidence, particularly in areas of maximum strain (Wardell, 1971, p. 209–211). Larger structures that have small deformation tolerances, such as large buildings (including apartments) (Thorburn and Reid, 1978, 87–99), schools (Stephenson and Aughenbaugh, 1978, 100–118), powerplants, or factories, are particularly susceptible to tensile and compressive strain, unless they are designed to withstand the effects of differential settlement.

Slope or tilt can damage structures or features sensitive to tilt (NCB, 1975, p. 45–58; Peng, 1978, p. 323–331). For example, slope changes can affect drainage by changing gradients in sewer pipes, streams, and lakes. Tilting can affect delicate equipment in factories and high structures, such as tall buildings and smoke stacks. A slope of 0.025, or 1 in 40, which is the maximum slope measured from subsidence profiles in the Somerset, Colo., and Salina, Utah, areas, could throw a structure 4 m high out of plumb by 0.1 m or a structure 40 m high out of plumb by 1 m, which in turn might cause local load redistribution and possible yielding of the foundation, footings, or bearing material beneath tall, massive buildings.

Differential vertical displacements in subsidence depressions can cause tensile, compressive, and shear failure in structures. Ground strains are transmitted to structures by vertical or frictional shear stresses acting on the foundation, footings, or pilings. Deformation commonly occurs at the weaker points of structures, such as near windows and doors of buildings, where tensile ground strains are greatest. Convex curvature of the ground surface, for example, can cause tensile failure near the top of a building and rotational failure within the structure where the stresses induced by curvature and strain exceed the strength of the structure (Fig. 4).

Results of studies of damage to buildings, caused by strain in the United Kingdom, led to a classification system whereby either damage to a structure of a given length can be estimated in terms of estimated ground strain or ground strain can be estimated in terms of severity of damage to buildings (Fig. 20). For example, changes in horizontal length of a structure (ΔL) of 0.03 to 0.06 m

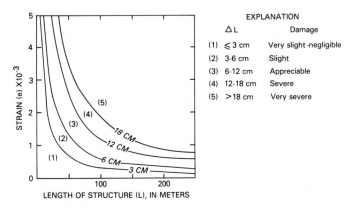

Figure 20. Relation between strain and length of structure. Damage classification as follows: (1) very slight to negligible—local hairline cracks inside of building; (2) slight—several hairline cracks inside building; windows and doors may stick slightly; (3) appreciable—local, small fractures showing on building exterior; doors and windows hard to open; local fracture of utility pipes; (4) severe—many utility pipes broken; open fractures in building that require patching; window and door frames distorted; floors and walls bulging or out of alignment; loss of support of local beams, columns, and joists; (5) very severe—same as (4) except damage is more extensive requiring partial or complete rebuilding—beams and columns loose; windows broken from distortion; floors and ceilings tilted and walls bulged and buckled noticeably. Modified from NCB (1975, p. 49).

(slight damage class) cause small cracks, sticking doors and windows, and many require minor cosmetic repairs inside; whereas changes in the length of a structure of 0.12 to 0.18 m (severe damage) commonly disrupt utilities, cause open cracks that require patching, distort windows and door frames, walls, floors, ceilings, and cause local loss of bearing support of beams. Damage to local structures in the Denver Metropolitan area (Myers and others, 1975, p. V-12; Fig. 21A) and in the Des Moines, Iowa, area (Fig. 21B, C) indicates strain in the appreciable to severe category.

Effects on Ground Water and Surface Water

The quality of ground water can be affected by subsidence cracks in bedrock. For example, both the acidity and proportion of magnesian salts increased in a potable water well in the United Kingdom, where subsidence cracks caused water from a sandstone aquifer to come in contact with carbonate rocks within the draw-down zone of the well (Wardell, 1971, p. 212–214). The quality and temperature of surface water was affected in the Kemmerer, Wyo., area in 1979, when the water was diverted by a subsidence pit into an underlying coal mine, filled the mine workings, and emerged again about 1.6 km (1 mi) downstream. Tests of the water by the Soil Conservation Service, Kemmerer, Wyo. (written communication, 1981), showed that the water at the outlet (1) was about 18°C (32°F) warmer, (2) had about three times more dissolved substances (Na, K, Mg, Cl, SO_4, Fe, and Mn), and (3) was more alkaline (pH of 6.8 to 7.3) than the water at the inlet.

Effects on Unstable Ground

Landslides also can be triggered by subsidence above underground mines, where depressions, cracks, or pits occur on unstable slopes. Landslides also can cause subsidence, where the increased surface loading from the slide promotes failure above unstable cavities located near the ground surface. The landslide, which covered part of Aberfan, Wales, and caused the deaths of 144 persons in October 1966, reportedly was caused partly by coal mine subsidence (Wardell and Piggott, 1969, p. 194; Bishop and others, 1969, p. 44–45). The large landslide at Frank, Alberta, Canada in 1903, which caused the death of 70 persons, covered part of the town and part of the Canadian Pacific Railway, was reported to have been triggered by coal mine subsidence (Briggs, 1929, p. 115).

Relation Between Hazards and Overburden Thickness and Lithology

Overburden thickness and strength commonly control the severity of subsidence hazards and the areal extent of potential subsidence. The subsidence area increases as the overburden thickness increases, but the rate of increase commonly is reduced where the overburden consists of strong sandstones compared with weak shales and mudstones. Vertical displacement often is greater where the overburden rocks consist of weak shales and mudstones than it is where the overburden contains strong, massive sandstones. Horizontal displacement, slope, and horizontal strain increase as the mining thickness increases and as the overburden thickness decreases (Table 2). Subsidence pits are limited to areas where the overburden thickness is less than the maximum stoping height, which commonly is less than 10 to 15 times the thickness of coal mined. The stoping height often is reduced in strong rocks because strong rocks may bridge stoped areas and because the caved fragments are larger and the bulking is greater.

SUBSIDENCE PREDICTION AND MITIGATION

Subsidence prediction involves estimating vertical and horizontal displacement, slope, strain, area, rate, and duration, based on current knowledge and on mining, geologic, geotechnical, and topographic factors existing at the mining site. Mitigation consists of reducing the risk to life, property, and (or) the environment by reducing the severity of subsidence or the consequences of subsidence. The capability to predict and mitigate potential coal mine subsidence hazards depends on whether the coal mines are currently active or are abandoned. Subsidence hazards commonly can be reduced more easily and cheaply in active mining areas than they can be in abandoned mining areas.

Subsidence Prediction

Active underground mining areas. In currently active or proposed underground mines in the western United States, subsi-

Figure 21. Photographs showing damage or potential damage to structures caused by subsidence. *A.* Depressions, pits, and cracks occur within a few meters of a state highway and utilities, including a high-pressure gasline and powerlines in the Marshall, Colo., area. *B.* X-shaped and en-echelon cracks in the middle and near the corners of a building in Des Moines, Iowa. *C.* Close-up view of cracks in brick building, which have been patched (located in *B* at arrow).

dence can be estimated by using existing information and relating mining and site factors of the area to subsidence amount, slope, strain, area, rate, and duration. Site factors often can be determined during the exploration and drilling phase of mining operations at costs that may be more than offset by increased safety and productivity of the mine. In proposed mining areas, subsidence estimates can be compared with actual measured results after mining begins, when mining and site factors can be further refined.

Results of subsidence studies in the Somerset, Colo., Salina, Utah, York Canyon, N. Mex., and Sheridan, Wyo., areas indicate that vertical displacement, limit angle and critical extraction angle, and slope and strain are related to mining thickness, areal extent of mining, and overburden thickness and strength as follows (Table 2):

1. Maximum vertical displacement ranges from about 0.45 to 0.90 t for critical and supercritical extraction panels. In the Salina, Utah, area, subsidence ratio is reduced by about 50 percent where a massive sandstone occurs in the upper part of the overburden.

2. Limit angle (ϕ) and critical extraction angle (γ) appear to decrease (become steeper) as the percentage of stronger rocks increases, particularly where these rocks occur near the surface. Limit angles range from 15° to 25° in the Somerset, Colo., and Salina, Utah, areas, where the overburden consists of about 25 percent to 35 percent of medium-strength sandstones and 65 percent to 75 percent low- to very low-strength shales and claystones, to 8° to 12° in the Salina, Utah, area, where the overburden consists of about 45 percent medium-strength sandstone and 55 percent low- to very low-strength shales and claystones. Critical extraction angles range from about 25° in the Salina, Utah, area, where a massive sandstone occurs at the surface, to 35° in the Somerset, Colo., area, where lenticular sandstones are interbedded with shale and mudstone.

3. Maximum slope and horizontal strain increase with decreasing overburden thickness but remain relatively constant in terms of the ratio of maximum vertical displacement to overburden thickness (S/d). Maximum slope ranges from 3 to 5 S/d (0.025 m/m) in the Somerset, Colo., and Salina, Utah, areas to 2.5 S/d (0.3 m/m) in the Sheridan, Wyo., area. Maximum horizontal strain is compressive in the Somerset, Colo., Salina, Utah, and York Canyon, N. Mex., areas and ranges from −0.7 to −4 S/d (−0.006 to −0.022 m/m); maximum strain in the Sheridan, Wyo., area is compressive (−2 S/d or −0.04 m/m) above panel boundaries but tensile above barrier pillars (2.3 S/d or 0.02 m/m).

Abandoned underground mining areas. Information needed to predict subsidence above abandoned mine workings is the same as for active mines; however, it is more difficult to obtain because the mine workings are rarely accessible to examine firsthand and information on the thickness of coal mined, the mine geometry, extraction percentage, and overburden thickness and strength commonly is incomplete or unavailable. The location of remaining pillars and their state of deformation commonly is unknown or not well known and often must be obtained by methods that usually are costly and that produce no coal. The available mining information also decreases with the age of the mine. Mining laws, which required that accurate and detailed mine maps be compiled, were passed in the late 1800's and early 1900's in the western United States.

Information gathering and reporting procedures consists of (1) superimposing mine maps on topographic maps, and compiling land-use and geologic maps of the area; (2) drilling to locate mine voids and to determine lithology, structure, and overburden thickness and strength; (3) conducting geophysical studies (seismic, gravity, electrical, etc.) to supplement drilling and reduce project costs (Burton, 1975, p. 75–102; Howell, 1975, p. 103–108); and (4) compiling maps that show (a) the location and geometry of the mine workings, (b) probable mining thickness, (c) overburden thickness and lithology, (d) subsidence depressions, pits, and cracks, and (e) potential subsidence hazards (for example, Ivey, 1978, p. 166–172; Myers and others, 1975, [95] p.).

Surface mining areas. Vertical and horizontal displacement and depressions, cracks, or pits in surface mining areas reportedly are controlled by (1) the amount and extent of lowering and raising of the water table during and after mining, (2) the amount of aquifer compaction around the mining area, (3) the nature and extent of compaction of spoil, and (4) the amount and rate of saturation of reclaimed spoil by ground water recharge or by direct surface moisture (for example, Charles and others, 1978, p. 229–251; Hutchings and others, 1978, p. 136–161; White and Farrow, 1982, p. 385–388). Variations in ground water levels during and after mining may be significantly different in arid and semiarid areas than they are in areas with higher precipitation levels. Saturation of reclaimed spoil in much of the western United States also may begin at the top and progress downward, during short, intense periods of precipitation.

Subsidence Mitigation

Mitigation consists of reducing the risk to life, property, and the environment by reducing the amount and severity of subsidence, building subsidence-resistant structures, restricting land use, or removing the features that are endangered. The severity of subsidence above underground mines can be reduced by increasing pillar sizes and decreasing the dimension of the mine openings, using harmonic extraction procedures (for example, Kaneshige, 1971, p. 169–199); by filling the mine openings (Colaizzi and others, 1981, 56 p.); or by designing mine extraction areas in harmony with surface development, where both mineral extraction and surface development are equally essential (for example, Dobson and Roberts, 1959, p. 691–697, 723–730). Subsidence in surface mining areas may be reduced by uniformly compacting reclaimed spoil and minimizing the dewatering of aquifers adjacent to the mine areas.

Underground mining areas. Subsidence above active underground mines can be reduced by (1) leaving coal pillars to support the overburden; (2) mechanically, pneumatically, or hydraulically filling the mine cavities with material; or (3) constructing support pillars from mine refuse (Rose and Howell, 1979, p. 290–298) or other material. In currently active coal mines in foreign countries, mine refuse (rock and bony coal) sometimes is placed mechanically, pneumatically, or hydraulically into mine openings in sequence with mining. Hydraulic backfilling, which was used first in the eastern United States in the late 1890's and later was used extensively in France, Belgium, Germany, and Poland, for example, can reduce subsidence from 50 percent to 90 percent to about 5 percent to 15 percent of the thickness of coal mined (Briggs, 1929, p. 81–86; Table 3; Fig. 8). Backfilling costs reportedly are 5 percent and 11.5 percent of coal production costs in Poland and England, respectively (Cochran, 1971, p. 15).

Subsidence effects above abandoned mines can be reduced by (1) filling the mine openings from the surface, (2) constructing grout columns from the mine to the surface, or (3) driving piling from the surface to the mine floor. Filling operations include placing mine refuse or other granular material, such as sand and silt, into the mine openings through drill holes by blind flushing, controlled flushing, and pumped-slurry injection (Colaizzi and others, 1981, p. 10–15). Costs are much higher than they are when filling is done concurrent with mining.

The pumped-slurry injection method, which is the most widely used of the three, was used by the U.S. Bureau of Mines in Rock Springs, Wyo., and in Scranton, Pa. This method consists of pumping a granular material and water down cased boreholes to mine openings to be filled, with a velocity sufficient to keep the solids in suspension, so that a number of openings can be filled from one borehole and mine openings beneath buildings can be filled from streets and alleys (Colaizzi and others, 1981, p. 18–20).

Structural damage in subsidence areas also can be decreased by reducing the frictional stresses between the ground and foundation or footings of buildings, by strengthening the foundation or the structures, trenching around buildings, and (or) by reducing the length of the structure or making longer structures flexible (for example, NCB, 1975, p. 65–73; Geddes, 1978b, p. 949–968). Specially reinforced, segmented, beam-column structures have been built on reinforced concrete slabs and founded on gravel to reduce the effects of strain and tilt (Bell, 1978, p. 562–578). These structures reportedly are unaffected by vertical displacements of as much as 0.6 m, slope changes of 0.004 m/m, and horizontal strain of 0.001 (Geddes, 1978a, p. 579–596).

Surface mining areas. Stabilization of reclaimed spoil has not been done in the western United States, to the writer's knowledge, primarily because most of the reclaimed land so far has been used for agricultural purposes. However, some of the reclaimed surface mining areas could become prime sites for residential and industrial development. The highest land-use priority may now change from one of coal mining to one of residential and industrial development, particularly where land is needed for development near expanding population centers.

Stabilization procedures involve reducing differential compaction of the spoil to levels that can be tolerated by the structures to be built. Subsidence in currently active surface mines can be reduced by grading dragline-dumped spoil piles before significant differential compaction occurs. Subsidence of dragline-dumped spoil in old, abandoned strip mines might be reduced by uniformly compacting the graded spoil or by allowing it to settle with time, wetting, and drying until it stabilizes and then regrading it. Use of compactors also could help reduce differential compaction in either currently active or old, abandoned mines.

Prediction and Mitigation Procedures—Underground Mining Areas

In currently active, proposed, or abandoned mining areas, subsidence might be estimated and mitigated by (1) determining the site factors (thickness, strength, and structure of the overburden, coal, and roof and floor rocks); (2) determining mining factors (mine geometry, mining methods, and method of goaf treatment); (3) estimating subsidence parameters (vertical and horizontal displacement, slope, curvature, strain) on the basis of site and mining factors; and (4) modifying mining or surface development plans that will minimize risk to life, property, and the environment, and that are in harmony with current or planned uses of the land (Fig. 22). Prediction and mitigation procedures might include the following:

1—Site factors:

a. Prepare topographic, geologic, geotechnical, and hydrologic maps from studies of outcrops, aerial photographs, and geologic and geophysical logs of the drill core and drill holes. Determine the geotechnical properties (point load index strength, slake durability, compressive and tensile strengths, density, porosity, water content, and elastic moduli) of the drill core. Obtain supplemental information from geophysical logs (resistivity, natural gamma, gamma density, caliper, sonic) of rotary drill holes.

b. Prepare (1) structural contour and isopach maps of existing coal beds; (2) thickness and strength of the overburden; (3) location and trends of lineaments, faults, major joint trends; and (4) location and amount of ground water and surface water.

2—Mining factors:

Make preliminary design of mine geometry (width (W) and length (L) of mining panels), mining thickness (t), mining rate (R_m) and method (room-and-pillar (R&P), longwall (LW), and shortwall (SW)), and goaf (gob) treatment (caving, mechanical stowing, backfilling). Superimpose existing or proposed mine layouts on maps (1-a) or derivative maps (1-b).

3—Subsidence parameters:

Determine maximum subsidence amount, area, slope, horizontal strain, rate, and duration by incorporating the foregoing site factors to mining factors, such as panel width and length, mining method, goaf treatment, and by using subsidence graphs and profiles as follows:

a. Estimate maximum vertical and horizontal displacement (S, S_h), subsidence area (A_s), slope (M), curvature (C), and horizontal strain (E) by using the subsidence graph (Fig. 8), the

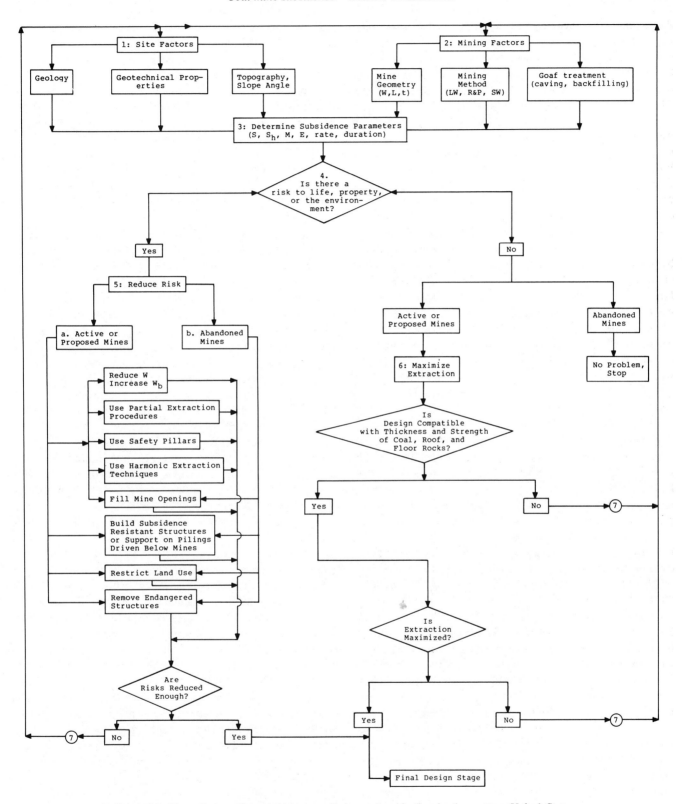

Figure 22. Flow diagram for subsidence prediction and mitigation in the western United States. Numbers are keyed to items in text under "Prediction and Mitigation Procedures—Underground Mining Areas."

subsidence parameter graphs (Fig. 11), the summary table (Table 2), and the graph showing maximum slope and curvature (Fig. 12). Maps of displacement, slope, curvature, and horizontal strain in relation to planned mining panels (Fig. 7B) can be made from subsidence parameter graphs (Fig. 11). Profile or distribution functions from finite element, analytic, and empirical model studies (for example, Berry, 1978, p. 781–811; Bräuner, 1973a, b; Salamon, 1978, p. 30–59; Peng, 1978, p. 291–307; Peng, 1980, p. 44–56; Savage, 1981, p. 195–198; Jones and Bellamy, 1973, p. 515–530; Marr, 1959, p. 693–696; Tandanand and Powell, 1982, 14 p.) may be used in conjunction with, supplemental to, or instead of this method.

b. Modify subsidence estimates for structural conditions, such as dip of bedrock (Fig. 14), faults (Figs. 15, 16), and major joints in the panel extraction areas. Estimate height of caving, fragmentation, and bulking based on mine geometry, method of mining, and lithology and structure of roof rocks. Caving, fragmentation, and bulking may increase where complete extraction, longwall mining, or straight pillar lines are used.

c. Estimate subsidence rate and duration by constructing subsidence development curves (Fig. 9A, B) for the mining method used. Modify for partial extraction procedures (Fig. 5); estimate the compressibility of caved rocks (for example, Peng, 1978, p. 223) and the effects that water may have on the caved material.

4—Is there a risk to life, property, environment?:

Determine risk, if any, to life, property, or environment. Evaluate current and potential use of the land surface and susceptibility of existing or potential features, such as (a) utilities (pipelines, aqueducts, etc.); (b) buildings (Figs. 20, 21); (c) springs, streams, and lakes; (d) croplands, orchards, forests, or any possible subsurface use, including (1) aquifers, (2) mineral deposits, and (3) underground structures (see, for example, Coal Mining Operating Regulations as implemented by the Surface Mining Reclamation and Enforcement Provisions in the Federal Register, 1977, Part II, 30 C.F.R. 700-725, December 13, 1977, p. 62639-62712 for more details).

5—Reduce risk:

A. Active or proposed mines. Where necessary, reduce subsidence amount, areal extent, rate, duration, slope, curvature, and strain where unacceptable in active or proposed mines by (1) reducing the width of the mining panels to subcritical width and increasing the width of the barrier pillars between panels (NCB, 1975, p. 90–93; Peng, 1978, p. 338–339; Figs. 5, 8); (2) using room-and-pillar mining methods and by designing for stable pillars (Fig. 5; Peng, 1978, p. 339); (3) leaving safety pillars (for example, Down and Stocks, 1978, p. 326); (4) using harmonic extraction procedures, where practical (for example, Kaneshige, 1971, p. 179–195; Dunrud, 1976, p. 7–8; Down and Stocks, 1978, p. 324–328); and (5) filling, or partially filling, the mine workings with mine refuse in sequence with mining (mechanical, hydraulic, or pneumatic backfilling).

B. Abandoned mines. Reduce risk above abandoned mines by filling the mine openings in problem areas (Colaizzi and others, 1981, p. 1–38), building subsidence-resistant structures, building structural support below the mine floor, restricting land use, or removing endangered structures.

6—Maximize extraction:

Where no risk is determined to exist and where mining conditions (thickness and strength of coal, roof, and floor rocks) are compatible, by using longwall, shortwall, or high recovery room-and-pillar methods (for example, Peng, 1978, p. 5–40, 208–280).

7—Iterate 1 through 6 to optimize mine design (Fig. 22).

LAND USE

A complete knowledge of the mining and geologic aspects of lands to be developed in coal mining areas is one of the more important aspects of land-use planning. Information about any mineral deposits, previous mining, and geologic and geotechnical conditions should be known before residential and industrial development begins. Case studies of subsidence in areas that were developed without considering the consequences of previous mining or possible future mining show that the time and money spent in determining the facts and making careful land-use plans before development were small, compared with the time and money spent in stabilizing the land or paying damage claims after development. The U.S. Bureau of Mines estimates that the average cost to control subsidence above abandoned underground mines is $30,000 per acre. On the basis of this estimate, it would cost about $1 billion to control subsidence in the estimated 13,300 hectares (33,000 acres) in urban areas alone in the 13 states west of the Mississippi River that are underlain by coal mines, that have not been affected by subsidence as of 1978, but that eventually may be affected by subsidence (Johnson and Miller, 1979, p. 6–7).

Implementation of recent laws in Colorado on land development is an example of how land use might be designed so that subsidence and other geologic hazards may not be a hazard to life and property in the future. State Senate Bill 35, which was passed in 1972, requires that a geological investigation and report be made on proposed subdivisions in unincorporated areas. The counties, since 1972, are responsible for approving or disapproving the proposal, based on potential geologic hazards. State House Bill 1041 (passed in 1974) provides that areas where potential geologic hazards or where economic mineral deposits occur may become a matter of state concern and cannot be legally developed unless the hazard is eliminated and (or) the mineral is extracted (Ivey, 1978, p. 164; for more details on laws related to coal mining, subsidence, and land use, see, for example, Zwartendyk, 1971, p. 213–241).

SUMMARY AND CONCLUSIONS

Subsidence is controlled by percentage of coal extracted and mine geometry, the ratio of mine area to critical extraction area, the caving height and amount of bulking near the mine openings, and lithology and structure of the overburden rocks. Subsidence depressions (troughs) commonly occur above mining areas where enough coal has been extracted to cause yielding of remaining pillars. Cracks locally are common near the margins of the subsidence depressions, where convex curvature causes lengthening of the ground surface. Pits can occur above mine openings where the overburden thickness is less than the maximum possible caving height—a thickness that usually is less than about 10 to 15 times the mining thickness.

Maximum vertical displacements commonly amount to about 45 percent to 90 percent of the thickness of coal mined above critical and supercritical room-and-pillar and longwall extraction panels in the western United States. Uplifts of as much as 5 percent of the thickness of coal mined also have been recorded and apparently are caused either by the swelling of clay minerals in caved or downwarped rocks as they become wet or by downwarping of overburden into adjacent extraction panels. The depth of subsidence pits varies from a few percent of the thickness of coal mined, where cavities are nearly full of bulked material when they reach the surface, to as much as about three times the thickness of coal mined, where caved material can move laterally into adjacent mine openings.

The limit or draw angle above panel extraction areas commonly ranges from about 5° to 25° in the western United States. The angle commonly is about 5° to 15° where massive sandstones are present in the overburden, particularly where they occur near the surface; but it commonly ranges from about 15° to 25° where the overburden consists of sequences of interbedded shale, mudstone, and sandstone. The area occupied by depressions above mining panels is greater than the mining area and increases with increasing overburden thickness, whereas the area occupied by subsidence pits commonly is less than, or equal to, the area of the mine openings.

Maximum changes in slope or tilt of the ground surface and horizontal strain are roughly proportional to maximum vertical displacement and inversely proportional to overburden thickness. A decrease of overburden thickness by a factor of 10 results in an increase in slope and horizontal strain by about a factor of 10. Maximum slope measured within subsidence depressions in the Somerset, Colo., and Salina, Utah, areas, where the overburden thickness ranges from 170 to 270 m is about 0.025 m/m, (3-5 S/d), whereas the maximum slope in the Sheridan, Wyo., area, where the overburden ranges from 20 to 30 m is 0.2 to 0.3 m/m (2.5 to 4 S/d). Maximum horizontal strain, estimated from curvature of subsidence profiles or measured directly, ranges from about −0.006 to −0.02 m/m (−0.7 to −4 S/d) in the Somerset, Colo., Salina, Utah, and York Canyon, N. Mex., areas to about 0.12 to −0.22 m/m (−2 to 2.3 S/d) in the Sheridan, Wyo., area (minus sign means that strain is compressive). Maximum strain commonly may be compressive above extraction panels bounded by large areas of unmined coal but is tensile above barrier pillars between two extraction areas.

Tension cracks are more common than compression features because rocks are stronger in compression. They also are more abundant in bedrock than they are in surficial material. Cracks can occur in bedrock along joints at angles as much as about 30° from the normal to the direction of maximum extension, whereas cracks in surficial material commonly are about normal to the direction of maximum extension. Cracking can occur in surficial material after cracks occur in the bedrock, where the material yields without cracking until weakened by such processes as cyclic wetting and drying.

The duration and rate of subsidence varies with mine geometry, method of mining, and overburden thickness and strength. Depressions commonly become stable above extraction panels in a few months to a few years, depending on mining method, extraction ratio, goaf (gob) treatment, and the behavior of the caved material and remaining coal, whereas pits may occur for many decades or even centuries.

Vertical displacement, slope, curvature, and strain can be analyzed by constructing a subsidence graph, a subsidence parameter graph, and a subsidence development curve for the area of interest, using information included herein (Table 2; Figs, 8, 9, 11). Computers could facilitate calculation of displacement, slope, curvature, and strain, particularly where the critical extraction area changes due to rugged topography, dipping coal beds, variable lithology in the overburden, or more than one vertically superimposed coal bed.

Subsidence depressions, cracks, and pits also can occur near or within surface mines, where differential compaction occurs in the reclaimed spoil or where differential settlement occurs near surface mines due to changes in loading of the earth's crust or to dewatering of aquifers. Subsidence is controlled by the thickness and geologic and geotechnical conditions of the overburden, the method of overburden replacement, grading, compaction, and hydrologic conditions. Methods of emplacement and grading and saturation of spoil, from rising water tables or from precipitation, appear to be major factors controlling subsidence in surface mining areas. Wetting of spoil replaced after the coal is mined increases the rate and amount of compaction. Dewatering of aquifers in thick coal deposits near deep pits may cause subsidence of areas greater than the mining areas.

Subsidence hazards to life and property vary from low, where narrow cracks or shallow pits occur, to high, where wide cracks and deep pits could severely damage structures and trap or injure persons or animals. The hazard can be high near areas of residential and industrial development above shallow underground mines, where large coal pillars occur adjacent to unstable mine openings. The land may appear to be suitable for development, particularly where the mines were abandoned many decades ago. Pits can occur suddenly and with little warning in these areas. The greatest hazard exists where underground mines are on fire, where cavities have moved up near the surface and are filled with fire or noxious and toxic chemicals, and where collapse occurs suddenly and without warning.

One of the more important aspects of land-use planning is to determine the depth, extent, geometry, and extraction ratio of any active and abandoned underground mines and any other economic coal resources that might be mined in the future in the area of interest. Detailed mapping and studies should be conducted to determine potential subsidence hazards before development begins. Experience in populated areas underlain by abandoned mines, such as Denver and Colorado Springs, Colo., Rock Springs, Wyo., Gallup, N. Mex., Black Diamond and Auburn, Wash., and Des Moines, Iowa, has shown that the time and money spent on site studies and land-use planning above abandoned underground coal mining areas, before development begins, are much less than will be spent on mitigation procedures after development has begun.

ACKNOWLEDGMENTS

The author wishes to thank the many individuals and organizations who contributed valuable maps, information, and suggestions, which materially improved the report. Officials of Colorado Westmoreland, Inc., U.S. Steel Corp., and Coastal States Energy Co. allowed access to their respective properties and leases and provided many valuable suggestions about mining and subsidence. L.R. Miller, superintendent, Somerset Mine; Art Garcia, manager, Orchard Valley Mine; Kerry Frame, chief engineer, Dall Dimick and Leonard Barney, engineers, Southern Utah Fuel Co. were particularly helpful.

Many colleagues provided information and advice which benefited the report. R. Edward McKinley made periodic subsidence surveys in the Somerset area. Jill M. Jappe skillfully prepared some of the figures. David J. Varnes, William Z. Savage, and Thomas L. Holzer, U.S.G.S., and James A. Pendleton, Colorado Department of Natural Resources, made many helpful suggestions on the technical contents of the report. Merline Van Dyke, U.S. Office of Surface Mining, provided local subsidence information for the states of New Mexico, Washington, and Wyoming. Gary Glass, Geological Survey of Wyoming, provided information for mines in Wyoming. Jerome DeGraff, U.S. Forest Service, provided information on subsidence cracks in the Salina, Utah, area.

REFERENCES CITED

Abel, J. F., Jr., and Lee, F. T., 1980, Lithologic controls on subsidence: Society of Mining Engineers Preprint 80-314, p. 1–16.

Avcin, M. J., Van Dorpe, P. E., Ravn, R. L., Swade, J. W., Bentzinger, K. C., and Watson, J. W., 1979, Coal resources program report: Iowa Geological Survey Program Report for 1979, 14 p.

Babcock, C., Morgan, T., Haramy, K., 1981, Review of pillar design equations including the effects of constraint, *in* Peng, S. S., ed., Conference on ground control in mining, West Virginia University, 1981: Proceedings, Department of Mining Engineering, Morgantown, W. Va., p. 23–34.

Bell, F. G., 1975, The character of the coal measures, *in* Bell, F. G., ed., Site investigations in areas of mining subsidence: London, Butterworth and Co., Ltd., p. 25–39.

Bell, S. E., 1978, Successful design for mining subsidence, *in* Geddes, J. D., ed., Large ground movements and structures, Conference at the University of Wales Institute of Science and Technology, Cardiff, 1977, Proceedings: New York, John Wiley, p. 562–596.

Berry, D. S., 1978, Progress in the analysis of ground movements due to mining, *in* Geddes, J. D., ed., Large ground movements and structures, Conference at the University of Wales Institute of Science and Technology, Cardiff, 1977, Proceedings: New York, John Wiley, p. 781–811.

Bieniawski, Z. T., 1981, Improved design of coal pillars for U.S. mining conditions, *in* Peng, S. S., ed., Conference on ground control in mining, West Virginia University, 1981: Proceedings, Department of Mining Engineering, Morgantown, W. Va., p. 13–22.

Bishop, A. W., Hutchinson, J. N., Penman, A.D.M., and Evans, H. E. , 1969, Geotechnical investigation into the causes and circumstances of the disaster of 21st October 1966: London, Her Majesty's Stationery Office, p. 1–80.

Braüner, Gerhard, 1973a, Subsidence due to underground mining, Part 1 *of* Theory and practices in predicting surface deformation: U.S. Bureau of Mines Information Circular 8571, 56 p.

—— 1973b, Ground movements and mining damage, Part 2 *of* Theory and practices in predicting surface deformation: U.S. Bureau of Mines Information Circular 8572, 53 p.

Brawner, C. O., 1978, Stability in open pit and strip mining coal projects, *in* Brawner, C. O., and Dorling, I.P.F., eds., Stability in coal mining, Proceedings, First International Symposium on Stability in Coal Mining, Vancouver, B.C., Canada: San Francisco, Miller Freeman Publications, p. 13–28.

Briggs, Henry, 1929, Mining subsidence: London, Edward Arnold and Co., 153 p.

Broch, E., and Franklin, J. A., 1972, The point-load strength test: International Journal of Rock Mechanics and Mining Sciences, v. 9, no. 6, p. 669–693.

Bruhn, R. W., Magnuson, M. O., and Gray, R. E., 1978, Subsidence over the mined-out Pittsburg coal, *in* Coal mine subsidence session, American Society of Civil Engineers Convention, Pittsburgh: American Society of Civil Engineers preprint, p. 26–55.

Burton, A. N., 1975, Geographical methods in site investigations in areas of mining subsidence, *in* Bell, F. G., ed., Site investigations in areas of mining subsidence: London, Butterworth and Co., Ltd., p. 75–102.

Capper, P. L., and Cassie, W. F., 1963, The mechanics of engineering soils: London, E. & F. N. Spon, 298 p.

Charles, J. A., Naismith, W. A., and Burford, D., 1978, Settlement of backfill at Horsley restored opencast coal mining site, *in* Geddes, J. D., ed., Large ground movements and structures, Conference at the University of Wales Institute of Science and Technology, Cardiff, 1977, Proceedings: New York, John Wiley, p. 229–251.

Cochran, William, 1971, Mine subsidence—Extent and cost of control in a selected area: U.S. Bureau of Mines Information Circular 8507, 32 p.

Colaizzi, G. J., Whaite, R. H., and Donner, D. L., 1981, Pumped-slurry backfilling of abandoned coal mine workings for subsidence control at Rock Springs, Wyo., *with an appendix on* Hydraulic model studies for backfilling mine cavities, by E. J. Carlson: U.S. Bureau of Mines Information Circular 8846, 56 p., appendix, 40 p.

Cook, N.G.W., and Hood, M., 1978, Stability of underground coal mine workings, *in* Brawner, C. O., and Dorling, I.P.F., eds., Stability in coal mining, Proceedings, First International Symposium on Stability in Coal Mining, Vancouver, B.C., Canada: San Francisco, Miller Freeman Publications, p. 135–147.

Corlett, A. V., and Emery, C. L., 1959, Prestress and stress redistribution around a mine opening: Canadian Mining and Metallurgical Bulletin, p. 372–384, June.

Current Science, 1981, Fire, *in* Earth burns for 20 years: Current Science, v. 67, no. 8, p. 6–7.

DeGraff, J. V., and Romesburg, H. Ch., 1981, Subsidence crack closure; rate, magnitude, and sequence: International Association of Engineering Geology Bulletin 23, p. 123–127.

Denver Post, 1976, Wyoming mine subsidence—Earth snatches at wary subdivision: Denver Post, Oct. 12, 1976.

Dobson, W. D., and Roberts, R.G.S., and Wilson, K., 1959, Surface and underground development at Peterlee, Co. Durham: Colliery Guardian, v. 148, nos. 5127, 5128, p. 691–697, 723–730.

Down, C. G., and Stocks, J., 1978, Environmental impact of mining: London, Applied Science Publishers, Ltd., 371 p.

Dunrud, C. R., 1976, Some engineering geologic factors controlling coal mine subsidence in Utah and Colorado: U.S. Geological Survey Professional Paper 969, 39 p.

Dunrud, C. R., and Nevins, B. B., 1981, Solution mining and subsidence in evaporite rocks in the United States: U.S. Geological Survey Miscellaneous Investigations Map I-1298, 2 sheets.

Dunrud, C. R., and Osterwald, F. W., 1965, Seismic study of coal mine bumps, Carbon and Emery Counties, Utah: Society of Mining Engineers Transactions, v. 232, p. 174–182.

Dunrud, C. Richard, Osterwald, Frank W., and Hernandez, Jerome, 1972, Summary of the seismic activity and its relation to geology and to mining in the Sunnyside mining district, Carbon and Emergy Counties, Utah during 1967 through 1970: U.S. Geological Survey open-file report, 61 p.

Dunrud, C. Richard, and Osterwald, Frank W., 1980, Effects of coal mine subsidence in the Sheridan, Wyoming, area: U.S. Geological Survey Professional Paper 1164, 49 p.

Foster, Dick, 1979, Mine shafts under hundreds of homes: Gazette Telegraph, Colorado Springs, Colo., p. 1A, March 31.

Franklin, J. A., and Chandra, R., 1972, The slake-durability test: International Journal of Rock Mechanics and Mining Sciences, v. 9, no. 3, p. 325–338.

Geddes, J. D., 1978a, The behaviour of a CLASP-system school subjected to mining movements, in Geddes, J. D., ed., Large ground movements and structures, Conference at the University of Wales Institute of Science and Technology, Cardiff, 1977, Proceedings: New York, John Wiley, p. 579–596.

——1978b, Construction in areas of large ground movement, in Geddes, J. D., ed., Large ground movements and structures, Conference at the University of Wales Institute of Science and Technology, Cardiff, 1977, Proceedings: New York, John Wiley, p. 949–974.

Gentry, D. W., and Abel, J. F., Jr., 1978, Surface response to longwall coal mining in mountainous terrain: Association of Engineering Geologists Bulletin, v. 15, no. 2, p. 191–220.

Grond, G.J.A., 1957, Ground movements due to mining: Colliery Engineering, p. 157–158, April, p. 197–205, May.

Handin, John, and Haeger, R. V., 1958, Experimental deformation of sedimentary rocks under confining pressure: tests at high temperature: American Association of Petroleum Geologists Bulletin, v. 42, no. 12, p. 2892–2934.

Handin, John, Haeger, R. V., Friedman, Melvin, and Feather, J. N., 1963, Experimental deformation of sedimentary rocks under confining pressure: pore pressure tests: American Association of Petroleum Geologists Bulletin, v. 47, no. 5, p. 717–755.

Hassani, F. P., Scoble, M. J., and Whittaker, B. N., 1980, Application of the point load index test to strength determination of rock and proposals for a new size-correction chart, in Summers, D. A., compiler, The state of the art in rock mechanics: Proceedings, U.S. Symposium on Rock Mechanics, 21st, Rolla, Missouri, p. 543–565.

Hertzberg, M., 1978, Mine fire detection, in [U.S.] Bureau of Mines, Pittsburg, Pa., compilers, Mining research; Coal mine fire and explosion prevention, Proceedings, Bureau of Mines Technology Transfer Seminars: U.S. Bureau of Mines Information Circular 8768, p. 38–50.

Hoek, E., and Brown, E. T., 1980, Underground excavations in rock: Institution of Mining and Metallurgy, London, 527 p.

Howell, M., 1975, Improved geophysical techniques for survey of disturbed ground, in Bell, F. G., ed., Site investigations in areas of mining subsidence: London, Butterworth and Co., Ltd., p. 103–108.

Hutchings, R., Fajdiga, M., and Raisbeck, D., 1978, The effect of large ground movements resulting from brown coal open cut excavations in the Latrobe Valley, Victoria, in Geddes, J. D., ed., Large ground movements and structures, Conference at the University of Wales Institute of Science and Technology, Cardiff, 1977, Proceedings: New York, John Wiley, p. 136–161.

Ivey, J. B., 1978, Guidelines for engineering geologic investigations in areas of coal mine subsidence—A response to land-use planning needs: Association of Engineering Geologists Bulletin, v. 15, no. 2, p. 163–174.

Jaeger, J. C., and Cook, N.G.W., 1979, Fundamentals of rock mechanics (third edition): London, Chapman and Hall, v. 18, 593 p.

Jeremic, M. L., 1981a, Effect of sub-coal strata on coal pillar stability: Coal Miner, v. 6, no. 1, p. 39–46.

——1981b, Coal mine roadway stability in relation to lateral tectonic stress; western Canada: Mining Engineering, v. 33, no. 6, p. 704–709.

Johnson, Wilton, and Miller, G. C., 1979, Abandoned coal mine lands—Nature, extent, and cost of reclamation: U.S. Bureau of Mines, 29 p.

Jones, C.J.F.P., and Bellamy, J. B., 1973, Computer prediction of ground movements due to mining subsidence: Geotechnique, v. 23, no. 4, p. 515–530.

Kaneshige, Osamu, 1971, The underground excavation to avoid subsidence damage to existing structures in Japan, in Symposium [on] geological and geographical problems of areas of high population density, Washington, D.C., 1970, Proceedings: Association of Engineering Geologists, p. 169–199.

Kanlybayeva, Zh. M., 1964, Dynamics of displacement of a stratum under the influence of working gently dipping coal seams, based on geophysical data; Fourth International Conference on Strata Control and Rock Mechanics [translated from Russian by H. Frisch]: Canada Department of Energy Mines and Resources, 18 p., 1965.

Kauffman, P. W., Hawkins, S. A., and Thompson, R. R., 1981, Room and pillar retreat mining: U.S. Bureau of Mines Information Circular 8849, 228 p.

Kim, A. G., 1977, Laboratory studies on spontaneous heating of coal: U.S. Bureau of Mines Information Circular 8756, 13 p.

Kuchta, J. M., Furno, A. L., Dalverny, L. E., Sapko, M. J, and Litton, C. D., 1982, Diagnostics of sealed coal mine fires: U.S. Bureau of Mines Report of Investigations 8625, 25 p.

Kunz, K. S., 1957, Numerical analysis: New York, McGraw-Hill, 381 p.

Kuti, Joseph, 1979, Longwall mining in America: Society of Mining Engineers Transactions, v. 266, p. 1593–1602.

Lee, K. L., and Shen, C. K., 1969, Horizontal movements related to subsidence: American Society of Civil Engineers Proceedings, Journal of the Soil Mechanics and Foundations Division, v. 95, p. 139–166.

Littlejohn, G. S., 1979, Surface stability in areas underlain by old coal workings: Ground Engineering, v. 12, no. 2, p. 22–30.

Marr, J. E., 1959, A new approach to the estimation of mining subsidence: Transactions of the Institution of Mining Engineers, London, v. 118, p. 692–707.

Mohr, H. F., 1956, Influence of mining on strata: Mine and Quarry Engineering, v. 22, no. 4, p. 140–152.

Myers, A. R., Hansen, J. B., Lindvall, R. A., Ivey, J. B., and Hynes, J. L., 1975, Coal mine subsidence and land use in the Boulder-Weld coalfield, Boulder and Weld Counties, Colorado: Colorado Geological Survey, Interim Text for Environmental Geology, no. 9, [95] p., includes 31 figs., 6 pls.

NCB, 1975, Subsidence engineers' handbook: National Coal Board [United Kingdom], Mining Department, 111 p.

Obert, Leonard, and Duvall, W. I., 1967, Rock mechanics and the design of structures in rock: New York, John Wiley, 650 p.

Olson, J. J., and Tandanand, Sathit, 1977, Mechanized longwall mining: U.S. Bureau of Mines Information Circular 8740, 201 p.

Osterwald, F. W., and Dunrud, C. R., 1965, Geology applied to the study of coal mine bumps at Sunnyside, Utah: Society of Mining Engineers Transactions, v. 232, no. 2, p. 168–174.

Pariseau, W. G., 1980, Inexpensive but technically sound mine pillar design analysis, in Summers, D. A., compiler, The state of the art in rock mechanics, Proceedings, U.S. Symposium on Rock Mechanics, Rolla, Missouri, 21st, p. 57–72.

Paul, R. A., 1978, The effects of geologic structures on slope stability at the Centralia coal mine, in Brawner, C. O., and Dorling, I.P.F., eds., Stability in coal mining, Proceedings, First International Symposium on Stability in Coal Mining, Vancouver, B.C., Canada: San Francisco, Miller Freeman Publications, p. 29–34.

Peng, S. S., 1978, Coal mine ground control: New York, John Wiley, 450 p.

——1980, 3-D structural analysis of longwall panels, in Summers, D. A., compiler, The state of the art of rock mechanics: Proceedings, U.S. Symposium on Rock Mechanics, 21st, Rolla, Missouri, p. 44–56.

Piggot, R. J., and Eynon, Peter, 1978, Ground movements arising from the presence of shallow abandoned mine workings, in Geddes, J. D., ed., Large ground movements and structures, Conference at the University of Wales Institute of Science and Technology, Cardiff, 1977, Proceedings: New York, John Wiley, p. 749–780.

Post, J. L., 1981, Expansive soils: California Geology, v. 34, no. 9, p. 197–203.

Pöttgens, J.J.E., 1979, Ground movements by coal mining in the Netherlands, in Saxena, S. K., ed., Evaluation and prediction of subsidence, International Engineering Foundation Conferences, Pensacola Beach, 1978: American Society of Civil Engineers, p. 267–282.

Rockaway, J. D., and Stephenson, R. W., 1982, Geotechnical evaluation of the support of coal pillars in underground coal mines: Association of Engineering Geologists Bulletin, v. 19, no. 1, p. 5–14.

Ropski, St., and Lama, R. D., 1973, Subsidence in the near-vicinity of a longwall face: International Journal of Rock Mechanics and Mining Sciences, v. 10, no. 2, p. 105–118.

Rose, J. G., and Howell, R. C., 1979, Proposed coal pillaring procedure using concrete containing coal refuse (coal-crete): Society of Mining Engineers Transactions, v. 266, p. 290–298.

Salamon, M.D.G., 1978, The role of linear models in the estimation of surface ground movements induced by mining tabular deposits, in Geddes, J. D., ed., Large ground movements and structures, Conference at the University of Wales Institute of Science and Technology, Cardiff, 1977, Proceedings: New York, John Wiley, p. 30–59.

Savage, W. Z., 1981, Prediction of vertical displacements in a subsiding elastic layer: Geophysical Research Letters, v. 8, no. 3, p. 195–198.

Schoemaker, R. P., 1948, A review of rock pressure problems: American Institute of Mining and Metallurgical Engineers Technical Publication 2495, 14 p.

Shadbolt, C. H., 1975, Mining subsidence, in Bell, F. G., ed., Site investigations in areas of mining subsidence: London, Butterworth and Co., Ltd., p. 109–124.

——1978, Mining subsidence—Historical review and state of the art, in Geddes, J. D., ed., Large ground movements and structures, Conference at the University of Wales Institute of Science and Technology, Cardiff, 1977, Proceedings: New York, John Wiley, p. 705–748.

Sherwood, G.E.F., and Taylor, A. E., 1957, Calculus: Englewood Cliffs, New Jersey, Prentice Hall, Inc., 579 p.

Shoemaker, H. D., Advani, S. H., and Gmeindl, F. D., 1979, Studies of thermomechanical subsidence associated with underground coal gasification, in Saxena, S. K., ed., Evaluation and prediction of subsidence, International Engineering Foundation Conferences, Pensacola, Florida, 1978: American Society of Civil Engineers, p. 140–153.

Singer, F. L., 1951, Strength of materials: New York, Harper and Brothers, 469 p.

Singh, M. M., 1979, Experience with subsidence due to mining, in Saxena, S. K., ed., Evaluation and prediction of subsidence, International Engineering Foundation Conferences, Pensacola, Florida, 1978: American Society of Civil Engineers, p. 92–112.

Smith, W. K., 1980, Long-term highwall stability in the northwestern Powder River Basin, Wyoming and Montana: U.S. Geological Survey Open-File Report 80-1229, 106 p.

Sprouls, M. W., 1982, Kemmerer creates largest coal pit: Coal Mining and Processing, v. 19, no. 5, p. 42–45.

Stephansson, O., 1971, Stability of single openings in horizontally bedded rock: Engineering Geology, v. 5, p. 5–71.

Stephenson, R. W., and Aughenbaugh, N. B., 1978, Analysis and prediction of ground subsidence due to coal mine entry collapse, in Geddes, J. D., ed., Large ground movements and structures, Conference at the University of Wales Institute of Science and Technology, Cardiff, 1977, Proceedings: New York, John Wiley, p. 100–118.

Stimpson, Brian, and Ross-Brown, D. M., 1979, Estimating the cohesive strength of randomly jointed rock masses: Mining Engineering, v. 31, no. 2, p. 182–188.

Tandanand, Sathit, and Powell, Larry, 1982, Assessment of subsidence data from the Northern Appalachian basin for subsidence prediction: U.S. Bureau of Mines Report of Investigations 8630, 14 p.

Taylor, R. K., 1975, Characteristics of shallow coal-mine workings and their implications in urban redevelopment areas, in Bell, F. G., ed., Site investigations in areas of mining subsidence: London, Butterworth and Co., Ltd., p. 125–148.

Thorburn, S., and Reid, W. M., 1978, Incipient failure and demolition of two-story dwellings due to large ground movements, in Geddes, J. D., ed., Large ground movements and structures, Conference at the University of Wales Institute of Science and Technology, Cardiff, 1977, Proceedings: New York, John Wiley, p. 87–99.

Voight, Barry, and Pariseau, William, 1970, State of predictive art in subsidence engineering: Proceedings of the American Society of Civil Engineers, Journal of the Soil Mechanics and Foundations Division, v. 96, no. SM2, p. 721–750.

Wardell, K., 1954, Some observations on the relationship between time and mining subsidence: Transactions of the Institution of Mining Engineers, London, v. 113, p. 471–482.

——1971, The effects of mineral and other underground excavations on the overlying ground surface, in Symposium [on] geological and geographical problems of areas of high population density, Washington, D.C., 1970, Proceedings: Association of Engineering Geologists, p. 201–217.

Wardell, K., and Piggott, R. J., 1969, Report on mining subsidence: Her Majesty's Stationery Office, p. 187–204.

White, J. M., and Farrow, R. A., 1982, Subsidence of reclaimed coal mine spoils, Colstrip, Montana: Association of Engineering Geologists Bulletin, v. 19, no. 4, p. 385–388.

Whittaker, B. N., and Pye, J. H., 1977, Design and layout aspects of longwall methods of coal mining, in Design methods in rock mechanics, Proceedings, Sixteenth Symposium on Rock Mechanics: American Society of Civil Engineers, p. 303–314.

Young, L. E., and Stoek, H. H., 1916, Subsidence resulting from mining: University of Illinois, Bulletin 91, 205 p.

Zwartendyk, Jan, 1971, Economic aspects of surface subsidence resulting from underground mineral exploitation [Ph.D. thesis]: Pennsylvania University, University Microfilms International, Ann Arbor, 411 p.

MANUSCRIPT ACCEPTED BY THE SOCIETY APRIL 18, 1984

Sinkholes resulting from ground-water withdrawals in carbonate terranes–an overview

J. G. Newton
U.S. Geological Survey
520 19th Avenue
Tuscaloosa, Alabama 35401

ABSTRACT

Numerous sinkholes resulting from declines in the water table due to ground-water withdrawals in carbonate terranes have occurred in the eastern United States and elsewhere. In Alabama alone, it is estimated that more than 4,000 of these sinkholes, areas of subsidence, or related features have formed since 1900. Almost all occur where cavities develop in residual or other unconsolidated deposits overlying openings in carbonate rocks. The downward migration of the deposits into underlying openings in bedrock and the formation and collapse of resulting cavities are caused or accelerated by a decline in the water table that results in (1) loss of buoyant support, (2) increase in the velocity of movement of water, (3) water-level fluctuations at the base of unconsolidated deposits, and (4) induced recharge.

Damage due to sinkhole activity related to ground-water withdrawals has resulted in a variety of studies utilizing available scientific methods. These studies indicate that identifying the terrane in which the activity most commonly occurs and limiting large withdrawals of water from it will eliminate or minimize the problem. The terrane, youthful in nature, exhibits little karstification, is usually a lowland area, has a water table above or near the top of bedrock, and contains perennial or near-perennial streams.

INTRODUCTION

Recent sinkholes resulting from ground-water withdrawals in carbonate terranes have resulted in a variety of problems related to the safety of man's structures and the pollution of existing and potential water supplies. Most of the sinkhole activity described here is restricted to that in Alabama; however, similar activity has occurred in Florida, Georgia, Maryland, Pennsylvania, South Carolina, and Tennessee. An inventory to determine the extent of this problem has not been made even though damage resulting from it has been significant. In Alabama alone, costly damage and numerous accidents have occurred or nearly occurred as a result of collapses beneath highways, streets, railroads, buildings, sewers, gas pipelines, vehicles, and people.

The purpose of this paper is to present a review of geologic and hydrologic mechanisms or forces involved in the subject subsidence process and any predictive capability (before the fact) relating to its occurrence. Because of limited work and findings in recent years, this paper is an excerpt from a similar report on man-related and natural sinkholes by the author (1976a). Limited excerpts have also been taken from subsequent, more condensed papers by the author in 1976 and 1980.

GEOLOGIC AND HYDROLOGIC SETTING

The development of sinkholes is primarily dependent on the presence of carbonate rocks such as limestone and dolomite; on the prehistoric, historical, and present relationships between these rocks and water, climatic conditions, vegetation, and topography; and on the presence or absence of residual or other unconsolidated deposits overlying carbonate bedrock. The source of water associated with the development of sinkholes is precipitation, of which part runs off directly into streams, part replenishes soil moisture but is returned to the atmosphere by evaporation and transpiration, and the remainder percolates downward below the soil zone to ground-water reservoirs.

Water is stored in and moves through interconnected openings in carbonate rocks. Most of the openings were created or existing openings along bedding planes, joints, fractures, and faults were enlarged by the solvent action of slightly acidic water coming in contact with the rocks. Water in the interconnected openings moves in response to gravity from higher to lower altitudes, generally toward a stream channel where it discharges and becomes a part of the streamflow. During dry weather, this discharge constitutes most of the streamflow in the channel.

Water in openings in carbonate rocks occurs under both water-table and artesian conditions; however, this paper is concerned primarily with that occurring under water-table conditions. The water table is the unconfined upper surface of a zone in which all openings are filled with water. The configuration of the water table conforms somewhat to that of the overlying topography but is influenced by geologic structure, withdrawal of water, and variations in recharge. The lowest altitude of the water table in a drainage basin containing a perennial stream occurs where the water table intersects the stream channel. A schematic diagram illustrating geologic and hydrologic conditions in the typical basin described is shown in Figure 1. In this area, openings in bedrock underlying lower parts of the basin are water-filled. This condition is maintained by recharge from precipitation in the basin. The water table underlying adjacent highland areas within the basin occurs at higher altitudes than the water near the perennial stream. Openings in bedrock between the land surface and the underlying water table in highland areas are air-filled (Fig. 1). The progressive enlargement of these openings by solution has resulted in the formation of caves in many areas.

The general movement of water through openings in bedrock underlying the basin, even though the route may be circuitous, is toward the stream channel and downstream under a gentle gradient approximating that of the stream. Some water moving from higher to lower altitudes is discharged through springs along flanks of the basin because of the intersection of the land surface and the water table. The velocity of movement of water in openings underlying most of the lowland area is probably sluggish when compared with that in openings at higher altitudes.

A mantle of unconsolidated deposits consisting chiefly of residual clay (residuum), the residue that has resulted from the solution of the underlying carbonate rocks, covers bedrock in most of the typical basin (Fig. 1). Alluvial or other unconsolidated deposits often overlie the residual clay. The residuum commonly contains varying amounts of chert debris that are insoluable remnants of the underlying bedrock. Some unconsolidated deposits are carried by water into openings in bedrock. These deposits commonly fill solutionally enlarged joints, fractures, or other openings underlying the lowland areas. The buried contact between the residuum and the underlying bedrock, because of differential solution, can be highly irregular (Fig. 1).

The hydrology and geology of the typical basin described

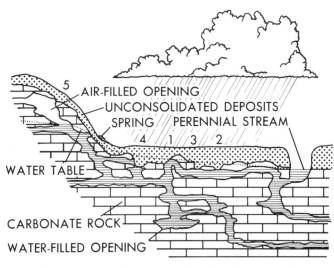

Figure 1. Schematic cross-sectional diagram of basin showing geologic and hydrologic conditions.

above illustrate conditions related to the active development of sinkholes. In actuality, geologic and hydrologic conditions in a particular carbonate-rock terrane are often quite different from those in other areas. The simple terrane described above was used because factors related to the development of sinkholes that have been observed in it are generally applicable to other carbonate-rock terranes. The terrane illustrated in Figure 1 differs from those examined only in that the inclination of beds is shown as horizontal for ease of illustration. Figure 1 was designed to show openings in highland areas that are interconnected with those in lowland areas. These types of openings in inclined strata are commonplace. They often trend at an angle almost 90° from the strike of geologic units, and the movement of water through such openings is sometimes opposite to the dip of strata.

NATURAL AND INDUCED SINKHOLES

Thousands of natural sinkholes are present in areas underlain by carbonate rocks in Alabama. These sinkholes, or depressions, are sometimes only a few meters in diameter, but others attain major dimensions of as much as 3.2 km. Similarly, their depth commonly ranges from a few meters to more than 30 m. The development of new collapses and the subsidence and drainage irregularities associated with their presence or development pose serious problems for existing or planned structures. These problems can be variable and depend largely on the type of sinkhole involved and its cause.

Sinkholes, as related to occurrence, can be separated into two categories, even though most factors involved in their development are the same. These categories are defined as "induced" and "natural." Induced sinkholes are those that can be related to man's activities, whereas natural ones cannot. A major difference in the categories is the time required for their development. Some

Figure 2. Sinkhole resulting from collapse near Calera in Shelby County, Alabama. Photograph by C. Frizzell.

induced sinkholes develop within minutes or hours after the effects of man's activities are exerted on existing geologic and hydrologic conditions. In contrast, the development of a natural sinkhole may require tens, hundreds, or even thousands of years. Both categories generally can be separated on the basis of their physical characteristics and environmental setting. The development of all sinkholes, regardless of their category, is dependent on some degree of solution of the underlying bedrock. The lack of solution eliminates their development.

The classification of sinkhole categories has one major advantage. The much larger number of induced collapses allows the investigator to observe, photograph, and record factors involved in initial stages of their development. Comparable information generally is not available to evaluate the initial stage of development of natural sinkholes. Because this paper is cause-oriented, the following text is devoted almost entirely to the description of the initial stage of development of induced sinkholes.

Induced sinkholes are divided into two types: those resulting from a decline in the water table due to ground-water withdrawals and those resulting from construction. For information concerning the latter, not included in this paper, the reader is referred to Newton (1976a). It is estimated that more than 4,000 sinkholes, areas of subsidence, or other related features have resulted from water withdrawals in Alabama since 1900. Most of them have occurred since 1950. Sufficient information is not available to make similar estimates for other areas.

Collapses forming sinkholes that result from water withdrawals are particularly dangerous and potentially catastrophic because (1) some form instantaneously; (2) they often occur in significant numbers during a short time span; and (3) because their relationship to man's activities, they often occur in populated areas. All of these collapses are potentially dangerous with the degree of danger depending more on where and how rapidly they form than on their size. A large sinkhole resulting from a

Figure 3. Sinkhole beneath railroad in Birmingham, Alabama. Photograph by B. Fitzgerald.

collapse in a wooded area in Shelby County, Alabama, that apparently occurred in a matter of seconds in December 1972, is shown in Figure 2. The collapse was about 90 m in diameter and 30 m deep but because of its location, the danger was minimal. In contrast, a much smaller collapse (Fig. 3) that occurred beneath a railroad in 1972 in Birmingham, Alabama, was very dangerous until discovered.

WATER WITHDRAWAL AND DECLINE OF WATER TABLE

A relationship between the formation of sinkholes and pumpage was recognized in Alabama as early as 1916 (Prouty, 1916). Subsequent investigations in Alabama (Johnston, 1933; Robinson and others, 1953; Powell and LaMoreaux, 1969; Newton and Hyde, 1971; and Newton and others, 1973) have verified this relationship. Sinkholes have developed at and near sites of single wells and in larger areas in which dewatering operations are located.

Documented collapses have occurred in the immediate vicinities of 36 wells tapping limestone and dolomite in Alabama. The actual number of wells related to collapses probably far exceeds this figure because no inventory has been attempted. Available information indicates that all of the collapses occurred in areas not associated with recent natural sinkhole development and that most occurred during initial stages of well development. Three collapses during a pumping test of a new well in Birmingham in 1959 are excellent examples of sinkholes resulting from man-created forces (Newton and Hyde, 1971, p. 17).

Dewatering or the continuous withdrawal of large quantities of water from carbonate rocks by wells, quarries, and mines in numerous other areas is associated with extremely active sinkhole development. Numerous collapses in these areas contrast sharply with their lack of occurrence in adjacent geologically and hydro-

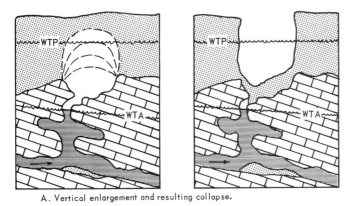

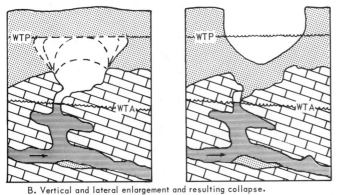

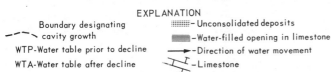

Figure 4. Development of cavities in unconsolidated deposits.

logically similar areas where withdrawals of water are minimal. For example, in five areas of sinkhole activity examined by the author, an estimated 1,700 collapses, areas of subsidence, or other associated features have formed in a total combined area of about 36 km². Recent collapses in adjacent areas underlain by the same geologic units are absent. This phenomenon is not unusual; the relationship of this type sinkhole occurrence to cones of depression created by water withdrawals in Pennsylvania and Africa has been well established by Foose (1953, p. 636; 1967, p. 1047).

Cause and Development

Two areas in Alabama in which intensive sinkhole development has occurred or is occurring have been studied in detail. Both areas were made prone to the development of sinkholes by major declines of the water table due to the withdrawals of ground water. Collapses forming sinkholes in both areas resulted from the creation or acceleration of growth of cavities in unconsolidated deposits caused by the declines (Newton and Hyde, 1971, p. 17–18; Newton and others, 1973, p. 20).

Cavities in unconsolidated deposits overlying carbonate rocks in areas where water-table declines have occurred have also been described and explored in Africa and Pennsylvania (Donaldson, 1963; Jennings and others, 1965; and Foose, 1967). The growth of one such cavity in Birmingham has been photographed through a small adjoining 460-mm opening (Newton, 1976a).

The enlargement and configuration of cavities in unconsolidated deposits varies. The growth and collapse of two cavities are illustrated in Figure 4. The vertical enlargement (Fig. 4A) and resulting collapse are one of the most common types. Vertical and lateral enlargement is illustrated in Figure 4B. The lateral enlargement occurs where the upward enlargement encounters a bed that is able to maintain its integrity. There is some evidence, though meager, that the upward migration of a roof stabilizes near the position occupied by the water table prior to its decline. This evidence is confined to the observation of two flat roofs of cavities in residual clay. When this stabilization occurs, the walls of the cavity sometimes expand outward until the roof fails. The strength of residual clays is surprising; the author unknowingly has been on a roof as thin as 0.5 m without falling through.

Excellent reports prior to 1971 associated the development of sinkholes and subsidence resulting from the downward migration of unconsolidated deposits with erosive forces caused by pumpage, the position of the water table, or a lowering of the water table due to withdrawals of ground water. Johnston (1933) noted that sinkholes appeared to be due to the removal, by moving ground water, of residual clay filling fissures in limestone. He described the stoping action, surmised that water would have to be moving fast enough to erode clay and, because of this, stated that there appeared to be a causal relationship between this type sinkhole and high pumpage from new wells. Robinson and others (1953) attributed the development of sinkholes in a cone of depression to the increased velocity of ground-water movement causing the collapse of clay- and rock-filled cavities in bedrock.

Foose (1953), in the first detailed investigation of the subject of this paper, associated the occurrence of recent sinkhole activity with pumpage and a subsequent decline in the water table. In this monumental study, Foose determined that collapses were confined to areas where a drastic lowering of water table had occurred, that their occurrence ceased when the water table recovered, and that the shape of recent collapses indicated a lowering of the water table and withdrawal of its support. Donaldson (1963, p. 123, 125) described the formation and enlargement of cavities in unconsolidated deposits and their eventual collapse. He attributed the phenomena to water percolating downward through the deposits and eroding them into openings in underlying bedrock. He also noted that where water is pumped out of a fissure in dolomite, the increased flow through it might cause erosion.

Jennings and others (1965, p. 51) have also associated development of sinkholes in the Transvaal, South Africa, with pumpage and the creation of cones of depression. They determined that sinkhole and subsidence problems increased where the water table was lowered and described the formation, enlargement, and collapse of cavities in unconsolidated deposits due to their downward migration. They also described geologic conditions necessary for the formation of the cavities. Foose (1967) also described the development of cavities in unconsolidated deposits in the same area and attributed them to shrinkage of desiccated debris and the downward migration of the debris into bedrock openings. He also outlined geologic conditions related to their development and stated that a lowering of the water table initiated their formation.

The cited reports associated or attributed the occurrence of sinkholes and subsidence to a lowering of the water table and described the related development of cavities in unconsolidated deposits. The reports, however, described only indirectly or in part the hydrologic forces or mechanisms resulting from a decline in the water table that create or accelerate the growth of cavities that collapse and form sinkholes. These forces or mechanisms, based on studies in Alabama (Newton and Hyde, 1971, p. 17–20; Newton and others, 1973, p. 20), are (1) a loss of buoyant support to roofs of cavities in bedrock previously filled with water and to residual clay or other unconsolidated deposits overlying openings in bedrock, (2) an increase in the velocity of movement of ground water, (3) an increase in the amplitude of water-table fluctuations, and (4) the movement of water from the land surface to openings in underlying bedrock where recharge had previously been rejected largely because the openings were water-filled.

The same forces creating cavities and subsequent collapses also result in subsidence. The movement of unconsolidated deposits into bedrock where the strength of the overlying material is not sufficient to maintain a cavity roof will result in subsidence at the surface (Donaldson, 1963, p. 124). Subsidence also can result from consolidation or compaction due to the draining of water from deposits previously located beneath the water table (Jennings, 1966, p. 48). The former determination applies more to observations made in Alabama although the latter may be a contributing factor in some instances. Recognizable subsidence sometimes precedes a collapse (Newton and Hyde, 1971, p. 24–25). This occurrence, where unconsolidated deposits are thin and consist chiefly of clay, indicates that the subsidence is due to a downward migration of the deposits rather than to compaction.

To demonstrate forces that result in the development of cavities and their eventual collapse, the schematic diagram shown in Figure 5 illustrates changes in natural geologic and hydrologic conditions previously described and shown in Figure 1. A description of the forces triggered by water withdrawals resulting in the creation of a cone of depression follows.

Loss of Buoyant Support. The loss of buoyant support following a decline in the water table can result in an immediate collapse of the roofs of openings in bedrock or can cause a downward migration of unconsolidated deposits overlying openings in bedrock. The buoyant support exerted by water on a solid and, hypothetically, unsaturated clay overlying an opening in bedrock, for instance, would be equal to about 40 percent of its

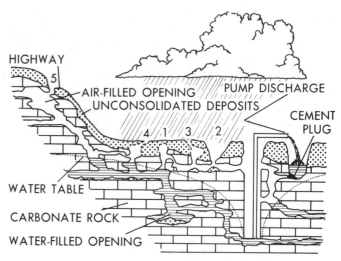

Figure 5. Schematic cross-sectional diagram of basin showing changes in geologic and hydrologic conditions resulting from water withdrawal.

weight. The support to a saturated clay containing chert debris would be less. The probability of a collapse resulting from loss of buoyant support depends primarily on the capacity of the overlying deposit to support itself.

The loss of support for a saturated, unconsolidated deposit overlying an opening in bedrock will, depending on its degree of saturation and strength, result in the deposit's downward movement into the opening. Site 1 in Figure 1 shows the unconsolidated deposit overlying a water-filled opening in bedrock. Site 1 in Figure 5 shows a decline in the water table and the resulting cavity in the deposit formed by the downward migration caused by the loss of support. The cavity may remain stable, or it may enlarge upward by the spalling of overlying debris until the roof collapses.

Increase in Velocity. The creation of a cone of depression in an area of water withdrawal results in an increased hydraulic gradient toward the point of discharge (Fig. 5) and a corresponding increase in the velocity of ground-water movement. This force can result in the flushing out of the finer grained, unconsolidated sediments that have accumulated in the interconnected, solutionally enlarged openings. This flushing action is a standard procedure used in the development of wells tapping carbonate rocks. This movement also transports unconsolidated deposits migrating downward into bedrock openings to the point of discharge or to a point of storage in openings at lower elevations. Without this velocity, residual clay entering openings eventually could result in blockage and a cessation of growth of cavities in unconsolidated deposits.

The increase in the velocity of ground-water movement also plays an important role in the development of cavities in unconsolidated deposits. Erosion caused by the movement of water through unobstructed openings and against joints, fractures, faults, or other openings filled with clay or other unconsolidated sediments results in the creation of cavities that can enlarge and

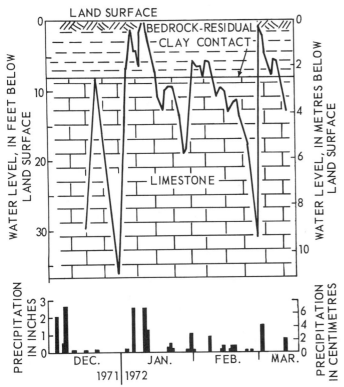

Figure 6. Relation of fluctuations of water level in a well at Greenwood, Alabama, to contact between bedrock and daily precipitation at Bessemer.

eventually collapse (Johnston, 1933, p. 40–41; Robinson and others, 1953, p. 44–45). Collapses and subsidence due to erosion of clay-filled, solutionally enlarged openings are occurring beneath and near Interstate Highway 59 in Birmingham, Alabama (Newton, 1976a, p. 18). Clay-filled openings underlying some of the areas of the highway extend more than 15 m below the top of bedrock adjacent to the openings.

Water-Level Fluctuations. Pumpage often results in fluctuations in ground-water levels that are of greater magnitude than those occurring under natural conditions. The repeated movement of water through openings in bedrock against overlying residuum or other unconsolidated sediments causes a repeated addition and subtraction of support to the sediments and repeated saturation and drying. This process might be best termed "erosion from below" because it results in the creation of cavities in unconsolidated deposits, their enlargement, and eventual collapse.

Fluctuations of the water table against the roof of a cavity in unconsolidated deposits near Greenwood, Alabama, have been observed and photographed through a small collapse in the center of the roof. These fluctuations, in conjunction with the movement of surface water into openings in the ground, resulted in the formation of the cavity and its collapse (Newton and others, 1973). The relation of these fluctuations in a nearby well to precipitation and to the contact between bedrock and residuum is shown in Figure 6. The well was located near a stream that was perennial prior to the decline in the water table. The rise and decline of the water against the base of residual clay that occurred three times during the three-month period illustrate the repeated addition and subtraction of support and the alternate wetting and drying of the clay.

Induced Recharge. A large decline of the water table in a lowland area (Fig. 5), in which all openings in the underlying carbonate rock were previously water-filled (Fig. 1), commonly results in induced recharge of surface water. This recharge was partly rejected prior to the decline because of the water-filled nature of the underlying openings. The quantity of surface water available as recharge to such an area generally would be significant because of the runoff moving to and through it from areas at higher elevations.

The inducement of surface water infiltration through openings in unconsolidated deposits interconnected with openings in underlying bedrock results in the creation of cavities where the material overlying the openings in bedrock is eroded to lower elevations. Repeated rains result in the progressive enlargement of this type cavity. A corresponding thinning of the cavity roof due to this enlargement eventually results in a collapse. The position of the water table below unconsolidated deposits and openings in bedrock that is favorable to induced recharge is illustrated in Figures 4 and 5. Sites 2, 3, and 4 on Figure 5 illustrate a collapse and cavities in unconsolidated deposits formed primarily or in part by induced recharge. The creation and eventual collapse of cavities in unconsolidated deposits by induced recharge is the same process described by many authors as "piping" or "subsurface mechanical erosion" where it has been applied mainly to collapses occurring on noncarbonate rocks (Allen, 1969).

Openings that will transmit water from the land surface to the subsurface are abundant in some areas. Numerous circular openings observed in walls of recent collapses apparently were formed by the rotting of old tree roots. The abundance of such openings at depths of 1.5 to 3 m in a carbonate terrane has been described by Walker (1956). Some cavities formed by the forces or processes described above undoubtedly would collapse when their roofs enlarge upward and encounter openings interconnected with the land surface. The joining of such openings allows the entrance of surface water that will aid in enlarging the cavities or in weakening their roofs sufficiently to collapse them.

Areas especially vulnerable to the formation of collapses due to induced recharge are those near streams that have ceased flowing because of a decline in the water table. Precipitation on such an area results in intermittent streamflow that forms sinkholes where it discharges from the streambed into the subsurface (Newton and others, 1973). Large numbers of collapses also occur in adjacent lowland areas subject to flooding.

Relationship of Forces. In an area of sinkhole development where a cone of depression is maintained by constant pumpage (Fig. 5), all of the forces described are in operation even though only one may be principally responsible for the creation of a cavity and its collapse. For instance, the inducement of recharge from the surface (site 2 on Fig. 5), where the water table is maintained at depths well below the base of unconsolidated deposits, can be solely responsible for the development of cavities and their collapse. In contrast, a cavity resulting from a loss of support (site 1 on Fig. 5) can be enlarged and collapsed by induced recharge if it has intersected openings interconnected with the surface. In an area near the outer margin of the cone (site 4 on Fig. 5), the creation of a cavity and its collapse can result from all forces. The cavity can originate from a loss of support, can be enlarged by the continual addition and subtraction of support and the alternate wetting and drying resulting from water-level fluctuations, can be enlarged by the increased velocity of movement of water, and can be enlarged and collapsed by water entering from the surface.

While some of the examples given are conjectural, an examination of conditions in an area of sinkhole development can sometimes determine the primary force or forces involved in the formation of a sinkhole. For instance, a sinkhole resulting from water-level fluctuations and induced recharge has been described by Newton and others (1973).

RECOGNITION PRIOR TO WITHDRAWAL

Damage due to sinkhole activity related to ground-water withdrawals has resulted in a variety of studies utilizing available scientific methods. These studies indicate that identifying the terrane in which the activity most commonly occurs is the only technique available that allows a predictive capability (before the fact) that will aid in recognizing a potential problem area. The terrane (Fig. 1), youthful in nature, exhibits little karstification, is usually a lowland area, has a water table above or near the top of bedrock, and contains perennial or near-perennial streams. Avoiding or limiting withdrawals of water from the terrane will eliminate or minimize the problem. Avoiding withdrawals eliminates the creation of forces that result in sinkhole activity. Limiting withdrawals, especially those that maintain a water level below the top of bedrock, eliminates the potential activity or minimizes the size of the area in which it can occur.

A variety of tools and techniques are available that will aid in assessing sites in an area of active sinkhole development resulting from ground-water withdrawals (after the fact). For an evaluation of remote sensing applications using multispectral aerial photography and thermal imagery, the reader is referred to Newton (1976a). For techniques utilizing surface and subsurface sensors, the reader is referred to Benson (1978) and to Fountain and others (1975).

REFERENCES CITED

Allen, Alice S., 1969, Geologic settings of subsidence, *in* Reviews in engineering geology: Geological Society of America, v. 2, p. 305–342.
Benson, Richard C., 1978, Assessment of localized subsidence (before the fact), *in* Proceedings, Engineering Foundation Conference on Land Subsidence, Pensacola: p. 47–57.
Donaldson, G. W., 1963, Sinkholes and subsidence caused by subsurface erosion, *in* Proceedings, Regional Conference for Africa on Soil Mechanics and Foundation Engineering, 3rd, Salisbury, Southern Rhodesia: p. 123–125.
Foose, R. M., 1953, Ground-water behavior in the Hershey Valley, Pennsylvania: Geological Society of America Bulletin, v. 64, p. 623–645.
—— 1967, Sinkhole formation by ground-water withdrawal, Far West Rand, South Africa: Science, v. 157, p. 1045–1048.
Fountain, L. S., Herzig, F. X., and Owen, T. E., 1975, Detection of subsurface cavities by surface remote sensing techniques: U.S. Department of Transportation, Report no. FHWA-RD-75-80, 125 p.
Jennings, J. E., 1966, Building on dolomites in the Transvaal: The Civil Engineer in South Africa, v. 8, no. 2, p. 41–62.
Jennings, J. E., Brink, A.B.A., Louw, A., and Gowan, G. D., 1965, Sinkholes and subsidences in the Transvaal dolomites of South Africa *in* Proceedings, International Conference on Soil Mechanics and Foundation Engineering, 6th, Montreal: p. 51–54.
Johnston, W. D., Jr., 1933, Ground water in the Paleozoic rocks of northern Alabama: Alabama Geological Survey Special Report 16, 414 p.
Newton, J. G., 1976a, Early detection and correction of sinkhole problems in Alabama, with a preliminary evaluation of remote sensing applications: Alabama Highway Department, Bureau of Research and Development, Research Report no. HPR-76, 83 p.
—— 1976b, Induced and natural sinkholes in Alabama—a continuing problem along highway corridors, *in* Subsidence over mines and caverns, moisture and frost actions, and classification: National Academy of Science, Transportation Research Record 612, p. 9–16.
—— 1976c, Induced sinkholes—a continuing problem along Alabama highways, *in* Proceedings, International Association of Hydrological Sciences, Anaheim: no. 121, p. 453–463.
—— 1980, Induced sinkholes: an engineering problem in carbonate terranes, *in* Proceedings, Applied Geography Conference, 3rd, Kent State University: v. 3, p. 185–193.
Newton, J. G., Copeland, C. W., and Scarbough, L. W., 1973, Sinkhole problem along proposed route of Interstate Highway 459 near Greenwood, Alabama: Alabama Geological Survey Circular 83, 53 p.
Newton, J. G., and Hyde, L. W., 1971, Sinkhole problem in and near Roberts Industrial Subdivision, Birmingham, Alabama—a reconnaissance: Alabama Geological Survey Circular 68, 42 p.
Powell, W. J., and LaMoreaux, P. E., 1969, A problem of subsidence in a limestone terrane at Columbiana, Alabama: Alabama Geological Survey Circular 56, 30 p.
Prouty, W. F., 1916, Preliminary report on the crystalline and other marbles of Alabama: Alabama Geological Survey Bulletin 18, 212 p.
Robinson, W. H., Ivey, J. B., and Billingsley, G. A., 1953, Water supply of the Birmingham area, Alabama: U.S. Geological Survey Circular 254, 53 p.
Walker, E. H., 1956, Ground-water resources in the Hopkinsville quadrangle, Kentucky: U.S. Geological Survey Water-Supply Paper 1328, 98 p.

MANUSCRIPT ACCEPTED BY THE SOCIETY APRIL 18, 1984

Mechanisms of surface subsidence resulting from solution extraction of salt

John R. Ege
U.S. Geological Survey
Box 25046, MS 903
Denver Federal Center
Denver, Colorado 80225

ABSTRACT

Extraction of soluble minerals, whether by natural or man-induced processes, can result in localized land-surface subsidence. The subsidence is caused by partial or total collapse of underground cavities resulting from dissolution of salt or other soluble evaporites. In many cases, subsidence is ultimately related to the strength limit of the overlying rocks that form the unsupported roof above the cavity. Downwarping results where strength of roof spans are exceeded. In other cases, collapse of the undermined roof causes stoping of the overburden rocks. If sufficient underground space is available for the loosely packed rock debris to collect, the void can migrate to the surface and produce surface subsidence, or in the extreme, catastrophic surface collapse. Another mechanism is subsurface erosion of susceptible layers (sandstone, silt, loess) overlying salt cavities. Ground water can erode and transport the loose material down subsidence-induced and natural cracks, or drill holes into the salt cavity. The voids formed in the higher eroded beds can then cause surface subsidence.

INTRODUCTION

This paper presents the distribution of major bedded-salt deposits and locates the major salt basins of North America; identifies both natural and man-induced processes related to solution subsidence; reviews a variety of ground-subsidence incidents above saline rocks; summarizes the results of a two-year study of subsidence over an active brine field in the Detroit area; and discusses detection and monitoring of areas susceptible to solution subsidence.

Approximately one-fourth of the world's continental areas may be underlain by evaporites ranging in age from Precambrian to Holocene (Kozary and others, 1968). The world-map projection (Fig. 1) shows the composite distribution of saline deposits.

In North America, the major salt deposits are found in former large evaporitic basins of different geological ages. In Canada, the principal known deposits are in three large basins (Fig. 2): a Carboniferous basin underlying the Maritime Provinces (Nova Scotia, New Brunswick, Prince Edward Island) and Newfoundland; the Michigan-Appalachian Basin (Silurian) in Ontario; and the Elk Point Basin (Devonian) extending southeast from Alberta into Saskatchewan and Maniboba (Pearson, 1963). Salt is also found in the Northwest Territories and British Columbia.

Salt deposits of the United States occur in three great basins and a number of isolated areas (Fig. 2). The three major basins are the eastern United States salt basin (Michigan-Appalachian Basin), in which the salt is of Silurian and Devonian age; the Gulf Coast Basin, in which the "mother lode" salt is probably Jurassic or older; and the southwestern or Permian Basin, in which the widespread salt deposits underlie parts of Kansas, Colorado, Oklahoma, Texas, and New Mexico. Salt bodies are also found in Virginia, Florida, North Dakota, South Dakota, Montana, Wyoming, Nebraska, Colorado, New Mexico, Utah, Nevada, and Arizona (Landes, 1960; Smith and others, 1973).

Rock-salt deposits (Fig. 2) are found in two areas of Mexico: in the northern part of the country (Chihuahua, Coahuila, Nuevo Leon, and Tamaulipas) which may represent extensions of the Permian and Gulf Coast Basins and in the Isthmus of Tehuantepec Salt Basin (Lefond, 1969).

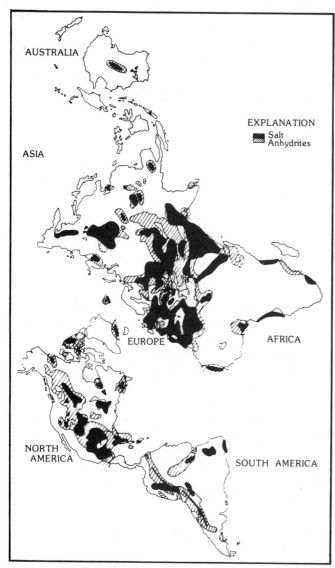

Figure 1. Worldwide distribution of salt and anhydrite deposits. (Modified from Kozary and others, 1968.)

MECHANISMS OF SOLUTION SUBSIDENCE

General Subsidence Processes

Subsidence is a local sinking of the ground surface. When an underground opening is created, whether by mining, dissolution, or subsurface mechanical erosion, the rock strata are disturbed and equilibrium conditions are altered. Downwarping or collapse of overlying rocks into an underground cavity can be reflected at the surface as subsidence. Rock when broken and deposited haphazardly in an opening will increase in volume, or bulk. This bulking action often will fill a void so that overlying strata will be supported, ceasing further lowering of the land surface. On the other hand, if a very large opening is created so that a great amount of space is available for storage of rock debris, the void can migrate to the surface and result in surface collapse.

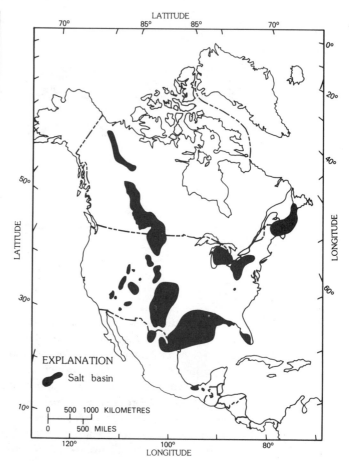

Figure 2. Major salt basins of North America. (Modified from Landes, 1963; Johnson and Gonzales, 1978.)

In discussing the many factors involved in producing subsidence, Stefanko (1973, sec. 13, p. 2) cites the span of the opening as one of the most important. If the width of the opening is small, the overlying rock strata can bridge across the void and little movement or convergence into the opening will take place. As the span increases in width, however, a point is reached where the stress in the overlying rock strata exceeds the strength of the rock and the roof ruptures. If the span of the opening is limited to some subcritical value of width and (or) is at great depth, a "pseudo-arch" (pressure arch) can form that, in many instances, will stabilize the region around the void before any rupturing reaches the surface. On the other hand, if the width of this opening exceeds some critical value, the overlying rock can progressively fail to the surface resulting in subsidence. Sowers (1976), referring to soluble rocks, specified solution channels or voids in these rocks as one of the focal points of surface subsidence. He perceived for solution cavities, that loss of support caused by enlargement of an underground opening beyond the ability of the materials above to bridge it, could result in continuing shear failure in the roof rock and eventual roof collapse and ground failure. Sowers attributes ultimate failure development to some alteration in the environment, such as ground water, ground stress, or change in the materials involved.

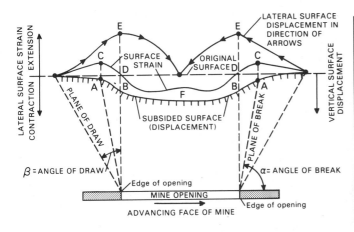

Figure 3. Idealized representation of trough subsidence showing shift of surface points with excavation (Modified from Rellensmann, 1957.)

Obert and Duvall (1967, p. 554–581) identified at least four processes, all applicable to solution cavities, that contribute to closure of underground openings: (1) trough subsidence, (2) subsurface caving, (3) plug caving, and (4) chimneying. These processes, according to the authors, depend on both the time-dependent (inelastic) and time-independent (elastic) characteristics of the rock, and on the stress conditions created in the rock by the geometry of the opening.

Trough subsidence, the most commonly observed process in underground mining, generally forms above openings in horizontal, thin-bedded deposits overlain by stratified sedimentary rocks. In solution mining, trough subsidence can occur above an elongate cavity, such as one formed by a row of coalesced brine wells. For example, in a mining operation if the opening is enlarged to a point where the roof and floor converge into the opening, the surface will subside almost immediately. An idealized profile of a subsidence trough and pattern of vertical and horizontal displacements and horizontal strains over an infinite opening was described by Rellensmann in 1957 (Fig. 3). Vertical displacement is maximum over the center of the opening (point F) and extends beyond the lateral limits (point E) of the opening. The lateral or horizontal surface strain is extensional outside the limits of the opening and contractional within the limits. Lateral strain is zero at point D, which coincides with the surface-displacement inflection point at B. Maximum horizontal strain occurs at C. If surface tension cracks form, they would be expected to appear at A, which is the surface location of point C. The angle of break, α, is the angle between the horizontal and the line connecting the edge of the opening with the point of maximum surface extensional strain, A. The plane of break passes through the point of maximum extensional strain on the surface of all subsurface strata and coincides with the line defining the angle of break. If the strains are large enough, fractures would tend to form in the rock along the plane of break.

Under certain conditions, caving or continuous failure of rock overlying an underground opening is sustained such that the broken zone progresses toward the surface. If the caving process

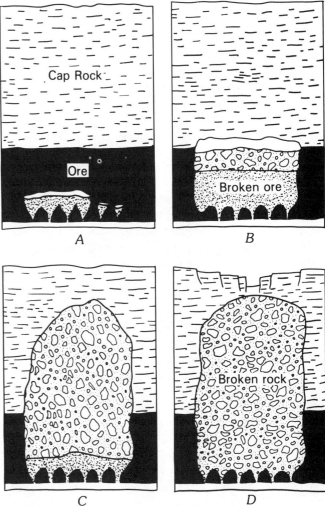

Figure 4. Progress of subsurface subsidence induced by the block caving method. Caving is initiated by removal of support (a) and exceeding the span (b). Removing ore creates void into which additional rock can fall (c), and if unchecked the broken zone can breach the surface (d). (Modified from Obert and Duvall, 1967.)

breaches the surface, a depression or sinkhole can form, sometimes catastrophically. In certain mining operations, rock failure of this type is deliberately induced and can serve as an illustration of the caving process for solution cavities where uncontrolled extraction of salt can cause excessive cavity enlargement. In the block-caving method of mining, a block of ore is undercut over its bottom area and the ore is removed leaving an opening. Caving is initiated by removing support and increasing the span in the ore (Fig. 4). At some width of span, the caving will sustain itself and continue until the void is filled with broken rock. As ore is withdrawn from the mined block, the overlying broken ore and rock will subside, creating a void into which additional overlying rock can fall. The cavity or void will thus migrate toward the surface at a rate determined largely by the rate at which the ore is removed.

Chimneying is a type of caving that is restricted to a relatively small area and progresses rapidly, sometimes in a matter of days, to the surface by a succession of failures or sloughs. The cross-sectional area over the length of the hole or chimney formed is usually constant and of small diameter relative to the height of the chimney. Although the mechanism that causes chimneying may be similar to that of progressive caving, its rapid and unpredictable formation and restriction to a small area distinguishes this process.

Plug caving is characterized by a sudden lowering en masse of the overburden above an unsupported opening and is usually accompanied at the surface by venting and a dust cloud. The subsidence plug seems to involve a unique mechanism for which Obert and Duvall (1967) offer no explanation.

Additional insight into the subsidence process is provided by Savage (1981), who investigated the effects of depth to the cavity and cavity size on resulting surface deformation. Savage's model (Figs. 5 and 6) differs from previous analytic models in permitting both the ground surface and cavity roof to be traction-free. Savage (1981) modeled subsidence over an underground cavity by assuming the subsiding region to be an infinitely long elastic layer that rests on a rigid base and deforms under its own weight into an opening under its lower surface. He determined an approximate analytic solution based on Fourier transform methods for vertical displacements of the ground surface and the roof of the opening, where layer thickness is much greater than the width of the opening. The layer material properties (density, Young's modulus and Poisson's ratio) and the layer and opening geometries (thickness of layer and width of opening) are required for the analytic solution. He found the shape of the surface-subsidence trough is controlled by the ratio of the layer thickness (h) to the width of the opening (2a); as the value of this ratio (h/2a) de-

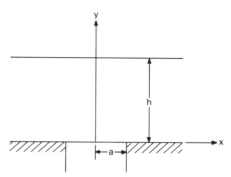

Figure 5. Elastic layer of thickness h overlying an opening of width 2a. (Modified from Savage, 1981.)

creases, the subsidence trough narrows and peripheral ridges form (Fig. 6). The present solution is best used as a model of subsidence in a terrain where a thick and relatively soft clay layer overlies an opening in much stiffer material.

An additional subsidence mechanism that is not directly related to solution but may contribute significantly to certain types of subsidence above soluble rocks is subsurface-mechanical erosion ("piping") of susceptible materials overlying salt beds. The creation of subsurface cavities by mechanical transportation of sediment below ground has been recognized as the cause for collapse of the ground surface at only a few localities, but the process may occur more frequently than has been recognized (Allen, 1969, p. 315). Three prerequisite conditions needed for underground transport of sediments are (1) an easily erodible, pervious bed, such as water-laid silt, loose sand, or weakly cemented sandstone, which is sufficiently competent to maintain a temporary roof over a cavity; (2) a source of water with sufficient head to transport silt or sand; (3) an outlet that serves both as a

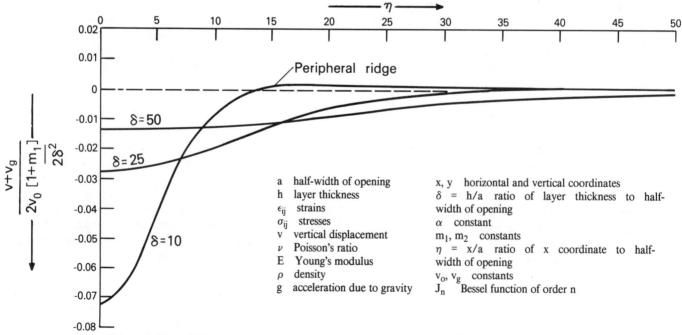

Figure 6. Vertical displacements on the free surface as a function on $\eta = x/a$ and $\delta = h/a$. (Modified from Savage, 1981.)

discharge point for the flowing water and as a disposal area for the transported silt and sand.

In the "piping" process, a cavity in erodible materials, overlying a deeper salt cavity, could be formed where sand or silt are removed by water and transported downward through cracks into voids in underlying rock, or through drill holes or fractures into a previously developed salt void. The surface subsidence activity could then be confined largely to the shallower cavity. Several examples of cavity formation in granular rocks by subsurface mechanical erosion have appeared in the literature.

Landes (1933) described caves and sinkholes found in Mitchell County, Kansas, that were formed in loess. A pervasive 5-cm bed of thick, flat discoidal limestone pebbles lay within the loess. Percolating ground water, using this permeable bed as a channelway to the local stream valley, had dissolved out the pebbles leaving a minute tunnel. Enlargement of the horizontal opening into caves occurred through falling of loess from the channel roof and transportation of loess out of the opening by underground streams. Sinkholes formed from cave collapse. Parks (1963, p. 16–19) reported on caves in Attala County, Mississippi, that are constructed in quartzitic rock. He attributes the origin to physical erosion. The highly fractured quartzitic rock overlies a leached silt. Surface water flowing through fractures in the quartzitic rock washed out tunnels and small cavities in the underlying silts by undercutting and leaving resistant rock as the roof. A number of sinkholes have formed where cave roofs have collapsed. Soper (1915) mapped a number of caves in sandstone underlying the city of Minneapolis. He reasoned that the caves formed when ground water, flowing beneath an impermeable shale above the sandstone, washed away loose sand creating openings. Soper also noted subsurface depressions under the city and suggested they represent slumping of overlying beds of limestone into the sandstone caves.

Wright (1964) investigated depressions formed in the flat crest of the Chuska Mountains of New Mexico. The depressions are located in the crossbedded siliceous eolian Chuska Sandstone (Tertiary) varying from 240 to 300 m in thickness and currently occupied by lakes. The Chuska Sandstone consists of alternating well-cemented and poorly cemented sandstone layers, meters to tens of meters thick. Openings in the sandstone were produced by piping of uncemented sand out of the steep escarpment bounding the mountains. Pipes originated at springs or seeps along the escarpment and extended headward under successive caprocks. Collapse occurred over the enlarged openings forming the depressions seen at the surface today.

Other examples of subsurface mechanical erosion are given by Buckham and Cockfield (1950) and Karl Terzaghi (1931, p. 90–92).

Subsidence Above Saline Rocks—Field occurrences

Natural subsidence. Many examples of natural dissolving of evaporite rocks and resultant subsidence structures are found throughout the world (Ege, 1979). In North America, incidents of solution-related subsidence have occurred in all major salt basins through geologic time and involve salt formations of differing ages. Salt deposits of the Middle Devonian Prairie Formation, which extend from North Dakota and Montana through Saskatchewan and Alberta to the Northwest Territories, contain structural lows that were formed through removal of salt by subsurface leaching while the salt bed was buried under hundreds of meters of sediments. Dissolution has been taking place from Late Devonian to present (De Mille and others, 1964). Gendzwill (1978) notes that in southern Saskatchewan, the salt-bearing Prairie Formation is underlain by reeflike carbonate mounds. Ground water moving through the mounds during Devonian times removed some of the salt above the carbonate mounds creating voids in the salt bed. Eventually, the salt-cavity roofs ruptured causing the overlying sediments to subside as much as 30 m.

A structural depression, the "Saskatoon low," south of Saskatoon, Saskatchewan, formed by collapse resulting from removal of salt from the Prairie Formation. It began forming during the Late Cretaceous and continued until at least late Pleistocene time. The continuity of the collapse mechanism suggests that the dissolution of salt has been a continuing process, and may be going on at the present time with future collapses possible (Christiansen, 1967). In the same area, Christiansen (1971) reported that a large water-filled depression, Crater Lake, is a surface expression of a collapse formed from removal of Prairie Formation salt by ground water. The depression is 244 m in diameter, 6 m deep, and comprises two main concentric fault zones. The inner cylinder was downfaulted periodically in Late Cretaceous–Tertiary–early Pleistocene time and the outer cylinder during the last deglaciation (13,600 years ago).

Parker (1967) reported that Middle Devonian and Permian salt beds underlying North Dakota, Montana, and Wyoming vary in thickness from a little over 1 m to 200 m. He attributes the salt thickness changes to postdepositional salt dissolution by ground water ascending along local and regional fractures from aquifers located below the salt beds.

Landes (1945) described limestone and dolomite breccias located in the Mackinac Straits region of Michigan. He attributed these breccias to collapse of solution cavities in salt beds of the Silurian Salina Formation. Stratigraphic evidence indicates some of the blocks fell as much as 200 m, suggesting the presence of very large caverns. Collapse took place during Silurian and Devonian times creating breccia thicknesses as great as 1,000 m.

Meade County, located in the southwestern corner of Kansas, has numerous hollows and sinks, some of which can be attributed to solution collapse of overlying strata into cavities dissolved from Permian salt beds. A large sink named the Meade Salt Well developed catastrophically in March 1879. Johnson (1901, p. 702–712) quotes an article taken from the May 15, 1879, issue of a local newspaper that describes how a water-filled sinkhole suddenly formed along a well-traveled cattle trail leading from northern Texas to Dodge City, Kansas (Fig. 7). The sinkhole measured 52 m in diameter and the water level was 4.3

Figure 7. View of the Meade Salt Well, a sinkhole formed from dissolution of underlying salt beds, Meade County Kansas. (Johnson, 1901.)

m below land surface. Soundings indicated that water depths ranged between 8.5 and 23 m in depth.

Frye and Schoff (1942), in postulating the origin of the Meade County solution sinkholes, noted that faults cut both Permian strata containing salt beds and overlying Pliocene-Pleistocene sediments. The latter sediments contain confined freshwater (Fig. 8). The local structure allows fresh artesian water to circulate down the fault zones into permeable Permian rocks at depth. The ground water flows down-gradient eastward dissolving pockets in the salt, causing collapse of the overlying strata. Salt springs emerge at lower elevations to the east, attesting to the dissolution.

A more recent collapse described by Bass (1931) and Landes (1931) took place in Hamilton County in western Kansas. A sinkhole about 30 m wide and 12 to 15 m deep formed on December 18, 1929. The cause was believed to be collapse of a cavern dissolved in salt or gypsum.

San Simon Swale, a large southeastward-trending depression in Lea County, southeastern New Mexico, covers an area of about 260 km^2. The lowest part of the swale contains a collapse feature called the San Simon Sink, which is about 30 m deep from the rim to the bottom and approximately 1.3 km^2 in area. Within the sink is a secondary collapse about 7 m deep. A great thickness of Permian salt beds underlies the southern Lea County area, and surface features such as the San Simon structures apparently were formed by the removal of salt by solution and collapse of the overlying beds. Triassic red beds are exposed at the surface both to the northeast and southwest of San Simon Swale; however, drilling over 120 m deep within the sink has not encountered these red beds. Bachman and Johnson (1973) proposed that the swale was initially formed by a very large collapse in the vicinity of the present San Simon Sink. On the basis of numerous ring fractures around San Simon Sink, the sink has had a long history of successive collapse events. Subsidence took place as recently as 1922, with the development of a fissure along the western side of the sink (Nicholson and Clebsch, 1961, p. 13–17, 46–47; Bachman and Johnson, 1973, p. 25–34).

Man-induced subsidence. Man-induced subsidence associated with evaporite rocks is usually the result of some form of mining or drilling operation or construction activity. Conventional mining of bedded salt and potash deposits is similar to coal mining, and the subsidence mechanisms of these mining methods are likewise similar (Obert and Duvall, 1967, p. 555). Solution mining or extraction by dissolving salt and other soluble evaporites is a specialized mining technique which can produce subsidence (Marsden and Lucas, 1973). Drilling through aquifers and salt beds in search of oil, gas, and water has occasionally resulted in induced salt dissolution and subsequent subsidence (Fader,

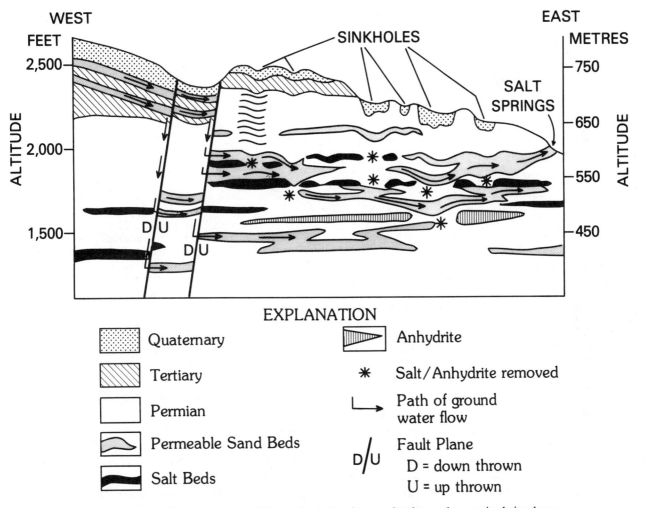

Figure 8. Generalized section through Meade Basin showing postulated ground-water circulation down fault planes and laterally along permeable beds where adjacent salt beds are dissolved causing development of solution subsidence features. (Modified from Frye and Schoff, 1942.)

1975). Construction of highways, dams, and reservoirs over saline or gypsiferous rock has resulted in subsidence, water loss, and dam failures (Burgot and Taylor, 1972; Sill and Baker, 1945).

The basic method of solution mining consists of drilling holes to the salt or evaporite deposit, injecting freshwater or unsaturated brine (salt-impregnated water still capable of dissolving solid salt) down the hole in order to dissolve the soluble minerals, and then removing the resulting brine through a return conduit to the surface. There are several variations of the solution-mining method which are employed according to local conditions and company policy (Querio, 1977). Three commonly used techniques are discussed below.

The top-injection single-well method (Fig. 9) pumps freshwater or unsaturated brine down the annular space between the well casing and central tube. The soluble minerals are dissolved and the heavier brine sinks to the bottom of the cavity and is drawn out through the central tube. This method tends to form a cone or "morning glory-shaped" cavity, and control of the

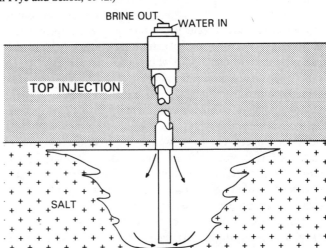

Figure 9. Solution mining by top- (annular-) injection single-well method. Freshwater is injected down annulus emplaced near top of salt bed. Heavier brine is pumped up tubing emplaced near base of salt bed. Top-injection method tends to form "morning-glory" cavity. (Modified from Querio, 1977.)

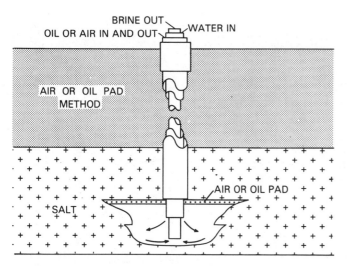
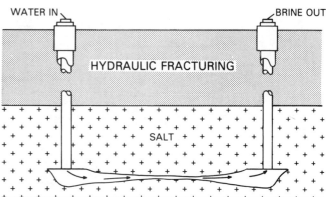

Figure 10. Solution mining by the air or hydrocarbon pad method. Gas or hydrocarbon "padding" material is carried down outer annulus. The lighter "pad" floats on the injected freshwater forming impermeable barrier between the salt and the dissolving water. This method is used to control the level at which salt solution occurs and a variety of techniques are used to determine the location and maintenance of the pad-solvent interface. Improved cavity shape is one advantage of the pad method. (Modified from Querio, 1977).

Figure 11. Hydrofracture technique between a water inlet and brine extraction. Hydrofracturing is used to coalesce a system of wells into a gallery. The rock is split at the desired depth by application of pressured water in a sealed interval of the boring. Once a fracture is initiated, pressure is maintained until it intersects an adjacent target well. Brine production begins with solutioning along the induced fracture. (Modified from Querio, 1977.)

geometry of the opening is limited. Top injection is an older technique and is the cause of many of the uncontrolled subsidences that have occurred in brine fields. It has largely been replaced by more controlled techniques. The air or oil pad method (Fig. 10) involves pumping air or hydrocarbon down the hole with water or through separate tubing. The less dense air or hydrocarbon floats on the water forming a barrier or "pad" which retards dissolution of material from the roof of the cavity. In this manner, control is maintained of the cavity geometry, particularly in keeping a predetermined thickness of salt or evaporite between the cavity roof and overlying geologic materials. In many cases, the evaporite is structurally more competent than the immediate overlying sedimentary rock which may be a weak shale or sandstone interbedded with soluble materials.

The hydrofracture technique (Fig. 11) is a relatively new method adapted from oilfield practice that prepares an area for solution mining. Two holes are drilled at a precalculated spacing. One hole is sealed off with packers within the evaporite deposit and water is injected under pressure until the rock fails in tension and a crack is formed, preferably horizontally. Continued application of pressurized water advances the crack until it intersects the companion boring. An open hydraulic system is thereby established, comprising an injection well and an extraction well. With the addition of an air or hydrocarbon pad, the solution mining can proceed with maximum control of extraction rates and cavity geometry. This technique allows better control of location, orientation, and size of the solution gallery (coalesced wells), and minimizes undesirable ground subsidence and collapse.

Walters (1977) discusses land subsidence occurring in central Kansas associated with conventional- and solution-salt mining, and oil and gas operations. He describes 13 subsidence areas, five induced by mining of salt and eight resulting from oil and gas activities. One illustrative example caused by solution mining involved dramatic subsidence that took place in October 1974, at a brine field in Hutchinson, Kansas. A sinkhole some 90 m in diameter formed over a period of three days and left railroad tracks suspended in air (Figs. 12 and 13). Salt is extracted locally from the approximately 105-m-thick Hutchinson Salt Member of the Permian Wellington Formation, which is encountered at a depth of about 120 m below ground surface. Unconsolidated Pleistocene sands, gravels, and loesslike soil overlie the salt and Permian shale. Salt has been produced at this location since 1888, and the locations of many of the earlier wells are not known. The older solution methods used were often uncontrolled, and the extent of many of the solution cavities in the area is unknown. The sinkhole formed in an active brine field included operating and abandoned wells, and cavities of unknown geometries.

Postsubsidence drilling of the collapse area as part of a Solution Mining Research Institute investigation (Walters, 1977) indicated that an elongate northeast-southwest cavity lay beneath the sinkhole, which paralleled a line of producing wells that were hydraulically connected (Fig. 14). The length of span of the gallery roof may have been more than 400 m, which apparently exceeded the span capabilities of the overlying rock layers. This, in turn, caused roof-rock failure which progressed by sequential collapse of the overlying rock layers until the uppermost rock layer was breached. Today the sinkhole is filled with freshwater and is apparently stable (Fig. 13).

Several examples of land subsidence associated with oil and gas operations in central Kansas are described by Walters (1977, p. 31–75). Oil in central Kansas is associated with gas- and water-driven brine-aquifer reservoirs. Unsaturated-waste salt-water is disposed of underground by brine-disposal wells that

Figure 12. Aerial oblique view of 1974 sinkhole near Hutchinson, Kansas. Sinkhole formed beneath railroad tracks seen here suspended in midair (Walters, 1977).

Figure 13. View of stabilized sinkhole shown in Figure 12 (1978), Hutchinson, Kansas.

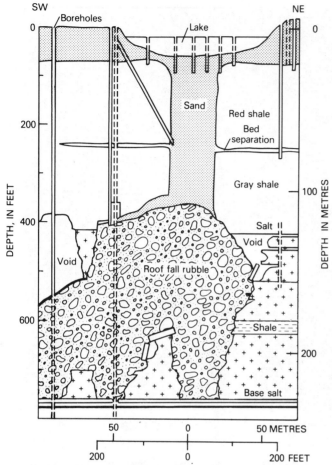

Figure 14. Cross section and interpretation of the sinkhole shown in Figure 12 based on post-collapse borings, Hutchinson, Kansas. (Modified from Walters, 1977).

penetrate a permeable dolomite of the Cambrian and Ordovician Arbuckle Group. In a few instances, particularly in salt-disposal wells, improperly sealed casings have been corroded and breached, allowing unsaturated saltwater to come in contact with salt strata of the Permian Hutchinson Salt Member overlying the dolomite, and form voids in the salt beds. Unsaturated-brine input extending over many years has permitted dissolution of sufficiently large quantities of salt to cause progressive upward caving of the salt layers, culminating in surface collapse.

On April 24, 1959, rapid subsidence developed around a saltwater-disposal well located in Barton County, Kansas. Continuing subsidence over a 12-hour period formed a water-filled circular sink nearly 90 m in diameter with the water level 15 to 18 m below the ground surface. Walters (1977) postulates that in 1938, during initial drilling of the well for oil extraction, freshwater drilling fluid dissolved salt to a diameter of 137 cm in a 35-m section that never had cement emplaced around the casing. The top of the salt section was at a depth of 297 m. In 1946, when boring was converted from an oil well to a saltwater-disposal well, brine was disposed of by gravity flow through tubing into the dolomite of the Arbuckle Group some 920 m below the surface. In 1949, the tubing was removed and brine was injected directly down the casing. Inspection showed that corrosion of the casing caused leaks which allowed unsaturated brine to circulate across the salt face, then downward into the dolomite aquifer. A huge cavern, larger than 90 m in diameter, was formed in the salt. The well was abandoned in January 1959. Successive roof falls of the cavity caused the void to migrate upward, eventually causing surface subsidence. On April 24, 1959, the upper rock layer failed and the void breached the surface, creating the 90-m-wide sink.

On April 27, 1976, a large sinkhole appeared suddenly in the city of Grand Saline, Texas (Grand Saline Sun, 1976). Failure occurred in two stages. First, a hole 4 to 6 m in diameter and more than 15 m deep formed on Tremont Street. Second, the hole widened rapidly as rim material moved down the hole by slabbing and toppling failure (Dunrud and Nevins, 1981). Grand Saline overlies the Grand Saline salt dome where salt was mined by solution methods between 1924 and 1949. A similar collapse took place in 1948, just east of the present sink.

GROSSE ILE SUBSIDENCE

Early in 1971, two sinkholes, designated North and Central sinkholes, formed within the Grosse Ile (Detroit, Michigan) brine field of BASF-Wyandotte Corp., where solution mining had been practiced for almost 20 years (Figs. 15 and 16). A similar collapse in 1954 took place at a nearby brine field operation in Windsor, Ontario (Fig. 16; Terzaghi, 1970). Subsidences, ranging from centimeters to decimeters at the surface above brine cavities, are expected as a normal consequence of brining operations. Although ground collapse is a rare occurrence, it is a matter of concern to the solution mining operator. Three investigations on causes and effects of the Grosse Ile events have been conducted and the results published. Landes and Piper (1972) reported on the environmental effects of the subsidence; Nieto-Pescetto and Hendron, Jr. (1977) advanced a stoping mechanism initiated by collapse of a salt cavity; and Stump, Nieto, and Ege (1982) proposed a subsurface mechanical-erosion mechanism for the Grosse Ile incidents.

Grosse Ile, an island in the Detroit River, is underlain by 18 m of morainal clay with scattered boulders. The unconsolidated surficial deposits are underlain by 150 m of nearly horizontal stratified rock consisting of impure dolomite (Detroit River Dolomite) underlain by friable, bedded sandstone (Sylvania Sandstone), and more impure dolomite (Bass Islands and Bois Blanc) (Fig. 17). The dolomites of Bass Islands and Bois Blanc contain zones of fractures, brecciation, vugs, and partings (Horvarth, 1957; Kalafatis, 1958). Beneath the sedimentary bedrock section is the salt which is about 220 m thick; the more massive salt beds that were solution-mined lie toward the base at depths between 340 and 400 m. The major salt exploited by BASF—Wyandotte lies in the B-unit (Fig. 17).

Landes and Piper (1972, p. 14) report that artesian-ground water first appears at the top of bedrock (Detroit River Dolomite,

Figure 15. Aerial view of the sinkholes formed at Grosse Ile, Michigan, on the BASF-Wyandotte Corp. property (Landes and Piper, 1972).

Fig. 17), and is confined by an overlying impermeable layer of clay at the base of the glacial material. A flow of 4,320,000 gallons of water per day coming from the Sylvania Sandstone was recorded at a well drilled in 1903-04 located about 12 km south of Grosse Ile. Today the well is flowing at an estimated rate of one million gallons per day.

In November 1969, concentric cracks were first observed in the ground above the North Gallery. One year later, a depression formed within the area bounded by the concentric cracks. Maximum measured subsidence was about 1 decimeter. On January 9, 1971, a new depression about 7 × 9 m appeared within the depressed area and progressively enlarged over a period of several months into a sink about 60 m in diameter, at which point it stabilized. On April 28, 1971, collapse started above the Central Gallery (Figs. 15, 18) which is about 800 m south of the North Gallery. Formation of the central sinkhole was not preceded by surficial subsidence or cracking as was the north sinkhole. The central sinkhole enlarged progressively over a period of months attaining a diameter of about 135 m before stabilizing. A smaller sink (62 m in diameter) formed adjacent to the central sinkhole but lies within the concentric cracks which delimit the central collapse zone.

Nieto-Pescetto and Hendron, Jr. (1977) proposed several factors that may have contributed to the formation of the Grosse Ile sinkholes. The initial uncontrolled development of the brine field, using older single-cavity injection wells, probably removed roof support in the salt section over large areas because this method commonly causes morning glory- or cone-shaped cavities. In addition, local thinning or pinch out of salt occurs in the salt bed along a narrow axis, concentrating extraction and contributing to formation of large cavities adjacent to the axis. The concentration occurred because the top of the salt sloped toward the axis and caused the less dense freshwater to dissolve more salt updip away from the pinchout axis, thereby enlarging the openings on either side of the axis (Fig. 19). Continued extraction of salt, and enlargement and coalescence of several cavities into a gallery, may have increased the span of the gallery beyond the strength of the roof rock and caused downwarping or failure of the overlying rock.

The preceding explanation fails to account for the relation between the size of the surface sinks, the volume of subsided material, and the amount of available space underground at the depths (345 m) of the salt cavities created by solution mining.

First, the large volume of material displaced at the surface would require that the entire 60-m-thick B-salt unit (345 m depth) be completely dissolved. However, four exploratory borings, drilled in 1969 over the Central Gallery prior to surface collapse, found the base of the B-salt to be intact and detected no open cavities within the B-unit or overlying section. Nieto-Pescetto and Hendron, Jr. (1977, p. 38) acknowledged this factor

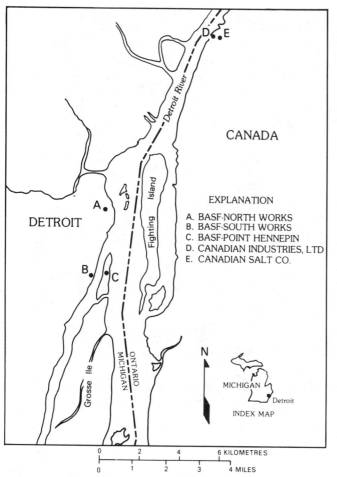

Figure 16. Map locating brine fields at Grosse Ile, Michigan, and Windsor, Ontario. (Modified from Landes and Piper, 1972.)

as an argument against large volume removal of salt from the B-unit as being the sole explanation for sinkhole formation.

Second, downhole TV camera logging indicated considerable bulking of rock fragments within collapsed zones above the B-salt where leaching of soluble minerals from thin salt layers had taken place (T. B. Piper, oral communication, 1979). Thus, any voids near the salt section that may have formed during brining operations would be filled with rubble and not be subjected to further collapse.

A. J. Robinson (Canadian Rock Salt Company, Ltd.) and T. B. Piper (BASF-Wyandotte Corp.) (oral communication, 1977) postulated that secondary mechanisms may have been active after salt-cavity enlargement and downwarping of overlying rock in these Michigan-Ontario brine fields. They proposed that the shallow, friable Sylvania Sandstone (Fig. 17) may have deformed in response to normal downwarping over the rubble-filled salt cavities. In addition, they suggested that the sandstone fractured during downwarping and was locally mobilized by ground water or introduced water percolating through the fractures and was transported as a sand slurry down cracks into the brine cavity. This would create fissures or voids in the sandstone bed.

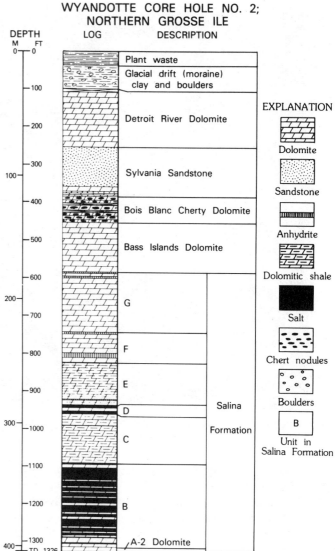

Figure 17. Stratigraphic column of rocks under Grosse Ile, Michigan. (Modified from Landes and Piper, 1972.)

In 1976, a separate investigation of subsidence associated with brining operations was conducted by BASF-Wyandotte Corp. at the Southwest Gallery site, located about 600 m southwest from the Central sinkhole (Fig. 18) (D. J. Dowhan, written communication, 1976). A benchmark network established in 1954 over the entire Grosse Ile brine field showed no significant settling (<5 cm) at the southwest locale until 1975. Surveys conducted over the Southwest Gallery in 1975, however, revealed accelerated rates of settling amounting to 2.5 cm/year.

A test-drilling program, centered in the zone of maximum surface subsidence, was initiated to investigate the increased local settling. Analysis of the borehole cores showed the overburden and Detroit River Dolomite were undisturbed. A void, however, was encountered in the underlying Sylvania Sandstone about 9.1 m from the top of the formation between depths of 85.3 and 87.2 m. A sonar survey indicated the opening extended 61 m north

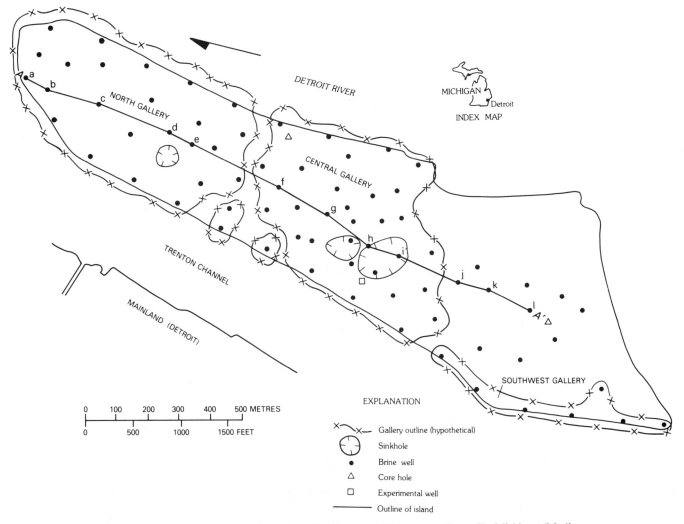

Figure 18. Map locating North, Central, and Southwest subsidence areas, Grosse Ile, Michigan. (Modified from Landes and Piper, 1972; D. J. Dowhan, written communication, 1976.)

and northeast, and 21 m east and west from the boring. No void was "detected" by the sonar device south or southwest of the hole, perhaps indicating collapse. Drilling also revealed zones of thin, leached layers of soluble rock material and accompanying broken rock (dolomite, shale, salt, anhydrite, gypsum) and lost circulation between depths of 234.7 and 313.9 m. The principal salt-production bed below 335.3-m depth, the B-salt of the Salina Formation, however, was intact to the bottom of the test hole at 378.0-m depth.

The sudden increase in surface settling over the Southwest Gallery, the discovery of a void in the Sylvania Sandstone, the observation that dissolution and accompanying broken rock were restricted to a zone containing thin, soluble layers above the main B-salt, and the absence of large cavities in the B-salt supported the concept of a disintegrating Sylvania Sandstone being a factor in Grosse Ile subsidence.

A field and laboratory investigation was begun in 1978 to examine subsidence mechanisms that would include the role of the Sylvania Sandstone in sink development in the Detroit area (Stump, Nieto, and Ege, 1982). Experience of local brine-field operators indicates that under normal conditions the Sylvania Sandstone will not flow, even in the presence of large injections of water from broken well casings, because the sandstone has some cohesion throughout its depth. Accordingly, Stump, Nieto, and Ege speculated that the Sylvania Sandstone might fail in compression and lose its cohesion if subjected to high horizontal stresses. These stresses could be regionally present or locally concentrated by downwarping of strata over a salt cavity.

Laboratory strength tests of Sylvania Sandstone cores taken at Grosse Ile showed variable unconfined strengths through the formation. The sandstone maintained fairly high (41.4 to 55.2 MPa; 6,000 to 8,000 lbf/in^2) dry, unconfined compressive strengths at the top and bottom of the unit, but low strengths in the middle of the section. Core samples taken between 9 and 18 m below the top of the bed were so weak that they did not remain intact and could not be tested. The lowest measured unconfined strength was approximately 8.3 MPa (1,200 lbf/in^2); the average unconfined compressive strength based on test values was about

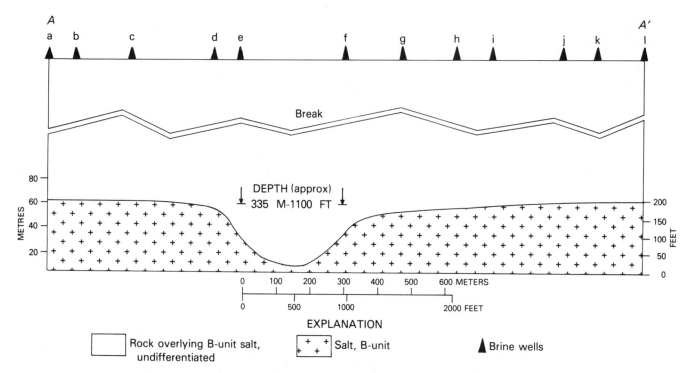

Figure 19. Cross section through Grosse Ile brine field. See Figure 18 for Line of section A-A'. (Modified from Nieto-Pescetto and Hendron, Jr., 1977).

20.7 MPa (3,000 lbf/in²). Furthermore, in more than 50 percent of the tests, sandstone cores failed explosively and were reduced to loose granular sand. The strength tests showed the Sylvania Sandstone can lose its cohesion when subjected to compressive stresses. Additionally, saturated-sandstone samples have been shown to be weaker where loaded parallel to bedding (as much as 40 percent) than similar dried samples that are loaded perpendicular to bedding (Boretti-Onysziewicz, 1966). In its natural state, the Sylvania Sandstone lies nearly horizontal within the water table and is saturated.

In situ stress measurements made in the vicinity of the Michigan Basin (Franklin and Hungr, 1978) indicate that horizontal stresses (S_h) in the range of 1.7 to 14.7 MPa (250 to 2,100 lbf/in²) exist at shallow depths between 2 and 37 m within 350 km of the Windsor-Detroit area. Herget (1973) and Herget, Pahl, and Oliver (1975) have proposed the following empirical equation for relating horizontal stress in MPa to depth below ground surface, based on measurements made at various locations in Canada, including southwestern Ontario.

$$S_h = 8.16 + 0.04\ H$$

where H = depth in meters.

The zone of low compressive strength 9.7 to 34.5 MPa (1,400 to 5,000 lbf/in²) of the Sylvania Sandstone is at an average depth of 84 m below ground surface at Grosse Ile (Figs. 17, 20). Using values obtained from the above equation, the Sylvania Sandstone in its undisturbed state (prior to solution-mining-induced deformation) could have horizontal stresses of at least 11.5 MPa (1,670 lbf/in²). Furthermore, any deformation that would increase horizontal stresses and decrease the confining effect of the overburden could bring the sandstone very close to failure.

A three-dimensional Vousoir arch model was used to examine the field behavior of the Sylvania Sandstone in terms of its arching action above a cavity (Fig. 21). Two modes of failure are predicted: (1) arch crushing, a compressive failure of the upper portion of arch due to compressive stresses exceeding the compressive strength of the material; and (2) arch collapse (buckling), a sagging of beds due to shortening strains which reduce the arch line to a length less than the original length (Fig. 22). The arch-

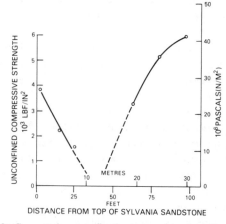

Figure 20. Corrected, unconfined compressive strength of saturated Sylvania Sandstone tested parallel to bedding versus depth within Sylvania Sandstone unit (Stump, Nieto, and Ege, 1982).

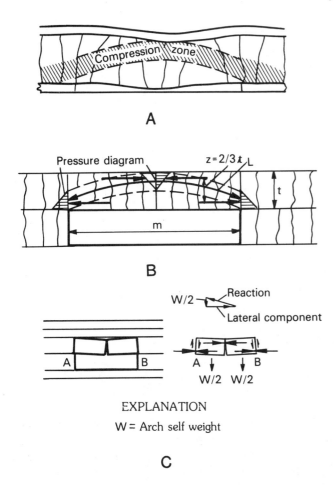

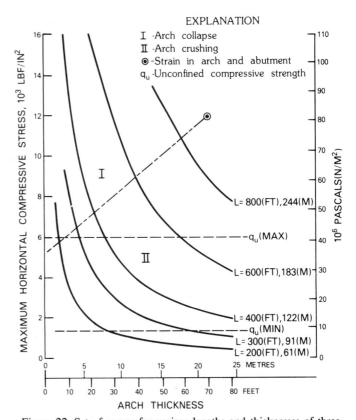

Figure 21. Diagrams of (a) compressive zone within a laterally restrained deflecting linear arch overlying an open cavity, (b) the comnpressive pressure fields developed in linear arch with elements shown, and (c) arching for jointed rock where arch reactions are transmitted through series of minor arch systems. (After Woodruff, 1966.)

Figure 22. Set of curves for various lengths and thicknesses of three-dimensional linear arches with maximum horizontal compressive strength (ordinate) and arch thickness (abscissa), showing zones of arch collapse and arch crushing with minimum and maximum measured uniaxial compressive strength for dry samples tested (Stump, Nieto, and Ege, 1982).

crushing mode of failure might yield the loose granular sand observed in the laboratory testing of Sylvania cores. Arch collapse would result in bed sagging without granulation of the sandstone (Evans, 1941; Woodruff, 1966).

A possible sequence of events leading to the shallow origin of the Detroit area sinks is conceptualized in Figures 23 through 26. Initially, general subsidence would occur resulting from normal downwarping of the overlying rocks into a brine cavity formed in the upper part of the B-salt, the lower salt still remaining intact (Fig. 23). The sagging of the saturated Sylvania Sandstone, under artesian conditions and high initial in situ horizontal stress field, causes separation along the weak zone observed in the formation (Fig. 20). The lower beds of Sylvania Sandstone (Fig. 17) sag with the lower dolomite (Bois Blanc Dolomite), and the upper part of the Sylvania Sandstone remains with the overlying dolomite (Detroit River Dolomite). As sagging continues, induced higher horizontal stresses cause increased failure of the lower-strength Sylvania Sandstone, and the area of the zone of separation within the sandstone bed increases. The loose sand

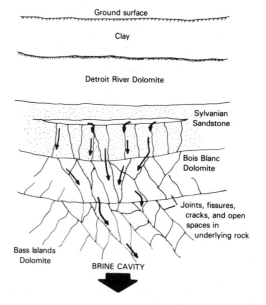

Figure 23. Sketch showing initial sagging of Sylvania Sandstone as responding to sagging of lower dolomite with separation of weak zone (Stump, Nieto, and Ege, 1982).

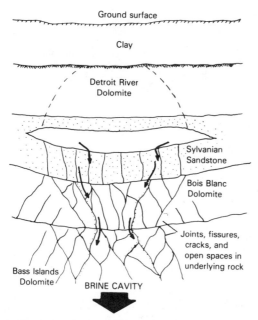

Figure 24. Sketch showing increased sagging of Sylvania Sandstone with greater crushing of weaker Sylvania Sandstone and migration of loose sand vertically downward. Incipient failure of roof of cavity within Sylvania Sandstone (Stump, Nieto, and Ege, 1982).

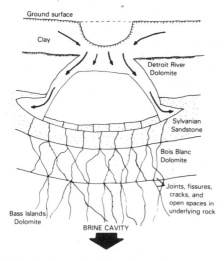

Figure 25. Sketch showing failure of overlying karstic dolomite and development of sink at surface with unconsolidated clay flowing into voids in Sylvania Sandstone and underlying Bois Blanc Dolomite (Stump, Nieto, and Ege, 1982).

mixed with ground water forms a sand slurry. In the dolomite below the Sylvania Sandstone, formerly tight joint sets open and provide paths for the sand slurry to flow into voids in the dolomite and upper B-salt. Further sagging induces higher horizontal stress fields, which causes more sandstone failure and cavity growth (Fig. 24). Eventually, the span of the cavity roof reaches a point where the weight of the overlying dolomite cannot be supported and a typical parabolic roof failure occurs (Fig. 25). The roof failure above the Sylvania Sandstone creates a path for migration of the overlying clay vertically downward, allowing the

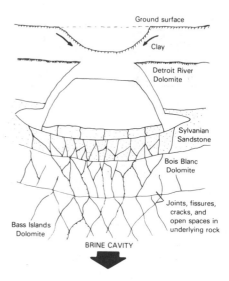

Figure 26. Sketch showing final phase of sink episode with effects of weathering reducing depth of sink with material removed from walls (Stump, Nieto, and Ege, 1982).

remaining cavity within the Sylvania Sandstone and the Detroit River Dolomite to be filled. The final step is the formation of a sink at the surface which enlarges as the unstable walls collapse and decrease in depth as the wall material fills the sink (Fig. 26).

SUBSIDENCE DETECTION AND MONITORING TECHNIQUES

A number of techniques are available to detect underground cavities and measure surface displacements resulting from subsidence. Because surface subsidence is a result of subsurface deformation above underground voids, locating cavities provides useful information for the prediction of possible subsidence or ground failure. On the other hand, the measurement of surface displacement caused by subsidence into underground voids requires that movement is occurring or has already taken place. Monitoring amounts and rates of surface displacements, therefore, can provide quantitative values useful for designing engineered structures and controlling future ground subsidence.

Subsidence detection techniques may be classified into three categories (Benson, 1979): remote sensors, surface sensors, and subsurface sensors. Remote sensing, which includes black-and-white, color, infrared-thermal, multispectral, and radar imagery gathered from aircraft or satellites, can be used to identify geometric or vegetation patterns associated with subsidence. Trained interpreters and the establishment of ground truth must comprise an integral part of the procedure to be effective.

Newton (1976), in an Alabama limestone-kart study, describes remote sensing as a tool to locate potential and active areas of sinkhole development. He considered only the visible and infrared parts of the spectrum and used color, color-infrared and black-and-white infrared film in the investigation. Newton concluded that remote sensing is helpful when used with other exploration methods in detecting subsidence, particularly in ac-

TABLE 1. DESCRIPTION AND EFFECTIVENESS OF SOME SURFACE GEOPHYSICAL-SENSOR METHODS*

Method	Areal coverage capability		Resolution	Comparative Traverse Speed
	Regional	Local cavities		
Seismic	yes	moderate	low	medium
Reisistivity	yes	good	high	medium
Gravity	yes	good	medium-high	slow
Ground penetrating radar	yes-shallow depth penetration	good	high	rapid
Electro-magnetics	yes-very shallow depth penetration	poor	low	rapid
Magnetics	yes	moderate	medium	rapid

*Modified from Benson (1977).

tive areas. It is most effective when repeated imagery is taken over time to record seasonal or time-dependent changes in structure, vegetation, and water conditions related to subsidence.

Surface-sensor methods and tools include (Benson, 1977 and 1979): the trained observer; seismic refraction and reflection; electrical resistivity and induced polarization; low-frequency electromagnetic waves (conductive); high-frequency electromagnetic waves (ground-penetrating radar); precise gravity, magnetic, thermal, and geochemical methods; vegetation patterns and stressed vegetation; and various displacement sensors. These surface methods are conducted on the ground, respond to different sets of physical parameters, and vary in depth of penetration. Many of the techniques can detect cavities directly, or can recognize changes in ground-water conditions that may be associated with subsidence (Table 1).

Subsurface methods and tools comprise (Benson, 1979): exploratory drilling, acoustic-scanner scope, borehole-logging tools (electrical, gamma, gamma-gamma, caliper), displacement transducers, thermal piezometers, flowmeters, dyes and tracers, and optical cameras and television. Most subsurface techniques are conducted from drill holes, mines, and tunnels, and generally provide a higher level of resolution than surface methods because the sensor is closer to the void.

Myers (1963) describes the "Sonar Caliper" subsurface-logging technique that has been used successfully to determine the size and shape of an underground cavity. A special acoustic tool is lowered into a cavity through a borehole, where it emits directional-sound impulses, and measures and records the elapsed time between emission and reflection from the cavern wall. A profile of the cavity is thus made by repeated soundings at successive depth intervals. The sonar method is widely used by the solution-mining industry to determine the size and shape of open salt cavities and galleries.

Panek (1970) outlined methods and reviewed equipment used to measure and monitor horizontal and vertical components of displacement at the surface, and to monitor strain, tilt, and curvature over a subsiding area. The measurement of subsidence is essentially a problem of applied geometry and begins with the installation of monuments, casings, or other permanent reference marks in a predetermined pattern. Measurements of horizontal and vertical distances, angles, and inclinations are repeated at intervals using various mechanical, optical, and electronic devices. Some of the combinations of instruments used to survey surface monuments include the theodolite, tape, level, electronic-distance-measuring device (EDM) and tiltmeter.

SUMMARY

Great thicknesses of bedded-salt deposits occur in many sedimentary basins. Voids or cavities in salt layers can develop through leaching of soluble minerals by continuous flow of water or unsaturated brine in contact with a salt bed. Continued uncontrolled dissolving of the salt minerals can enlarge the roof span of the opening to the strength limit of the overlying rocks and cause downwarping or failure of the overlying beds. Where disintegration of overlying rock is sufficient and space is available to accept falling rock, stoping or chimneying can progress upward and produce ground subsidence or, in the extreme case, ground collapse (sinkhole).

Natural dissolution of salt beds over long periods of time has locally produced very large-diameter (100 to 1,000 m) and sometimes spectacular subsidence features. The process leading to subsidence generally involves a source of fresh ground water or unsaturated brine in continuous contact with a salt bed either by flow through enclosing permeable beds along faults or other discontinuities in surrounding rocks, or by leaching of caves in exposed salt beds adjacent to lakes and streams.

Man-induced subsidence usually is caused by mining, drilling, or construction activity that generally involves the introduction of freshwater or unsaturated brine into a salt bed. Examples of man-induced subsidence are those produced from industrial solution-mining operations; oil- and gas-field activities in salt basins; and construction of dams, reservoirs, and highways over saline rock.

Subsurface-mechanical erosion of overlying beds also may be an important component of the subsidence process. "Piping" or mechanical transport of sediment from granular beds into deeper salt cavities may form voids in the overlying beds. Ground failure occurs if the roof span of the void formed in the shallower beds reaches the strength limit of the rock.

Many detection techniques have been employed for locating underground voids which are the ultimate cause of solution subsidence. The methods, including remote sensing, surface and subsurface sensors, are most efficient when combined with topographic, geologic, and hydrologic investigations.

Monitoring of surface displacements caused by subsidence is done by periodic measurement of monuments installed over a subsiding area so that they reflect only movements in the underlying rock. Magnitudes, directions, slope, curvature, strain, and rates of movement can be calculated from the repeated measurements.

REFERENCES CITED

Allen, A. S., 1969, Geologic settings of subsidence, *in* Varnes, D. J., and Kiersch, George, eds., Reviews in engineering geology, Volume II: Geological Society of America, p. 305–342.

Bachman, G. O., and Johnson, R. B., 1973, Stability of salt in the Permian Salt Basin of Kansas, Oklahoma, Texas, and New Mexico, *with a section on* Dissolved salts in surface water, by F. A. Swenson: U.S. Geological Survey Open-File Report 73-14, 62 p.

Bass, N. W., 1931, Recent subsidence in Hamilton County, Kansas: American Association of Petroleum Geologists Bulletin, v. 15, no. 2, p. 201–205.

Benson, R. C., 1977, An overview of cavity detection methods, *in* Symposium on detection of subsurface cavities: Vicksburg, Mississippi, U.S. Army Engineer Waterways Experiment Station, p. 44–79.

—— 1979, Assessment of localized subsidence (before the fact), *in* Saxena, S. K., ed., Evaluation and prediction of subsidence: Pensacola, Florida, Engineering Foundations Conference, p. 47–57.

Boretti-Onyszkiewicz, W., 1966, Joints in the Flysch Sandstones on the ground of strength examinations: Congress of International Society of Rock Mechanics, 1st, Lisbon, Proceedings, v. 1, p. 153–157.

Buckham, A. F., and Cockfield, W. E., 1950, Gullies formed by sinking of the ground (British Columbia): American Journal of Science, v. 248, no. 2, p. 137–141.

Burgot, V. A., and Taylor, W. K., 1972, Highway subsidence caused by salt solutioning: Association of Engineering Geologists, Program and Abstracts, Annual Meeting, Kansas City, Missouri, 1972, no. 15, p. 20.

Christiansen, E. A., 1967, Collapse structures near Saskatoon, Saskatchewan, Canada: Canadian Journal of Earth Sciences, v. 4, no. 5, p. 757–767.

—— 1971, Geology of the Crater Lake collapse structure in southeastern Saskatchewan: Canadian Journal of Earth Sciences, v. 8, no. 12, p. 1505–1513.

DeMille, G., Shouldice, J. R., and Nelson, H. W., 1964, Collapse structures related to evaporites of the Prairie Formation, Saskatchewan: Geological Society of America Bulletin, v. 75, p. 307–316.

Dunrud, C. R., and Nevins, B. B., 1981, Solution mining and subsidence in evaporite rocks in the United States: U.S. Geological Survey Miscellaneous Investigations Map I-1298, scale 1:5,000,000.

Ege, J. R., 1979, Selected bibliography on ground subsidence caused by dissolution and removal of salt and other soluble evaporites: U.S. Geological Survey Open-File Report 79-1133, 26 p.

Evans, W. H., 1941, The strength of undermined strata: Institution of Mining and Metallurgy Transactions, 1940–1941, p. 475–500.

Fader, S. W., 1975 [1976], Land subsidence caused by dissolution of salt near four oil and gas wells in central Kansas: U.S. Geological Survey Water Resources Investigation WRI 27-75, 33 p.; available *only* from U.S. Department of Commerce, National Technical Information Service, Springfield, Va. 22161, as Report PB-250 979/AS.

Franklin, J. A., and Hungr, O., 1978, Rock stresses in Canada, their relevance to engineering project: Rock Mechanics, Supplementum 6, p. 25–46.

Frye, J. C., and Schoff, S. L., 1942, Deep-seated solution in the Meade Basin and vicinity, Kansas and Oklahoma: American Geophysical Union Transactions, v. 23, Part 1, p. 35–39.

Gendzwill, D. L., 1978, Winnipegosis mounds and Prairie Evaporite Formation of Saskatchewan—seismic study: American Association of Petroleum Geologists Bulletin, v. 62, no. 1, p. 73–86.

Grand Saline Sun, 1976, Salt well cave-in attracts wide interest: Grand Saline, Texas, Grand Saline Sun, May 6.

Herget, G., 1973, Variations of rock stresses with depth at a Canadian iron mine: International Journal of Rock Mechanics and Mining Science, v. 10, p. 37–51.

Herget, G., Pahl, A., and Olive, P., 1975, Ground stresses below 3,000 feet: Canadian Rock Mechanics Symposium, 10th, Kingston, Ontario, Proceedings, p. 281–307.

Horvath, A. L., 1957, The stratigraphy and paleontology of drill hole H-1A, Taylor Township, Wayne County, Michigan [Master's thesis]: Ann Arbor, Michigan University, 33 p.

Johnson, K. S., and Gonzales, Serge, 1978, Salt deposits in the United States and regional characteristics important for storage of radioactive waste: U.S. Department of Energy, Office of Waste Isolation Report Y/OWI/SUB-7414/1, 188 p.

Johnson, W. D., 1901, The high plains and their utilization: U.S. Geological Survey 21st Annual Report, p. 601–741.

Kalafatis, C. A., 1958, The stratigraphy and paleontology of core H-1A of Upper Silurian strata from Taylor Township, Wayne County, Michigan [Master's thesis]: Ann Arbor, Michigan University, p. 1–45.

Kozary, M. T., Dunlap, J. C., and Humphrey, W. E., 1968, Incidence of saline deposits in geologic time, *in* Saline deposits—International Conference Saline Deposits, Houston, Tex., Symposium, 1962: Geological Society of America Special Paper 88, p. 44–57.

Landes, K. K., 1931, Recent subsidence, Hamilton County, Kansas: American Association of Petroleum Geologists Bulletin, v. 15, no. 6, p. 708.

—— 1933, Caverns in loess: American Journal of Science, 5th ser., v. 25, no. 146, p. 137–139.

—— 1945, Mackinac breccia, *in* Geology of the Mackinac Straits region: Michigan Geological Survey Publication 44, p. 123–154.

—— 1960, Salt deposits of the United States, *in* Sodium chloride—the production and properties of salt and brine: American Chemical Society Monograph Series 145, p. 70–95.

—— 1963a, Effects of solution of bedrock salt in the Earth's crust, *in* Bersticker, A. C., ed., Symposium on salt, 1st, Cleveland, 1962, Proceedings: Northern Ohio Geological Society, p. 64–73.

—— 1963b, Origin of salt deposits, *in* Bersticker, A. C., ed., Symposium on salt, 1st, Cleveland, 1962, Proceedings: Northern Ohio Geological Society, p. 3–9.

Landes, K. K., and Piper, T. B., 1972, Effect upon environment of brine cavity subsidence at Grosse Ile, Michigan—1971: Solution Mining Research Institute open-file report, 812 Muriel Street, Woodstock, Ill. 60098, 52 p.

Lefond, S. J., 1969, Handbook of World salt resources: New York, Plenum Press, 384 p.

Marsden, R. W., and Lucas, J. R., 1973, Specialized underground extraction systems, *in* Cummins, A. B., and Given, I. A., eds., SME mining engineering handbook: Society of Mining Engineers of American Institute of Mining, Metallurgical, and Petroleum Engineers, v. 2, sec. 21, p. 1–118.

Myers, A. J., 1963, Sonar measurements of brine cavity shapes, *in* Bersticker, A. C., ed., Symposium on salt, 1st, Cleveland, 1962, Proceedings: Northern Ohio Geological Society, p. 546–554.

Newton, J. G., 1976, Early detection and correction of sinkhole problems in Alabama, with a preliminary evaluation of remote sensing applications: Alabama Highway Research Report 76, 83 p.

Nicholson, Alexander, Jr., and Clebsch, Alfred, Jr., 1961, Geology and groundwater conditions in southern Lea County, New Mexico: New Mexico Bureau of Mines and Mineral Resources Report 6, 123 p.

Nieto-Pescetto, A. S., and Hendron, A. J., Jr., 1977, Study of sinkholes related to salt production in the area of Detroit, Michigan: Solution Mining Research Institute open-file report 812 Muriel Street, Woodstock, Ill. 60098, 49 p.

Obert, Leonard, and Duvall, W. I., 1967, Rock mechanics and the design of structures in rocks: New York, John Wiley, 650 p.

Panek, L. A., 1970, Methods and equipment for measuring subsidence, *in* Rau, J. L., and Dellwig, L. F., eds., Symposium on salt, 3rd, Cleveland, 1969, Proceedings: Northern Ohio Geological Society, p. 321–338.

Parker, J. H., 1967, Salt solution and subsidence structures, Wyoming, North Dakota and Montana: American Association of Petroleum Geologists Bulletin, v. 51, no. 10, p. 1929–1947.

Parks, W. S., 1963, Attala County mineral resources: Mississippi Geological Economic and Topographical Survey Bulletin 99, 191 p.

Pearson, W. J., 1963, Salt deposits of Canada, *in* Bersticker, A. C., ed., Symposium on salt, 1st, Cleveland, 1962, Proceedings: Northern Ohio Geological Society, p. 197–239.

Querio, C. W., 1977, Current practices in solution mining of salt, *in* Martinez, J. D., and Thoms, R. L., eds., Symposium on salt dome utilization and environmental considerations, Baton Rouge, 1976, Proceedings: Louisiana State Institute University Institute for Environmental Studies, p. 3–41.

Rellensmann, Otto, 1957, Rock mechanics in regard to static loading caused by mining excavation, *in* Behavior of materials of the Earth's crust, Symposium on rock mechanics, 2d Annual: Golden, Colo., 1957, Colorado School of Mines Quarterly, v. 52, no. 3, p. 35–49.

Savage, W. Z., 1981, Prediction of vertical displacements in a subsiding elastic layer: Geophysical Research Letter, v. 8, no. 3, p. 195–198.

Seely, F. B., and Smith, J. D., 1952, Advance mechanics of materials: New York, John Wiley, 680 p.

Sill, R. T., and Baker, D. M., 1945, Prevention of seepage in foundations for dams: American Geophysical Union Transactions, v. 26, no. 2, p. 169–273.

Smith, G. I., Jones, C. L., Culbertson, W. C., Ericksen, G. E., and Dyni, J. R., 1973, Evaporites and brines, *in* Brobst, D. A., and Pratt, W. P., eds., United States mineral resources: U.S. Geological Survey Professional Paper 820, p. 197–216.

Soper, E. K., 1915, The buried rock surface and pre-glacial river valleys of Minneapolis and vicinity: Journal of Geology, v. 23, p. 444–460.

Sowers, G. F., 1976, Mechanisms of subsidence due to underground openings, *in* Subsidence over mines and caverns, moisture and frost action, and classification: Washington, D.C., Transportation Research Board Record 612, p. 2–8.

Stefanko, Robert, 1973, Roof and ground control—subsidence and ground movement, *in* Cummins, A. B., and Given, I. A., eds., SME mining engineering handbook: Society of Mining Engineers of American Institute of Mining, Metallurgical, and Petroleum Engineers, v. 2, sec. 13, p. 2–9.

Stump, Daniel, Nieto, A. S., and Ege, J. R., 1982, An alternative hypothesis for sink development above solution-mine cavities in the Detroit area: U.S. Geological Survey Open-File Report 82-297, 61 p.

Terzaghi, Karl, 1931, Earth slips and subsidences from underground erosion: Engineering News-Record, v. 107, no. 3, p. 90–92.

Terzaghi, R. D., 1970, Brine field subsidence at Windsor, Ontario, *in* Rau, J. L., and Dellwig, L. F., eds., Symposium on salt, 3rd, Cleveland, 1969, Proceedings: Northern Ohio Geological Society, v. 2, p. 298–307.

Walters, R. F., 1977, Land subsidence in central Kansas related to salt dissolution: Kansas Geological Survey Bulletin 214, 82 p.

Woodruff, S. D., 1966, Methods of working coal and metal mines, Volume 1: London, Pergamon Press, p. 282–305.

Wright, H. E., Jr., 1964, Origin of the lakes in the Chuska Mountains, northwestern New Mexico: Geological Society of America Bulletin, v. 75, p. 589–598.

MANUSCRIPT ACCEPTED BY THE SOCIETY APRIL 18, 1984

Typeset by WESType Publishing Services, Inc., Boulder, Colorado
Printed in U.S.A. by Malloy Lithographing, Inc., Ann Arbor, Michigan